AIR POLLUTION

AIR POLLUTION

Edited by

David H.F. Liu
Béla G. Lipták

Paul A. Bouis
Special Consultant

CRC Press
Taylor & Francis Group
Boca Raton London New York

CRC Press is an imprint of the
Taylor & Francis Group, an **informa** business

CRC Press
Taylor & Francis Group
6000 Broken Sound Parkway NW, Suite 300
Boca Raton, FL 33487-2742

First issued in paperback 2019

© 2000 by Taylor & Francis Group, LLC
CRC Press is an imprint of Taylor & Francis Group, an Informa business

No claim to original U.S. Government works

ISBN-13: 978-1-56670-513-4 (hbk)
ISBN-13: 978-0-367-39913-9 (pbk)

Library of Congress Cataloging-in-Publication Data

Catalog record is available from the Library of Congress

Visit the Taylor & Francis Web site at
http://www.taylorandfrancis.com

and the CRC Press Web site at
http://www.crcpress.com

Preface

Dr. David H.F. Liu passed away prior to the preparation of this book.
He will be long remembered by his coworkers,
and the readers of this book will carry his memory into the 21st Century.

Engineers respond to the needs of society with technical innovations. Their tools are the basic sciences. Some engineers might end up working *on* these tools instead of working *with* them. Environmental engineers are in a privileged and challenging position, because their tools are the totality of man's scientific knowledge, and their target is nothing less than human survival through making man's peace with nature.

The contributors to this book came from all continents and their backgrounds cover not only engineering, but also legal, medical, agricultural, meteorological, biological and other fields of training. In addition to discussing the causes, effects, and remedies of pollution, this book also emphasizes reuse and recovery. Nature does not cause pollution; by total recycling, *nature makes resources out of all wastes*. Our goal should be to learn from nature in this respect.

To the best of our knowledge today, life in the universe exists only in a ten-mile-thick layer on the 200-million-square-mile surface of this planet. During the 5 million years of human existence, we lived in this thin crust of earth, air, and water. Initially man relied only on inexhaustible resources. The planet appeared to be without limits and the laws of nature directed our evolution. Later we started to supplement our muscle power with exhaustible energy sources (coal, oil, uranium) and to substitute the routine functions of our brains by machines. As a result, in some respects we have "conquered nature" and today we are directing our own evolution. Today, our children grow up in man-made environments; virtual reality or cyberspace is more familiar to them than the open spaces of meadows.

While our role and power have changed, our consciousness did not. Subconsciously we still consider the planet inexhaustible and we are still incapable of thinking in time-frames which exceed a few lifetimes. These human limitations hold risks, not only for the planet, nor even for life

on this planet, but for our species. Therefore, it is necessary to pay attention not only to our physical environment but also to our cultural and spiritual environment.

It is absolutely necessary to bring up a new generation which no longer shares our deeply rooted subconscious belief in continuous growth: A new generation which no longer desires the forever increasing consumption of space, raw materials, and energy.

It is also necessary to realize that, while as individuals we might not be able to think in longer terms than centuries, as a society we must. This can and must be achieved by developing rules and regulations which are appropriate to the time-frame of the processes which we control or influence. The half-life of plutonium is 24,000 years, the replacement of the water in the deep oceans takes 1000 years. For us it is difficult to be concerned about the consequences of our actions, if those consequences will take centuries or millennia to evolve. Therefore, it is essential that we develop both an educational system and a body of law which would protect our descendants from our own shortsightedness.

Protecting life on this planet will give the coming generations a unifying common purpose. The healing of environmental ills will necessitate changes in our subconscious and in our value system. Once these changes have occurred, they will not only guarantee human survival, but will also help in overcoming human divisions and thereby change human history.

The Condition of the Air

There is little question about the harmful effects of ozone depletion, acid rain, or the greenhouse effect. One might debate if the prime cause of desertification is acid rain, excessive lumbering, soil erosion, or changes in the weather, but it is a fact that the rain forests are diminishing and the deserts are spreading (Figure 1). We do not know what quantity of acid fumes, fluorinated hydrocarbons, or carbon dioxide gases can be released before climatic changes become irreversible. But we do know that the carbon dioxide content of the atmosphere has substantially increased, that each automobile releases 5 tons of carbon dioxide every year, and that the number of gas-burning oil platforms in the oceans is approaching 10,000.

Conditions on the land and in the waters are determined by complex biosystems. The nonbiological nature of air makes the setting of emission standards and their enforcement somewhat easier. As discussed in Chapter 5 of this handbook, the United States has air quality and emission standards for particulates, carbon monoxide, sulfur and nitrogen oxides, hydrocarbons, photochemical oxidants, asbestos, beryllium, and mercury.

For other materials, such as the "possible human carcinogens," the furans and dioxins (PCDD and PCDF), there are no firm emission or air quality standards yet. These materials are the byproducts of paper bleaching, wood preservative and pesticide manufacturing, and the incineration of plastics. Because typical municipal solid

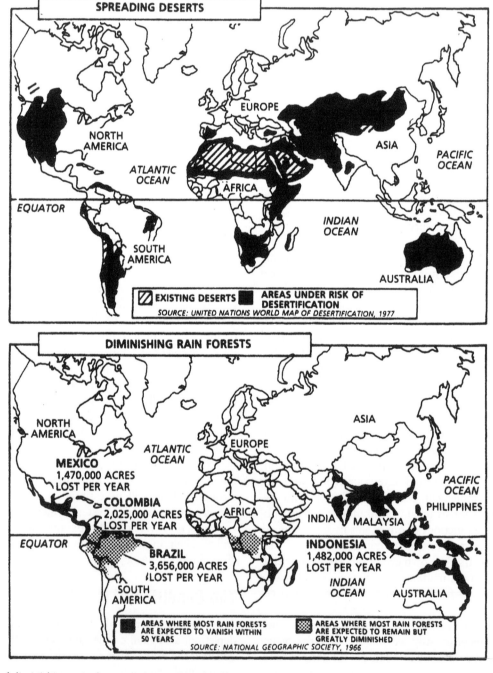

FIG. 1 Areas of diminishing rain forests and spreading deserts.

waste (MSW) in the U.S. contains some 8% plastics, incineration is probably the prime source of dioxin emissions. Dioxins are formed on incinerator fly ash and end up either in landfills or are released into the atmosphere. Dioxin is suspected to be not only a carcinogen but also a cause of birth defects. It is concentrated through the food chain, is deposited in human fat tissues, and in some cases dioxin concentrations of 1.0 ppb have already been found in mother's milk.

Although in the last decades the air quality in the U.S. improved and the newer standards (such as the Clean Air Act of 1990) became stricter, lately we have seen misguided attempts to reverse this progress. Regulations protecting wetlands, forbidding clear-cutting of forests, and mandating use of electric cars have all been relaxed or reversed. In the rest of the world, the overall trend is continued deterioration of air quality. In the U.S., part of the improvement in air quality is due not to pollution abatement but to the exporting of manufacturing industries; part of the improvement is made possible by relatively low population density, not the result of conservation efforts.

On a per capita basis the American contribution to worldwide pollutant emissions is high. For example, the yearly per capita generation of carbon dioxide in the U.S. is about 20 tons. This is twentyfold the per capita CO_2 generation of India. Therefore, even if the emission levels in the West are stabilized or reduced, the global generation of pollutants is likely to continue to rise as worldwide living standards slowly equalize.

Protecting the global environment, protecting life on this planet, must become a single-minded, unifying goal for all of us. The struggle will overshadow our differences, will give meaning and purpose to our lives and, if we succeed, it will mean survival for our children and the generations to come.

Béla G.
Lipták

Contributors

Elmar R. Altwicker

Larry W. Canter
BE, MS, PhD, PE;
Sun Company Chair of Ground Water Hydrology,
University of Oklahoma

Samuel Shih-hsien Cha
BS, MS; Consulting Chemist, TRC Environmental Corp.

Karl T. Chuang
PhD, ChE; Professor, Department of Chemical
Engineering, University of Alberta

Béla G. Lipták
ME, MME, PE; Process Control and Safety Consultant,
President, Liptak Associates, P.C.

David H.F. Liu
PhD, ChE; Principal Scientist, J.T. Baker, Inc. a division
of Procter & Gamble

Gurumurthy Ramachandran
BSEE, PhD; Assistant Professor,
Division of Environmental and Occupational Health,
University of Minnesota

R. A. Herrick

Roger K. Raufer
BSChE, MSCE, MA, PhD, PE; Associate Director,
Environmental Studies,
Center for Energy and the Environment,
University of Pennsylvania

Parker C. Reist
ScD, PE; Professor of Air and Industrial Hygiene
Engineering, University of North Carolina

Alan R. Sanger
BSc, MSc, DPhil; Consultant and Professor,
Department of Chemical Engineering,
University of Alberta

Amos Turk
BS, MA, PhD; Professor Emeritus,
Department of Chemistry,
The City College of New York

Curtis P. Wagner
BA, MS; Senior Project Manager, TRC Environmental,
Inc.

Contents

1

Pollutants: Sources, Effects, and Dispersion Modeling

Larry W. Canter | David H.F. Liu | Roger K. Raufer | Curtis P. Wagner

1.1
SOURCES, EFFECTS, AND FATE OF POLLUTANTS

Air pollution is defined as the presence in the outdoor atmosphere of one or more contaminants (pollutants) in quantities and duration that can injure human, plant, or animal life or property (materials) or which unreasonably interferes with the enjoyment of life or the conduct of business. Examples of traditional contaminants include sulfur dioxide, nitrogen oxides, carbon monoxide, hydrocarbons, volatile organic compounds (VOCs), hydrogen sulfide, particulate matter, smoke, and haze. This list of air pollutants can be subdivided into pollutants that are gases or particulates. Gases, such as sulfur dioxide and nitrogen oxides exhibit diffusion properties and are normally formless fluids that change to the liquid or solid state only by a combined effect of increased pressure and decreased temperature. Particulates represent any dispersed matter, solid or liquid, in which the individual aggregates are larger than single small molecules (about 0.0002 μm in diameter) but smaller than about 500 micrometers (μm). Of recent attention is particulate matter equal to or less than 10 μm in size, with this size range of concern relative to potential human health effects. (One μm is 10^{-4} cm).

Currently the focus is on air toxics (or hazardous air pollutants [HAPs]). Air toxics refer to compounds that are present in the atmosphere and exhibit potentially toxic effects not only to humans but also to the overall ecosystem. In the 1990 Clean Air Act Amendments (CAAAs), the air toxics category includes 189 specific chemicals. These chemicals represent typical compounds of concern in the industrial air environment adjusted from workplace standards and associated quality standards to outdoor atmospheric conditions.

The preceding definition includes the quantity or concentration of the contaminant in the atmosphere and its associated duration or time period of occurrence. This concept is important in that pollutants that are present at low concentrations for short time periods can be insignificant in terms of ambient air quality concerns.

Additional air pollutants or atmospheric effects that have become of concern include photochemical smog, acid rain, and global warming. Photochemical smog refers to the formation of oxidizing constituents such as ozone in the atmosphere as a result of the photo-induced reaction of hydrocarbons (or VOCs) and nitrogen oxides. This phenomenon was first recognized in Los Angeles, California, following World War II, and ozone has become a major air pollutant of concern throughout the United States.

Acid rain refers to atmospheric reactions that lead to precipitation which exhibits a pH value less than the normal pH of rainfall (the normal pH is approximately 5.7

when the carbon dioxide equilibrium is considered). Recently, researchers in central Europe, several Scandinavian countries, Canada, and the northeastern United States, have directed their attention to the potential environmental consequences of acid precipitation. Causative agents in acid rain formation are typically associated with sulfur dioxide emissions and nitrogen oxide emissions, along with gaseous hydrogen chloride. From a worldwide perspective, sulfur dioxide emissions are the dominant precursor of acid rain formation.

Another global issue is the influence of air pollution on atmospheric heat balances and associated absorption or reflection of incoming solar radiation. As a result of increasing levels of carbon dioxide and other carbon-containing compounds in the atmosphere, concern is growing that the earth's surface is exhibiting increased temperature levels, and this increase has major implications in shifting climatic conditions throughout the world.

Sources of Air Pollution

Air pollutant sources can be categorized according to the type of source, their number and spatial distribution, and the type of emissions. Categorization by type includes natural and manmade sources. Natural air pollutant sources include plant pollens, wind-blown dust, volcanic eruptions, and lightning-generated forest fires. Manmade sources include transportation vehicles, industrial processes, power plants, municipal incinerators, and others.

POINT, AREA, AND LINE SOURCES

Source categorization according to number and spatial distribution includes single or point sources (stationary), area or multiple sources (stationary or mobile), and line sources. Point sources characterize pollutant emissions from industrial process stacks and fuel combustion facility stacks. Area sources include vehicular traffic in a geographical area as well as fugitive dust emissions from open-air stock piles of resource materials at industrial plants. Figure 1.1.1 shows point and area sources of air pollution. Included in these categories are transportation sources, fuel combustion in stationary sources, industrial process losses, solid waste disposal, and miscellaneous items. This organization of source categories is basic to the development of emission inventories. Line sources include heavily travelled highway facilities and the leading edges of uncontrolled forest fires.

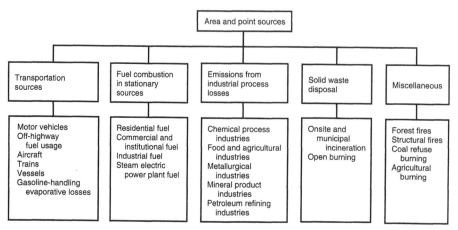

FIG. 1.1.1 Source categories for emission inventories.

GASEOUS AND PARTICULATE EMISSIONS

As stated earlier, air pollution sources can also be categorized according to whether the emissions are gaseous or particulates. Examples of gaseous pollutant emissions include carbon monoxide, hydrocarbons, sulfur dioxide, and nitrogen oxides. Examples of particulate emissions include smoke and dust emissions from a variety of sources. Often, an air pollution source emits both gases and particulates into the ambient air.

PRIMARY AND SECONDARY AIR POLLUTANTS

An additional source concept is that of primary and secondary air pollutants. This terminology does not refer to the National Ambient Air Quality Standards (NAAQSs), nor is it related to primary and secondary impacts on air quality that result from project construction and operation. Primary air pollutants are pollutants in the atmosphere that exist in the same form as in source emissions. Examples of primary air pollutants include carbon monoxide, sulfur dioxide, and total suspended particulates. Secondary air pollutants are pollutants formed in the atmosphere as a result of reactions such as hydrolysis, oxidation, and photochemical oxidation. Secondary air pollutants include acidic mists and photochemical oxidants. In terms of air quality management, the main strategies are directed toward source control of primary air pollutants. The most effective means of controlling secondary air pollutants is to achieve source control of the primary air pollutant; primary pollutants react in the atmosphere to form secondary pollutants.

EMISSION FACTORS

In evaluating air quality levels in a geographical area, an environmental engineer must have accurate information on the quantity and characteristics of the emissions from numerous sources contributing pollutant emissions into the ambient air. One approach for identifying the types and estimating the quantities of emissions is to use emission factors. An emission factor is the average rate at which a pollutant is released into the atmosphere as a result of an activity, such as combustion or industrial production, divided by the level of that activity. Emission factors relate the types and quantities of pollutants emitted to an indicator such as production capacity, quantity of fuel burned, or miles traveled by an automobile.

EMISSION INVENTORIES

An emission inventory is a compilation of all air pollution quantities entering the atmosphere from all sources in a geographical area for a time period. The emission inventory is an important planning tool in air quality management. A properly developed inventory provides information concerning source emissions and defines the location, magnitude, frequency, duration, and relative contribution of these emissions. It can be used to measure past successes and anticipate future problems. The emission inventory is also a useful tool in designing air sampling networks. In many cases, the inventory is the basis for identifying air quality management strategies such as transportation control plans, and it is useful for examining the long-term effectiveness of selected strategies.

NATIONWIDE AIR POLLUTION TRENDS

Based on source emission factors and geographically based emission inventories, nationwide information can be developed. Figure 1.1.2 summarizes nationwide air pollution emission trends from 1970 to 1991 for six key pollutants. The figure shows significant emission reductions for total suspended particulates, VOCs, carbon monoxide, and lead. The greatest reduction from 1982–1991 was an 89% reduction in lead levels in the air resulting primarily from the removal of lead from most gasoline. In addition, the gradual phase in of cleaner automobiles and powerplants

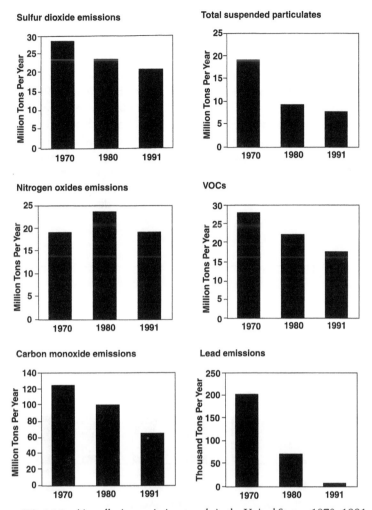

FIG. 1.1.2 Air pollution emission trends in the United States, 1970–1991. (Reprinted from Council on Environmental Quality, 1993, *Environmental quality,* 23rd Annual Report, Washington, D.C.: U.S. Government Printing Office [January].)

reduced atmospheric levels of carbon monoxide by 30%, nitrogen oxides by 6%, ozone by 8%, and sulfur dioxide by 20%. Levels of fine particulate matter (PM-10, otherwise known as dust and soot) dropped 10% since the PM-10 standard was set in 1987 (Council on Environmental Quality 1993).

Despite this progress, 86 million people live in U.S. counties where the pollution levels in 1991 exceeded at least one national air quality standard, based on data for a single year. Figure 1.1.3 shows this data. Urban smog continues to be the most prevalent problem; 70 million people live in U.S. counties where the 1991 pollution levels exceeded the standard for ozone.

Many areas release toxic pollutants into the air. The latest EPA toxics release inventory shows a total of 2.2 billion lb of air toxics released nationwide in 1990 (Council on Environmental Quality 1993).

The primary sources of major air pollutants in the

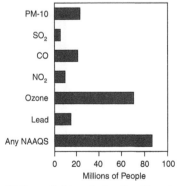

PM-10 = particulate matter less than 10 μm in diameter (dust and soot).

FIG. 1.1.3 People residing in counties that fail to meet NAAQS. Numbers are for 1991 based on 1990 U.S. county population data. Sensitivity to air pollutants can vary from individual to individual. (Reprinted from Council on Environmental Quality 1993.)

United States are transportation, fuel combustion, industrial processes, and solid waste disposal. Figures 1.1.4 through 1.1.9 show the relative contribution of these sources on a nationwide basis for particulates, sulfur oxides, nitrogen oxides, VOCs, carbon monoxide, and lead. Table 1.1.1 contains statistics on the emissions from key sources of these six major pollutants.

Figure 1.1.10 shows anthropogenic sources of carbon dioxide emissions, mainly fuel combustion, from 1950–1990. Table 1.1.2 contains information on the source contributions. Solid and liquid fuel combustion have been the major contributors.

Effects of Air Pollution

Manifold potential effects result from air pollution in an area. These effects are manifested in humans, animals, plants, materials, or climatological variations.

The potential effects of air pollution can be categorized in many ways. One approach is to consider the type of effect and identify the potential air pollutants causing that effect. Another approach is to select an air pollutant such as sulfur dioxide and list all potential effects caused by sulfur dioxide. The types of potential air pollutant effects include aesthetic losses, economic losses, safety hazards, personal discomfort, and health effects. Aesthetic effects

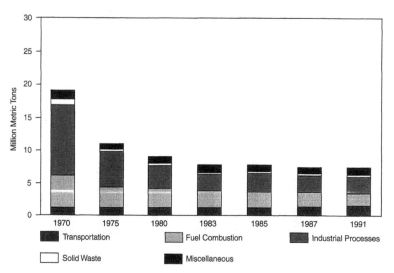

FIG. 1.1.4 U.S. emissions of particulates by source, 1970–1991. (Reprinted from Council on Environmental Quality 1993.)

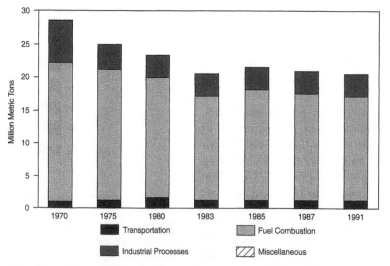

FIG. 1.1.5 U.S. emissions of sulfur oxides by source, 1970–1991. (Reprinted from Council on Environmental Quality 1993.)

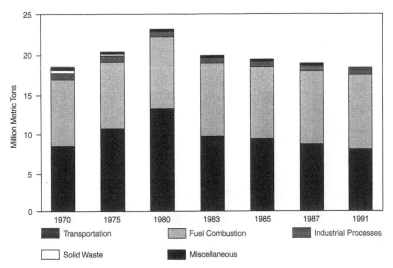

FIG. 1.1.6 U.S. emissions of nitrogen oxides by source, 1970–1991. (Reprinted from Council on Environmental Quality 1993.)

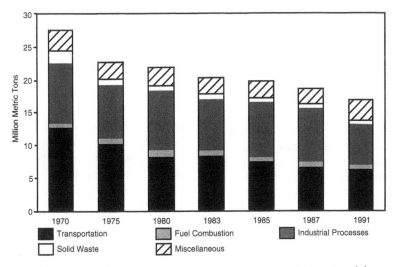

FIG. 1.1.7 U.S. emissions of VOCs by source, 1970–1991. (Reprinted from Council on Environmental Quality 1993.)

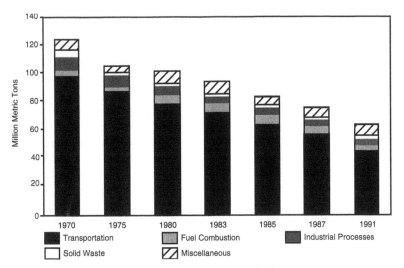

FIG. 1.1.8 U.S. emissions of carbon monoxide by source, 1970–1991. (Reprinted from Council on Environmental Quality 1993.)

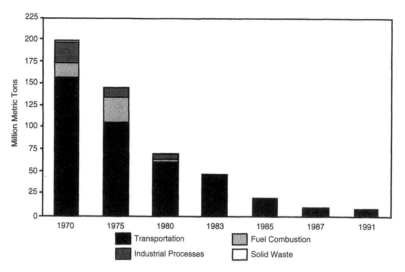

FIG. 1.1.9 U.S. emissions of lead by source, 1970–1991. (Reprinted from Council on Environmental Quality 1993.)

include loss of clarity of the atmosphere as well as the presence of objectionable odors. Atmospheric clarity loss can be caused by particulates and smog as well as by visibility reductions due to nitrate and sulfate particles. Objectionable odors encompass a range of potential air pollutants; the majority are associated with the gaseous form. Examples of odorous air pollutants include hydrogen sulfide, ammonia, and mercaptans. Mercaptans are thio alcohols which are characterized by strong odors often associated with sulfur.

ECONOMIC LOSSES

Economic losses resulting from air pollutants include soiling, damage to vegetation, damage to livestock, and deterioration of exposed materials. Soiling represents the general dirtiness of the environment that necessitates more frequent cleaning. Examples include more frequent cleaning of clothes, washing of automobiles, and repainting of structures. Soiling is typically due to particulate matter being deposited, with the key component being settleable particulates or dustfall.

Examples of damage to vegetation are numerous and include both commercial crops and vegetation in scenic areas. Most vegetation damage is due to excessive exposure to gaseous air pollutants, including sulfur dioxide and nitrogen oxides. Oxidants formed in the atmosphere due to photochemically induced reactions also cause damage to vegetation. Some studies indicate that settleable particulates also disrupt normal functional processes within vegetation and thus undesirable effects take place. An example is the deposit of settleable particulates around a cement plant.

VISIBLE AND QUANTIFIABLE EFFECTS

The visible and quantifiable effects of air pollution include tree injury and crop damage, with examples occurring nationwide (Mackenzie and El-Ashry 1989). Many influences shape the overall health and growth of trees and crops. Some of these influences are natural: competition among species, changes in precipitation, temperature fluctuations, insects, and disease. Others result from air pollution, use of pesticides and herbicides, logging, land-use practices, and other human activities. With so many possible stresses, determining which are responsible when trees or crops are damaged is difficult. Crop failures are usually easier to diagnose than widespread tree declines. By nature, agricultural systems are highly managed and ecologically simpler than forests. Also, larger resources have been devoted to developing and understanding agricultural systems than natural forests. Figure 1.1.11 shows the states in the contiguous United States where air pollution can affect trees or crops (Mackenzie and El-Ashry 1989).

The air pollutants of greatest national concern to agriculture are ozone (O_3), sulfur dioxide (SO_2), nitrogen dioxide (NO_2), sulfates, and nitrates. Of these, ozone is of greatest concern; the potential role of acid deposition at ambient levels has not been determined. At present deposition rates, most studies indicate that acid deposition does no identifiable harm to foliage. However, at lower-than-ambient pH levels, various impacts include leaf spotting, acceleration of epicuticular wax weathering, and changes in foliar leaching rates. When applied simultaneously with ozone, acid deposition also reduces a plant's dry weight (Mackenzie and El-Ashry 1989).

TABLE 1.1.1   U.S. EMISSIONS OF SIX MAJOR AIR POLLUTANTS BY SOURCE, 1970–1991

			Sulfur Oxides			
Year	*Transpor-tation*	*Fuel Com-bustion*	*Industrial Processes*	*Solid Waste*	*Miscel-laneous*	*Total*
			(million metric tons)			
1970	0.61	21.29	6.43	0.00	0.10	28.42
1975	0.64	20.23	4.57	0.04	0.02	25.51
1980	0.90	19.10	3.74	0.03	0.01	23.78
1981	0.89	17.78	3.80	0.03	0.01	22.51
1982	0.83	17.27	3.08	0.03	0.01	21.21
1983	0.79	16.69	3.11	0.02	0.01	20.62
1984	0.82	17.41	3.20	0.02	0.01	21.47
1985	0.88	17.58	3.17	0.02	0.01	21.67
1986	0.87	17.09	3.16	0.02	0.01	21.15
1987	0.89	17.04	3.01	0.02	0.01	20.97
1988	0.94	17.25	3.08	0.02	0.01	21.30
1989	0.96	17.42	3.10	0.02	0.01	21.51
1990	0.99	16.98	3.05	0.02	0.01	21.05
1991	0.99	16.55	3.16	0.02	0.01	20.73

			Nitrogen Oxides			
Year	*Transpor-tation*	*Fuel Com-bustion*	*Industrial Processes*	*Solid Waste*	*Miscel-laneous*	*Total*
			(million metric tons)			
1970	8.45	9.11	0.70	0.40	0.3	18.96
1975	10.02	9.33	0.68	0.14	0.15	20.33
1980	12.46	10.10	0.68	0.10	0.23	23.56
1981	10.42	10.01	0.64	0.10	0.19	21.35
1982	9.74	9.84	0.55	0.09	0.15	20.37
1983	9.35	9.60	0.55	0.08	0.23	19.80
1984	9.10	10.16	0.58	0.08	0.19	20.11
1985	9.15	9.38	0.56	0.08	0.21	19.39
1986	8.49	9.55	0.56	0.08	0.16	18.83
1987	8.14	10.05	0.56	0.08	0.19	19.03
1988	8.19	10.52	0.58	0.08	0.28	19.65
1989	7.85	10.59	0.59	0.08	0.19	19.29
1990	7.83	10.63	0.59	0.08	0.26	19.38
1991	7.26	10.59	0.60	0.01	0.21	18.76

			Reactive VOCs			
Year	*Transpor-tation*	*Fuel Com-bustion*	*Industrial Processes*	*Solid Waste*	*Miscel-laneous*	*Total*
			(million metric tons)			
1970	12.76	0.61	8.93	1.80	3.30	27.40
1975	10.32	0.60	8.19	0.88	2.54	22.53
1980	8.10	0.95	9.13	0.67	2.90	21.75
1981	8.94	0.95	8.24	0.65	2.44	21.22
1982	8.32	1.01	7.41	0.63	2.13	19.50
1983	8.19	1.00	7.80	0.60	2.65	20.26
1984	8.07	1.01	8.68	0.60	2.64	20.99
1985	7.47	0.90	8.35	0.60	2.49	19.80
1986	6.88	0.89	7.92	0.58	2.19	18.45
1987	6.59	0.90	8.17	0.58	2.40	18.64
1988	6.26	0.89	8.00	0.58	2.88	18.61
1989	5.45	0.91	7.97	0.58	2.44	17.35
1990	5.54	0.62	8.02	0.58	2.82	17.58
1991	5.08	0.67	7.86	0.69	2.59	16.88

Continued on next page

TABLE 1.1.1 *Continued*

Year	Carbon Monoxide					
	Transpor-tation	Fuel Com-bustion	Industrial Processes	Solid Waste	Miscel-laneous	Total
	(million metric tons)					
1970	96.85	4.21	8.95	6.40	7.20	123.61
1975	86.15	4.03	6.88	2.93	4.77	104.76
1980	77.38	6.59	6.34	2.09	7.57	99.97
1981	77.08	6.65	5.87	2.01	6.43	98.04
1982	72.26	7.07	4.35	1.94	4.91	90.53
1983	71.40	6.97	4.34	1.84	7.76	92.31
1984	67.68	7.05	4.66	1.84	6.36	87.60
1985	63.52	6.29	4.38	1.85	7.09	83.12
1986	58.71	6.27	4.20	1.70	5.15	76.03
1987	56.24	6.34	4.33	1.70	6.44	75.05
1988	53.45	6.27	4.60	1.70	9.51	75.53
1989	49.30	6.40	4.58	1.70	6.34	68.32
1990	48.48	4.30	4.64	1.70	8.62	67.74
1991	43.49	4.68	4.69	2.06	7.18	62.10

Year	National Total Suspended Particulates					
	Transpor-tation	Fuel Com-bustion	Industrial Processes	Solid Waste	Miscel-laneous	Total
	(million metric tons)					
1970	1.18	5.07	10.54	1.10	1.10	18.99
1975	1.30	3.28	5.19	0.44	0.75	10.96
1980	1.31	3.04	3.31	0.33	1.08	9.06
1981	1.33	2.96	3.03	0.32	0.94	8.58
1982	1.30	2.75	2.57	0.31	0.75	7.67
1983	1.28	2.72	2.39	0.29	1.09	7.77
1984	1.31	2.76	2.80	0.29	0.93	8.08
1985	1.38	2.47	2.70	0.29	1.01	7.85
1986	1.36	2.46	2.43	0.28	0.78	7.31
1987	1.39	2.44	2.38	0.28	0.93	7.42
1988	1.48	2.40	2.48	0.28	1.30	7.94
1989	1.52	2.41	2.46	0.27	0.92	7.57
1990	1.54	1.87	2.53	0.28	1.19	7.40
1991	1.57	1.94	2.55	0.34	1.01	7.41

Year	National PM-10 Particulates					
	Transpor-tation	Fuel Com-bustion	Industrial Processes	Solid Waste	Miscel-laneous	Total
	(million metric tons)					
1985	1.32	1.46	1.90	0.21	0.73	5.61
1986	1.31	1.48	1.74	0.20	0.54	5.27
1987	1.35	1.49	1.70	0.20	0.66	5.40
1988	1.43	1.45	1.73	0.20	0.96	5.76
1989	1.47	1.49	1.77	0.20	0.65	5.59
1990	1.48	1.04	1.81	0.20	0.87	5.42
1991	1.51	1.10	1.84	0.26	0.73	5.45

Year	National PM-10 Fugitive Particulates					
	Agricultural Tilling	Con-struction	Mining & Quarrying	Paved Roads	Unpaved Roads	Wind Erosion
	(million metric tons)					
1985	6.20	11.49	0.31	5.95	13.34	3.23
1986	6.26	10.73	0.28	6.18	13.30	8.52

Continued on next page

TABLE 1.1.1 *Continued*

| Year | National PM-10 Fugitive Particulates | | | | | |
	Agricultural Tilling	*Con-struction*	*Mining & Quarrying*	*Paved Roads*	*Unpaved Roads*	*Wind Erosion*
1987	6.36	11.00	0.34	6.47	12.65	1.32
1988	6.43	10.58	0.31	6.91	14.17	15.88
1989	6.29	10.22	0.35	6.72	13.91	10.73
1990	6.35	9.11	0.34	6.83	14.20	3.80
1991	6.32	8.77	0.36	7.39	14.36	9.19

| Year | Lead | | | | | |
	Transpor-tation	*Fuel Com-bustion*	*Industrial Processes*	*Solid Waste*	*Miscel-laneous*	*Total*
	(thousand metric tons)					
1970	163.60	9.60	23.86	2.00	0.00	199.06
1975	122.67	9.39	10.32	1.45	0.00	143.83
1980	59.43	3.90	3.57	1.10	0.00	68.00
1981	46.46	2.81	3.05	1.10	0.00	53.42
1982	46.96	1.70	2.71	0.94	0.00	52.31
1983	40.80	0.60	2.44	0.82	0.00	44.66
1984	34.69	0.49	2.30	0.82	0.00	38.30
1985	14.70	0.47	2.30	0.79	0.00	18.26
1986	3.45	0.47	1.93	0.77	0.00	6.62
1987	3.03	0.46	1.94	0.77	0.00	6.21
1988	2.64	0.46	2.02	0.74	0.00	5.86
1989	2.15	0.46	2.23	0.69	0.00	5.53
1990	1.71	0.46	2.23	0.73	0.00	5.13
1991	1.62	0.45	2.21	0.69	0.00	4.97

Source: Council on Environmental Quality, 1993.

Notes: Estimates of emissions from transportation sources have been recalculated using a revised model. These estimates supersede those reported in 1992's report and are not directly comparable to historical estimates calculated using different models. PM-10 refers to particulates with an aerodynamic diameter smaller than 10 μm. These smaller particles are likely responsible for most adverse health effects of particulates because of their ability to reach the thoracic or lower regions of the respiratory tract. Detail may not agree with totals because of independent rounding.

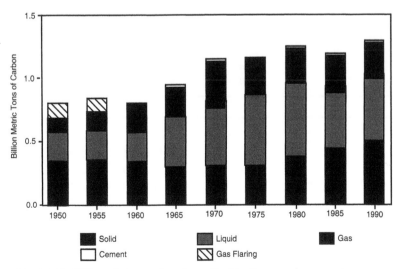

FIG. 1.1.10 U.S. emissions of carbon dioxide from anthropogenic sources, 1950–1990. (Reprinted from Council on Environmental Quality, 1993.)

TABLE 1.1.2 U.S. EMISSIONS OF CARBON DIOXIDE FROM ANTHROPOGENIC SOURCES 1950–1990

Year	Solid	Liquid	Gas	Cement	Gas Flaring	Total	Per Capita
			(million metric tons of carbon)				(metric tons)
1950	347.1	244.8	87.1	5.3	11.8	696.1	4.57
1951	334.5	262.2	102.7	5.7	11.7	716.7	4.63
1952	296.6	273.2	109.9	5.8	12.5	697.9	4.44
1953	294.3	286.6	115.5	6.1	11.9	714.5	4.46
1954	252.2	290.2	121.2	6.3	10.6	680.5	4.18
1955	283.3	313.3	130.8	7.2	11.4	746.0	4.50
1956	295.0	328.5	138.1	7.6	12.7	781.9	4.63
1957	282.7	325.8	147.6	7.1	1.9	775.1	4.51
1958	245.3	333.0	155.8	7.5	9.3	750.8	4.29
1959	251.5	343.5	169.9	8.1	8.4	781.4	4.40
1960	253.4	349.8	180.4	7.6	8.3	799.5	4.43
1961	245.0	354.1	187.4	7.7	7.7	801.9	4.37
1962	254.2	364.3	198.7	8.0	6.3	831.5	4.46
1963	272.5	378.8	210.3	8.4	5.6	875.6	4.63
1964	289.7	389.7	219.8	8.8	5.0	912.9	4.76
1965	301.1	405.6	228.0	8.9	4.7	948.3	4.88
1966	312.7	425.9	246.4	9.1	5.5	990.7	0.08
1967	321.1	443.6	258.5	8.8	7.2	1039.2	5.23
1968	314.8	471.9	277.4	9.4	7.6	1081.0	5.38
1969	319.7	497.4	297.8	9.5	7.7	1132.0	5.58
1970	322.4	514.8	312.1	9.0	7.2	1165.5	5.68
1971	305.7	530.5	323.3	9.7	4.2	1173.2	5.66
1972	310.4	575.5	327.6	10.2	3.6	1227.3	5.86
1973	334.0	605.4	321.7	10.6	3.6	1275.4	6.03
1974	330.1	580.7	307.9	10.0	2.4	1231.1	5.76
1975	317.6	565.1	286.0	8.4	1.9	1179.0	5.46
1976	351.6	608.1	291.3	9.0	2.0	1262.0	5.78
1977	355.6	641.9	260.5	9.7	2.0	1269.7	5.76
1978	361.2	655.0	264.7	10.4	2.2	1293.4	5.80
1979	378.7	634.6	274.8	10.4	2.4	1300.9	5.77
1980	394.6	581.0	272.5	9.3	1.8	1259.3	5.53
1981	403.0	533.1	264.2	8.8	1.4	1210.6	5.26
1982	390.1	502.2	245.4	7.8	1.4	1146.9	4.93
1983	405.5	500.1	233.8	8.7	1.4	1149.4	4.89
1984	427.8	507.1	241.5	9.6	1.6	1187.5	5.01
1985	448.0	505.6	236.7	9.6	1.4	1201.3	5.02
1986	439.7	531.1	222.6	9.7	1.4	1204.5	4.99
1987	463.3	545.3	233.8	9.6	1.8	1253.8	5.15
1988	491.4	566.3	244.6	9.5	2.1	1313.8	5.35
1989	498.4	566.5	252.4	9.5	2.1	1328.9	5.37
1990	508.1	542.9	247.9	9.5	1.9	1310.3	5.26

Source: Council on Environmental Quality, 1993.

BIODIVERSITY

Air pollution can effect biodiversity. For example, prolonged exposure of the vegetation in the San Bernardino Mountains in southern California to photochemical oxidants has shifted the vegetation dominance from ozone-sensitive pines to ozone-tolerant oaks and deciduous shrubs (Barker and Tingey 1992). The fundamental influencing factors include the pollutant's environmental partitioning, exposure pattern, and toxicity and the sensitiv-

ity of the affected species. Biodiversity impacts occur on local, regional, and global scales. Local plume effects reduce vegetation cover, diversity, and ecosystem stability. Regional impacts occur via exposure to photochemical oxidants, wet or dry acid or metal deposition, and the long-range transport of toxic chemicals.

Air pollution effects on biodiversity are difficult to document. Unlike habitat destruction, which results in a pronounced and rapid environmental change, the effects of air pollution on biota are usually subtle and elusive be-

FIG. 1.1.11 Areas where air pollution affects forest trees and agricultural crops. (Reprinted, with permission, from J.J. Mackenzie and M.T. El-Ashry, 1989, Tree and crop injury: A summary of the evidence, chap. 1 in *Air pollution's toll on forests and crops,* edited by J.J. Mackenzie and M.T. El-Ashry, New Haven, Conn.: Yale University Press.)

cause of their interactions with natural stressors. Years can be required before the ecological changes or damage within ecosystems become evident due to continuous or episodic exposure to toxic airborne contaminants or global climate changes (Barker and Tingey 1992).

A number of domestic animals are subject to air pollutant effects. The most frequently cited example is the effects of fluoride on cattle. Other air pollutants also affect animals, including ammonia, carbon monoxide, dust, hydrogen sulfide, sulfur dioxide, and nitrogen oxides.

DETERIORATION OF EXPOSED MATERIALS

The deterioration of exposed materials includes the corrosion of metals, weathering of stone, darkening of lead-based white paint, accelerated cracking of rubber, and deterioration of various manmade fabrics. Sulfur dioxide accelerates the corrosion of metals, necessitating more frequent repainting of metal structures and bridges. The weathering of stone is attributed to the effects of acidic mists formed in the atmosphere as a result of oxidative processes combined with water vapor. Some types of acidic mists include sulfuric acid, carbonic acid, and nitric acid.

HEALTH EFFECTS

The category of health effects ranges from personal discomfort to actual health hazards. Personal discomfort is characterized by eye irritation and irritation to individuals with respiratory difficulties. Eye irritation is associated with oxidants and the components within the oxidant pool such as ozone, proxyacetylnitrate, and others. The burning sensation experienced routinely in many large urban areas is due to high oxidant concentrations. Individuals with respiratory difficulties associated with asthma, bronchitis, and sinusitis experience increased discomfort as a result of oxidants, nitrogen oxides, and particulates.

Health effects result from either acute or chronic exposures. Acute exposures result from accidental releases of pollutants or air pollution episodes. Episodes with documented illness or death are typically caused by persistent (three to six days) thermal inversions with poor atmospheric dispersion and high air pollutant concentrations (Godish 1991). Exposures to lower concentrations for extended periods of time have resulted in chronic respiratory and cardiovascular disease; alterations of body functions such as lung ventilation and oxygen transport; impairment of performance of work and athletic activities; sensory irritation of the eyes, nose, and throat; and aggravation of existing respiratory conditions such as asthma (Godish 1991).

An overview of ambient air quality indicates the potential health effects. Table 1.1.3 shows ambient air quality trends in major urban areas in the United States. The table uses the pollutants standard index (PSI) to depict trends for fifteen of the largest urban areas.

Table 1.1.4 summarizes the effects attributed to specific air pollutants. Many of these effects are described in previous examples, thus this table is a composite of the range of effects of these air pollutants. Table 1.1.5 contains information on the effects of sulfur dioxide. The effects are arranged in terms of health, visibility, materials, and vegetation. Many health effects and visibility are related to the combination of sulfur dioxide and particulates in the atmosphere.

Numerous acute air pollution episodes have caused dra-

TABLE 1.1.3   AIR QUALITY TRENDS IN MAJOR URBAN AREAS, 1980–1991

PMSA	1980	1981	1982	1983	1984	1985	1986	1987	1988	1989	1990	1991
					(number of PSI days greater than 100)							
Atlanta	7	9	5	23	8	9	17	19	15	3	16	5
Boston	8	2	5	16	6	2	0	5	11	1	1	5
Chicago	na	3	31	14	8	4	5	9	18	2	3	8
Dallas	10	12	11	17	10	12	5	6	3	3	5	0
Denver	35	51	52	67	59	37	43	34	18	11	7	7
Detroit	na	18	19	18	7	2	6	9	17	12	3	7
Houston	10	32	25	43	30	30	28	31	31	19	35	39
Kansas City	13	7	0	4	12	4	8	5	3	2	2	1
Los Angeles	220	228	195	184	208	196	210	187	226	212	163	156
New York	119	100	69	65	53	21	16	16	35	9	10	16
Philadelphia	52	29	44	56	31	25	21	36	34	19	11	24
Pittsburgh	20	17	14	36	24	6	9	15	31	11	12	3
San Francisco	2	1	2	4	2	5	4	1	1	0	1	0
Seattle	33	42	19	19	4	26	18	13	8	4	2	0
Washington	38	23	25	53	30	15	11	23	34	7	5	16
Total	567	576	488	619	492	396	400	409	484	315	276	285

Source: Council on Environmental Quality, 1993, *Environmental quality,* 23rd Annual Report (Washington, D.C.: U.S. Government Printing Office [January]).

Notes: PMSA = Primary Metropolitan Statistical Area. PSI = Pollutant Standards Index. na = not applicable. The PSI index integrates information from many pollutants across an entire monitoring network into a single number which represents the worst daily air quality experienced in the urban area. Only carbon monoxide and ozone monitoring sites with adequate historical data are included in the PSI trend analysis above, except for Pittsburgh, where sulfur dioxide contributes a significant number of days in the PSI high range. PSI index ranges and health effect descriptor words are as follows: 0 to 50 (good); 51 to 100 (moderate); 101 to 199 (unhealthful); 200 to 299 (very unhealthful); and 300 and above (hazardous). The table shows the number of days when the PSI was greater than 100 (= unhealthy or worse days).

TABLE 1.1.4   QUALITATIVE SUMMARY OF THE EFFECTS ATTRIBUTED TO SPECIFIC POLLUTANTS

Air Pollutant	Effects
Particulates	Speeds chemical reactions; obscures vision; corrodes metals; causes grime on belongings and buildings; aggravates lung illness
Sulfur oxides	Causes acute and chronic leaf injury; attacks a wide variety of trees; irritates upper respiratory tract; destroys paint pigments; erodes statuary; corrodes metals; ruins hosiery; harms textiles; disintegrates book pages and leather
Hydrocarbons (in solid and gaseous states)	May be cancer-producing (carcinogenic); retards plant growth; causes abnormal leaf and bud development
Carbon monoxide	Causes headaches, dizziness, and nausea; absorbs into blood; reduces oxygen content; impairs mental processes
Nitrogen oxides	Causes visible leaf damage; irritates eyes and nose; stunts plant growth even when not causing visible damage; creates brown haze; corrodes metals
Oxidants: Ozone	Discolors the upper surface of leaves of many crops, trees, and shrubs; damages and fades textiles; reduces athletic performance; hastens cracking of rubber; disturbs lung function; irritates eyes, nose, and throat; induces coughing
Peroxyacetyl nitrate (PAN)	Discolors the lower leaf surface; irritates eyes; disturbs lung function

matic health effects to the human population. One of the first occurred in the Meuse Valley in Belgium in 1930 and was characterized by sixty deaths and thousands of ill people. In Donoro, Pennsylvania in 1948, seventeen people died, and 6000 of the population of 14,000 were reported ill. In Poza Rica, Mexico in 1950, twenty-two people died, and 320 people were hospitalized as a result of an episode. Several episodes with excess deaths have been recorded in London, England, with the most famous being in 1952 when 3500 to 4000 excess deaths occurred over a one-week time period. Other episodes occurred recently in locations throughout the United States, and others are anticipated in subsequent years. Generally, the individuals most affected by these episodes are older people already experiencing difficulties with their respiratory systems. Common characteristics of these episodes include pollu-

TABLE 1.1.5 EFFECTS ATTRIBUTED TO SULFUR DIOXIDE

Category of Effect	Comments
Health	a. At concentrations of about 1500 $\mu g/m^3$ (0.52 ppm) of sulfur dioxide (24-hr average) and suspended particulate matter measured as a soiling index of 6 cohs or greater, mortality can increase.
	b. At concentrations of about 715 $\mu g/m^3$ (0.25 ppm) of sulfur dioxide and higher (24-hr mean), accompanied by smoke at a concentration of 750 $\mu g/m^3$, the daily death rate can increase.
	c. At concentrations of about 500 $\mu g/m^3$ (0.19 ppm) of sulfur dioxide (24-hr mean), with low particulate levels, mortality rates can increase.
	d. At concentrations ranging from 300 to 500 $\mu g/m^3$ (0.11 to 0.19 ppm) of sulfur dioxide (24-hr mean) with low particulate levels, increase hospital admissions of older people for respiratory disease can increase; absenteeism from work, particularly with older people, can also occur.
	e. At concentrations of about 715 $\mu g/m^3$ (0.25 ppm) of sulfur dioxide (24-hr mean) accompanied by particulate matter, illness rates for patients over age 54 with severe bronchitis can rise sharply.
	f. At concentrations of about 600 $\mu g/m^3$ (about 0.21 ppm) of sulfur dioxide (24-hr mean) with smoke concentrations of about 300 $\mu g/m^3$, patients with chronic lung disease can experience accentuation of symptoms.
	g. At concentrations ranging from 105 to 265 $\mu g/m^3$ (0.037 to 0.092 ppm) of sulfur dioxide (annual mean) accompanied by smoke concentrations of about 185 $\mu g/m^3$, the frequency of respiratory symptoms and lung disease can increase.
	h. At concentrations of about 120 $\mu g/m^3$ (0.046 ppm) of sulfur dioxide (annual mean) accompanied by smoke concentrations of about 100 $\mu g/m^3$, the frequency and severity of respiratory diseases in school children can increase.
	i. At concentrations of about 115 $\mu g/m^3$ (0.040 ppm) of sulfur dioxide (annual mean) accompanied by smoke concentrations of about 160 $\mu g/m^3$, mortality from bronchitis and lung cancer can increase.
Visibility	At a concentration of 285 $\mu g/m^3$ (0.10 ppm) of sulfur dioxide with a comparable concentration of particulate matter and relative humidity of 50%, visibility can be reduced to about 5 mi.
Materials	At a mean sulfur dioxide level of 345 $\mu g/m^3$ (0.12 ppm) accompanied by high particulate levels, the corrosion rate for steel panels can increase by 50%.
Vegetation	a. At a concentration of about 85 $\mu g/m^3$ (0.03 ppm) of sulfur dioxide (annual mean), chronic plant injury and excessive leaf drop can occur.
	b. After exposure to about 860 $\mu g/m^3$ (0.3 ppm) of sulfur dioxide for 8 hr, some species of trees and shrubs show injury.
	c. At concentrations of about 145 to 715 $\mu g/m^3$ (0.05 to 0.25 ppm), sulfur dioxide can react synergistically with either ozone or nitrogen dioxide in short-term exposures (e.g., 4 hr) to produce moderate to severe injury to sensitive plants.

Source: National Air Pollution Control Administration, 1969, *Air quality criteria for sulfur oxides*, Pub. No. AP-50 (Washington, D.C. [January]: 161–162).

tant releases from many sources, including industry, and limiting atmospheric dispersion conditions.

ATMOSPHERIC EFFECTS

Air pollution causes atmospheric effects including reductions in visibility, changes in urban climatological characteristics, increased frequency of rainfall and attendant meteorological phenomena, changes in the chemical characteristics of precipitation, reductions in stratospheric ozone levels, and global warming (Godish 1991). The latter three effects can be considered from a macro (large-scale) perspective and are addressed in Section 1.5.

Particulate matter can reduce visibility and increase atmospheric turbidity. Visibility is defined as the greatest distance in any direction at which a person can see and iden-

tify with the unaided eye (1) a prominent dark object against the sky at the horizon in the daytime, and (2) a known, preferably unfocused, moderately intense light source at night. In general, visibility decreases as the concentration of particulate matter in the atmosphere increases. Particle size is important in terms of visibility reduction, with sizes in the micron and submicron range of greatest importance. Turbidity in ambient air describes the phenomena of back scattering of direct sunlight by particles in the air, thus reducing the amount of direct sunlight reaching the earth. As an illustration of the effect of turbidity increases in the atmosphere, the total sunshine in urban areas is approximately 80% of that in nearby rural areas. The ultraviolet (UV) component of sunlight in the winter in urban areas is only 70% of that in nearby rural areas; in the summer the UV component in urban areas is

95% of the rural areas' value.

Table 1.1.6 summarizes the quality factors of urban air in ratio to those of rural air when rural air is a factor of 1. The quantity of urban air pollutants and some of the results of the effects of cloudiness and fog are evident in urban areas more than rural areas. Urban areas and the associated air pollutants also influence certain climatological features such as temperature, relative humidity, cloudiness, windspeed and precipitation.

RAINFALL QUALITY

One issue related to the general effects of air pollution is the physical and chemical quality of rainfall. Air pollution can cause the pH of rainfall to decrease, while the suspended dissolved solids and total solids in rainfall increase. Nitrogen and phosphorus concentrations in rainfall can also increase as a result of the atmospheric releases of pollutants containing these nutrients. Finally, increases in lead and cadmium in rainfall are also a result of air pollutant emissions.

An important issue related to air pollution effects is acid rainfall and the resultant effects on aquatic ecosystems. Acid rainfall is any rainfall with a pH less than 5.7. The natural pH of rainfall is 5.7 and reflects the presence of weak carbonic acid (H_2CO_3) resulting from the reaction of water and carbon dioxide from green plants. Rainfall becomes more acid as a result of acidic mists such as H_2SO_4 and HNO_3. Atmospheric emissions of carbon monoxide also add to the carbonic acid mist in the atmosphere and cause the pH of rainfall to be less than 5.7. Numerous locations in the United States have rainfall with the pH values around 4.0. Some of the lowest recorded pH values of rainfall are 2.0 to 3.0.

The chief concerns related to acid rainfall are the potential adverse effects. For example, the pH of the soil can be changed and this change can have unfavorable implications. Changes in the pH in soil can cause changes in

adsorption and desorption patterns and lead to differences in nonpoint source water pollution as well as changes in nutrients in both surface runoff as well as from infiltration to groundwater. Acid rain can decrease plant growth, crop growth, and growth in forested areas. Acid rainfall can accelerate the weathering and erosion of metals, stone buildings, and monuments. One concern is related to changes in the quality of surface water and the resultant potential toxicity to aquatic species.

Tropospheric Ozone—A Special Problem

The most widespread air quality problem in the United States is exceedances of the ozone standard (0.12 ppm for 1 hr per year) in urban areas. The ozone standard is based on protecting public health. Ozone is produced when its precursors, VOCs and nitrogen oxides (NO_x), combine in the presence of sunlight (Office of Technology Assessment 1989). VOCs, a broad class of pollutants encompassing hundreds of specific compounds, come from manmade sources including automobile and truck exhaust, evaporation of solvents and gasoline, chemical manufacturing, and petroleum refining. In most urban areas, such manmade sources account for the majority of VOC emissions, but in the summer in some regions, natural vegetation produces an almost equal quantity. NO_x arises primarily from fossil fuel combustion. Major sources include highway vehicles and utility and industrial boilers.

About 100 nonattainment areas dot the country from coast to coast, with *design values* (a measure of peak ozone concentrations) ranging from 0.13 ppm to as high as 0.36 ppm. Figure 1.1.12 summarizes the data for the 3-year period 1983–85 (Office of Technology Assessment 1989). Generally, the higher the design value, the stricter the emission controls needed to meet the standard.

From one-third to one-half of all Americans live in areas that exceed the standard at least once a year. As shown in Figure 1.1.13, 130 of the 317 urban and rural areas exceeded 0.12 ppm for at least 1 hr between 1983 and 1985 (Office of Technology Assessment 1989). Sixty had concentrations that high for at least 6 hr per year. A number of areas topped the standard for 20 or more hr, with the worst, Los Angeles, averaging 275 hr per year.

Ozone's most perceptible short-term effects on human health are respiratory symptoms such as coughing and painful deep breathing (Office of Technology Assessment 1989). It also reduces people's ability to inhale and exhale normally, affecting the most commonly used measures of lung function (e.g., the maximum amount of air a person can exhale in 1 sec or the maximum a person can exhale after taking a deep breath). As the intensity of exercise rises so does the amount of air drawn into the lungs and thus the dose of ozone. The more heavily a person exercises at

TABLE 1.1.6   QUALITY FACTORS OF URBAN AIR IN RATIO TO THOSE OF RURAL AIR EXPRESSED AS 1

Urban	Quality Factor
10	Dust particles
5	Sulfur dioxide
10	Carbon dioxide
25	Carbon monoxide
0.8	Total sunshine
0.7	Ultraviolet, winter
0.95	Ultraviolet, summer
1.1	Cloudiness
2	Fog, winter
1.3	Fog, summer

FIG. 1.1.12 Areas classified as nonattainment for ozone based on 1983–85 data. The shading indicates the fourth highest daily maximum one-hour average ozone concentration, or design value, for each area. (Reprinted from Office of Technology Assessment, 1989, *Catching our breath—Next steps for reducing urban ozone*, OTA-0-412, Washington, D.C.: U.S. Congress [July].)

FIG. 1.1.13 Areas where ozone concentrations exceeded 0.12 ppm at least one hour per year on average, from 1983–85. Data from all monitors located in each area were averaged in the map construction. The shading indicates the number of hours that a concentration of 0.12 ppm was exceeded. The areas shown have 130 million residents. (Reprinted from Office of Technology Assessment, 1989.)

a level of ozone concentration and the longer the exercise lasts, the larger the potential effect on lung function.

The U.S. Environmental Protection Agency (EPA) has identified two subgroups of people who may be at special risk for adverse effects: athletes and workers who exercise heavily outdoors and people with preexisting respiratory problems (Office of Technology Assessment 1989). Also problematic are children, who appear to be less suscepti-

ble to (or at least less aware of) acute symptoms and thus spend more time outdoors in high ozone concentrations. Most laboratory studies show no special effects in asthmatics, but epidemiologic evidence suggests that they suffer more frequent attacks, respiratory symptoms, and hospital admissions during periods of high ozone. In addition, about 5 to 20% of the healthy adult population appear to be *responders,* who for no apparent reason are more sensitive than average to a dose of ozone.

At the summertime ozone levels in many cities, some people who engage in moderate exercise for extended periods can experience adverse effects. For example, as shown in Figure 1.1.14, on a summer day when ozone concentrations average 0.14 ppm, a construction worker on an 8-hr shift can experience a temporary decrease in lung function that most scientists consider harmful (Office of Technology Assessment 1989). On those same summer days, children playing outdoors for half the day also risk the effects on lung function that some scientists consider adverse. And some heavy exercisers, such as runners and bicyclists, notice adverse effects in about 2 hr. Even higher levels of ozone, which prevail in a number of areas, have swifter and more severe impacts on health.

Brief Synopsis of Fate of Air Pollutants

Atmospheric dispersion of air pollutants from point or area sources is influenced by wind speed and direction, atmospheric turbulence, and atmospheric stability (Godish 1991).

EFFECTS OF WIND SPEED AND DIRECTION

Horizontal winds play a significant role in the transport and dilution of pollutants. As wind speed increases, the volume of air moving by a source in a period of time also increases. If the emission rate is relatively constant, a doubling of the wind speed halves the pollutant concentration, as the concentration is an inverse function of the wind speed.

Pollutant dispersion is also affected by the variability in wind direction (Godish 1991). If the wind direction is rel-

atively constant, the same area is continuously exposed to high pollutant levels. If the wind direction is constantly shifting, pollutants are dispersed over a larger area, and concentrations over any exposed area are lower. Large changes in wind direction can occur over short periods of time.

EFFECTS OF ATMOSPHERIC TURBULENCE

Air does not flow smoothly near the earth's surface; rather, it follows patterns of three-dimensional movement which are called turbulence. Turbulent eddies are produced by two specific processes: (1) thermal turbulence, resulting from atmospheric heating, and (2) mechanical turbulence caused by the movement of air past an obstruction in a windstream. Usually both types of turbulence occur in any atmospheric situation, although sometimes one prevails. Thermal turbulence is dominant on clear, sunny days with light winds. Although mechanical turbulence occurs under a variety of atmospheric conditions, it is dominant on windy nights with neutral atmospheric stability. Turbulence enhances the dispersion process although in mechanical turbulence, downwash from the pollution source can result in high pollution levels immediately downstream (Godish 1991).

EFFECTS OF ATMOSPHERIC STABILITY

In the troposphere, temperature decreases with height to an elevation of approximately 10 km. This decrease is due to reduced heating processes with height and radiative cooling of air and reaches its maximum in the upper levels of the troposphere. Temperature decrease with height is described by the lapse rate. On the average, tempera-

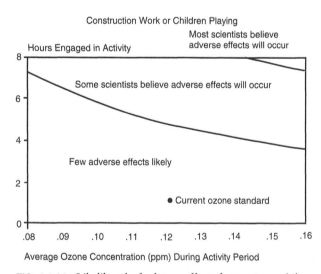

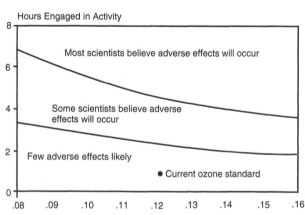

FIG. 1.1.14 Likelihood of adverse effects from ozone while exercising. The likelihood of experiencing adverse effects depends on 1) the ozone concentration, 2) the vigorousness of the activity, and 3) the number of hours engaged in that activity. The figure on the left shows the number of hours to reach an adverse effect under moderate exercise conditions (e.g., construction work or children playing). The figure on the right shows that fewer hours are needed under heavy exercise (e.g., competitive sports or bicycling). The current 1-hr ozone standard is shown for comparison. (Reprinted from Office of Technology Assessment 1989.)

ture decreases −0.65°C/100 m or −6.5°C/km. This decrease is the normal lapse rate. If warm dry air is lifted in a dry environment, it undergoes adiabatic expansion and cooling. This adiabatic cooling results in a lapse rate of −1°C/100 m or −10°C/km, the dry adiabatic lapse rate.

Individual vertical temperature measurements vary from either the normal or dry adiabatic lapse rate. This change of temperature with height for measurement is the *environmental lapse rate.* Values for the environmental lapse rates characterize the stability of the atmosphere and profoundly affect vertical air motion and the dispersion of pollutants (Godish 1991).

If the environmental lapse rate is greater than the dry adiabatic lapse rate, dispersion characteristics are good to excellent. The greater the difference, the more unstable the atmosphere and the more enhanced the dispersion. If the environmental lapse rate is less than the dry adiabatic lapse rate, the atmosphere becomes stable, and dispersion becomes more limited. The greater the difference from the adiabatic lapse rate, the more stable the atmosphere and the poorer the dispersion potential (Godish 1991).

EFFECTS OF TOPOGRAPHY ON AIR MOTION

Topography can affect micro- and mesoscale air motion near point and area sources. Most large urban centers in this country are located along sea (New York City and Los Angeles) and lake (Chicago and Detroit) coastal areas, and heavy industry is often located in river valleys, e.g., the Ohio River Valley. Local air flow patterns in these regions have a significant impact on pollution dispersion processes. For example, land–water mesoscale air circulation patterns develop from the differential heating and cooling of land and water surfaces. During the summer when skies are clear and prevailing winds are light, land surfaces heat more rapidly than water. The warm air rises and moves toward water. Because of the differences of temperature and pressure, air flows in from the water, and a sea or lake breeze forms. Over water, the warm air from the land cools and subsides to produce a weak circulation cell. At night, the more rapid radiational cooling of land surfaces results in a horizontal flow toward water, and a land breeze forms (Godish 1991).

Air flows downhill into valley floors, and the winds produced are called *slope winds.* As the air reaches the valley floor, it flows with the path of the river. This air movement is called the *valley wind.* The formation of valley wind lags several hours after slope winds. Because of a smaller vertical gradient, downriver valley winds are lighter and because of the large volume, cool dense air accumulates, flooding the valley floor and intensifying the surface

inversion that is normally produced by radiative cooling (Godish 1991). The inversion deepens over the course of the night and reaches its maximum depth just before sunrise. The height of the inversion layer depends on the depth of the valley and the intensity of the radiative cooling process.

Mountains affect local air flow by increasing surface roughness and thereby decreasing wind speed. In addition, mountains and hills form physical barriers to air movement.

In summary, the atmospheric dispersion of air pollution emissions depends on the interplay of a number of factors which include (1) the physical and chemical nature of the pollutants, (2) meteorological parameters, (3) the location of the source relative to obstructions, and (4) downwind topography (Godish 1991).

OTHER FACTORS

In addition to dispersion, wet and dry removal processes as well as atmospheric reactions affect the concentrations of air pollutants in the atmosphere. Atmospheric reactions include ozone or acid rain formation. In dry removal, particles are removed by gravity or impaction, and gases diffuse to surfaces where they are absorbed or adsorbed. Wet removal is the major removal process for most particles and can be a factor in the removal of gaseous contaminants as well. Wet removal can involve the in-cloud capture of gases or particles (rainout) or the below-cloud capture (washout). In washout, raindrops or snowflakes strike particles and carry them to the surface; gases are removed by absorption (Godish 1991).

—Larry W. Canter

References

Barker, J.R., and D.T. Tingey. 1992. The effects of air pollution on biodiversity: A synopsis. Chap. 1 in *Air pollution effects on biodiversity,* edited by J.R. Barker and D.T. Tingey, 3–8. New York: Van Nostrand Reinhold.

Council on Environmental Quality. 1993. *Environmental quality.* 23rd Annual Report. Washington, D.C.: U.S. Government Printing Office. (January): 7–9, 14–16, and 326–340.

Godish, T. 1991. *Air quality,* 2d ed., 65–85, 89, 131–133, and 173. Chelsea, Mich.: Lewis Publishers, Inc.

Mackenzie, J.J., and M.T. El-Ashry. 1989. Tree and crop injury: A summary of the evidence. Chap. 1 in *Air pollution's toll on forests and crops,* edited by J.J. Mackenzie and M.T. El-Ashry, 1–19. New Haven, Conn.: Yale University Press.

Office of Technology Assessment. 1989. *Catching our breath—Next steps for reducing urban ozone.* OTA-0-412. Washington, D.C.: U.S. Congress. (July): 4–9.

1.2
VOCs AND HAPs EMISSION FROM CHEMICAL PLANTS

Emission Points

Emission sources (or points) of volatile organic chemicals (VOCs) and hazardous air pollutants (HAPs) in a chemical plant can be classified into three groups: (1) process point sources, (2) process fugitive sources, and (3) area fugitive sources (U.S. EPA 1991). VOCs refer to compounds which produce vapors at room temperature and pressure; whereas, HAPs include VOCs as well as nonvolatile organics and inorganics present as vapors or particulates.

PROCESS POINT SOURCES

Process point sources of VOCs and HAPs can be individually defined for a chemical plant. Chemical reactors, distillation columns, catalytic cracking units, condensers, strippers, furnaces, and boilers are examples of point sources that discharge both air toxics and criteria pollutants through vent pipes or stacks. Emission reductions or control are achieved through process changes focused on pollution prevention and the use of add-on control devices such as adsorbers, absorbers, thermal or catalytic incinerators, fabric filters, or electrostatic precipitators (ESPs).

PROCESS FUGITIVE SOURCES

Although typically more numerous than process point sources, process fugitive sources can also be individually defined for a chemical plant. Inadvertent emissions from or through pumps, valves, compressors, access ports, storage tank vents, and feed or discharge openings to a process classify such units or equipment as process fugitive sources. Vent fans from rooms or enclosures containing an emissions source can also be classified this way (U.S. EPA 1991). Once process fugitive emissions are captured by hooding, enclosures, or closed-vent systems, they can often be controlled by add-on devices used for process point sources.

AREA FUGITIVE SOURCES

Large surface areas characterize area fugitive sources. Examples of such sources include waste storage ponds and raw material storage piles at many chemical plants. VOC and HAP control measures for area fugitive sources typically focus on release prevention measures such as the use of covers or chemical adjustments in terms of the pH and oxidation state for liquid wastes.

Classification of VOCs and HAPs

The HAPs described in this manual are not limited to the specific compounds listed in current laws such as the CAAAs of 1990, the Resource Conservation and Recovery Act (RCRA), or the Toxic Substances Control Act. HAPs can be classified relative to the type of compounds (i.e., organic or inorganic) and the form in which they are emitted from process point, process fugitive, or area fugitive sources (i.e., vapor or particulate).

This section discusses two examples of VOC and HAP emissions from chemical plant classes. Table 1.2.1 summarizes emissions from the inorganic chemical manufacturing industry. This industry produces basic inorganic chemicals for either direct use or use in manufacturing other chemical products. Although the potential for emissions is high, in many cases they are recovered due to economic reasons. As shown in Table 1.2.1, the chemical types of inorganic emissions depend on the source category, while the emission sources vary with the processes used to produce the inorganic chemical.

The second example is from petroleum-related industries, including the oil and gas production industry, the petroleum refining industry, and the basic petrochemicals industry. Table 1.2.2 summarizes the emission sources within these three categories. Sources of emissions from the oil and gas production industry include blowouts during drilling operations; storage tank breathing and filling losses; wastewater treatment processes; and fugitive leaks in valves, pumps, pipes, and vessels. In the petroleum refining industry, emission sources include distillation and fractionating columns, catalytic cracking units, sulfur recovery processes, storage tanks, fugitives, and combustion units (e.g., process heaters). Fugitive emissions are a major source in this industry. Emission sources in the basic petrochemicals industry are similar to those from the petroleum refining segments (U.S. EPA 1991).

Table 1.2.3 summarizes the potential HAP emissions from the petroleum refining segment of the petroleum industries. A large proportion of the emissions occur as organic vapors; for example, benzene, toluene, and xylenes are the principal organic vapor emissions. These organic vapors are due to the chemical composition of the two starting materials used in these industries: crude oil and natural gas. Crude oil is composed chiefly of hydrocarbons (paraffins, napthalenes, and aromatics) with small amounts of trace elements and organic compounds containing sulfur, nitrogen, and oxygen. Natural gas is largely

TABLE 1.2.1 POTENTIAL HAPS FOR INORGANIC CHEMICAL MANUFACTURING INDUSTRY

Source Category	Potential HAPs		Potential Emission Sources		
	Inorganic		Process Point	Process Fugitive	Area Fugitive
	Vapor	Particulate			
Aluminum chloride	4,10		X	X	
Aluminum fluoride	17		X	X	
Ammonia	1		B,D,E	K	J,S
Ammonium acetate	1		X	X	
Ammonium-nitrate, sulfate, thiocyanate, formate, tartrate	1		C,F,I,L	Q	
Ammonium phosphate	1,17		X	X	
Antimony oxide	5		X	X	
Arsenic-disulfide, iodide, pentafluoride, thioarsenate, tribromide, trichloride, trifluoride, trioxide, orthoarsenic acid	2	2	H,U	K,Q,T	J,S
Barium-carbonate, chloride, hydroxide, sulfate, sulfide		6	C,E,G,I,L,U	N,P,Q,T	
Beryllium-oxide, hydroxide		7	X	X	
Boric acid and borax		9	X	X	
Bromine	8,10		X	X	
Cadmium (pigment)—sulfide, sulfoselenide, lithopone		15	X	X	
Calcium-carbide, arsenate, phosphate	3,17	2	H	K,P	
Chlorine	10	25	H,C	K,R	J
Chlorosulfonic acid	19,34		X	X	
Chromic acid	12	11,12	H	K,N,O,Q	J,S
Chromium-acetate, borides, halides, etc.		11	X	X	
Chromium (pigment)—oxide		11	X	X	
Cobalt—acetate, carbonate, halides, etc.		13	X	X	
Copper sulfate	14		X	X	
Fluorine	17		X	X	
Hydrazine	1,39		X	X	
Hydrochloric acid	10,20	20	B		
Hydrofluoric acid	17		B,G	K,R	
Iodine (crude)	10	38	X	X	
Iron chloride	10,20	20	X	X	
Iron (pigment)—oxide	40		X	X	
Lead-arsenate, halides, hydroxides, dioxide, nitrate	3	2,21	G,L	P,Q	
Lead chromate	22		G,R	P,Q	
Lead (pigments)—oxide, carbonate, sulfate		21	G,R	P,Q	
Manganese dioxide (Potassium permanganate)	24	23	G,L	Q,P,T	
Manganese sulfate		23	G,L	Q,P,T	
Mercury-halides, nitrates, oxides		25	X	X	
Nickel-halides, nitrates, oxides		26		P,Q	
Nickel sulfate	27	26	L	Q,T	
Nitric acid	28	28	B,H	K,N,R	J,S
Phosphoric acid					
Wet process	10,17,18,30	30	H,C,W	K,N,P,T	J,S
Thermal process			B,G	K,N,R,T	J,S
Phosphorus	17		X	X	
Phosphorus oxychloride	10		X	X	
Phosphorus pentasulfide	29,31	29	X	X	

Continued on next page

TABLE 1.2.1 *Continued*

Source Category	Potential HAPs Inorganic Vapor	Potential HAPs Inorganic Particulate	Process Point	Process Fugitive	Area Fugitive
Phosphorus trichloride	32,10,29	29	X	X	
Potassium-bichromate, chromate	16	16	I		
Potassium hydroxide	10	25	X	X	
Sodium arsenate		2	H	K,P	
Sodium carbonate	1		I,L,V	P	
Sodium chlorate	10		X	X	
Sodium chromate-dichromate	16	16	G,I,L,M	P,Q	
Sodium hydrosulfide	18		X	X	
Sodium-siliconfluoride, fluoride	17	16	X	X	
Sulfuric acid	33,34	33	A,B,C,H	K,R	J,S
Sulfur monochloride-dichloride	10		X	X	
Zinc chloride	36,21	21	X	X	
Zinc chromate (pigment)	35		X	X	
Zinc oxide (pigment)	37		X	X	

Source: U.S. Environmental Protection Agency, 1991, *Handbook: Control technologies for hazardous air pollutants.* EPA/625/6-91/014 (Cincinnati, Ohio [June]).

Pollutant Key
1. ammonia
2. arsenic
3. arsenic trioxide
4. aluminum chloride
5. antimony trioxide
6. barium salts
7. beryllium
8. bromine
9. boron salts
10. chlorine
11. chromium salts
12. chromic acid mist
13. cobalt metal fumes
14. copper sulfate
15. cadmium salts
16. chromates (chromium)
17. fluorine
18. hydrogen sulfide
19. hydrogen chloride
20. hydrochloric acid
21. lead
22. lead chromate
23. manganese salts
24. manganese dioxide
25. mercury
26. nickel
27. nickel sulfate
28. nitric acid mist
29. phosphorus
30. phosphoric acid mist
31. phosphorus pentasulfide
32. phosphorus trichloride
33. sulfuric acid mist
34. sulfur trioxide
35. zinc chromate
36. zinc chloride fumes
37. zinc oxide fumes
38. iodine
39. hydrazine
40. iron oxide

Source Key
A. converter
B. absorption tower
C. concentrator
D. desulfurizer
E. reformer
F. neutralizer
G. kiln
H. reactor
I. crystallizer
J. compressor and pump seals
K. storage tank vents
L. dryer
M. leaching tanks
N. filter
O. flakers
P. milling, grinding, and crushing
Q. product handling and packaging
R. cooler (cooling tower and condenser)
S. pressure relief valves
T. raw material unloading
U. purification
V. calciner
W. hot well
X. no information

saturated hydrocarbons (mainly methane). The remainder can include nitrogen, carbon dioxide, hydrogen sulfide, and helium. Organic and inorganic particulate emissions, such as coke fires or catalyst fires, can be generated from some processes (U.S. EPA 1991).

—*Larry W. Canter*

Reference

U.S. Environmental Protection Agency (EPA). 1991. *Handbook: Control technologies for hazardous air pollutants.* EPA/625/6-91/014. Cincinnati, Ohio. (June) 2-1 to 2-13.

TABLE 1.2.2 EMISSION SOURCES FOR THE PETROLEUM-RELATED INDUSTRIES

	Potential HAP Emission Sources		
Source Category	Process Point	Process Fugitive	Area Fugitive
Oil and Gas Production			
Exploration, site preparation and drilling	A	C	D,E
Crude processing	G	F,H	
Natural gas processing	G,J,K	H	I
Secondary and tertiary recovery techniques	G		I
Petroleum Refining Industry			
Crude separation	G,J,L	F,H,M,N	I
Light hydrocarbon processing	O,G	F,H	Q
Middle and heavy distillate processing	G,O,P,R	F,H	I
Residual hydrocarbon processing	B,G,K,O,R	H	I
Auxiliary processes	G	F,H	I
Basic Petrochemicals Industry			
Olefins production	G,K,O	F,H	I
Butadiene production	G,J,L,O,R	F,H,N	I
Benzene/toluene/xylene (BTX) production	G,K,O,R	F,Q	I
Naphthalene production	G,L,O	F,H	I
Cresol/cresylic acids production	G,L	F,H	I
Normal paraffin production	G,O	F,H	I

Source: U.S. EPA, 1991.

Source Key

A. blowout during drilling
B. visbreaker furnace
C. cuttings
D. drilling fluid
E. pipe leaks (due to corrosion)
F. wastewater disposal (process drain, blow-down, and cooling water)

G. flare, incinerator, process heater, and boiler
H. storage, transfer, and handling
I. pumps, valves, compressors, and fittings
J. absorber
K. process vent
L. distillation and fractionation
M. hotwells

N. steam ejectors
O. catalyst regeneration
P. evaporation
Q. catalytic cracker
R. stripper

TABLE 1.2.3 POTENTIAL HAPs FOR PETROLEUM REFINING INDUSTRIES (SPECIFIC LISTING FOR PETROLEUM REFINING SEGMENT)

	Potential HAPs			
	Organic		Inorganic	
Process	Vapor	Particulate	Vapor	Particulate
Crude separation	a,b,d,e,f,g,h,i,j,k,l,m,o, A,B,C,D,E,F,J	o	c,m,t,u,v,x,y,L	p,I,Q,R
Light hydrocarbon processing	g,h,i,n,N,O,P	R	t,v	G,H,Q
Middle and heavy distillate processing	a,d,e,f,g,h,i,j,k,l, F,J,K,O,P,S,T	o,R	m,t,u,v,x,y,L	p,q,G,H,I,Q,U
Residual hydrocarbon processing	a,d,e,f,g,h,i,j,k,l,n, F,J,M,N,P,S,T	o,R	m,s,t,u,v,x,y,L	p,q,G,H,I,Q,U
Auxiliary processes	a,b,d,e,f,g,h,i,j,k,l,n, A,B,C,D,J,K,M,T	o,R	c,m,s,u,y,L	p,q,r,z,I

Source: U.S. Environmental Protection Agency (EPA), 1991, *Handbook: Control technologies for hazardous air pollutants,* EPA/625/6-91/014, Cincinnati, Ohio, (June) 2-1 to 2-13.

Pollutant Key

a. maleic anhydride
b. benzoic acid
c. chlorides
d. ketones
e. aldehydes
f. heterocyclic compounds (e.g., pyridines)
g. benzene
h. toluene
i. xylene
j. phenols
k. organic compounds containing sulfur (sulfonates, sulfones)
l. cresols
m. inorganic sulfides
n. mercaptans
o. polynuclear compounds (benzopyrene, anthracene)

p. vanadium
q. nickel
r. lead
s. sulfuric acid
t. hydrogen sulfide
u. ammonia
v. carbon disulfide
x. carbonyl sulfide
y. cyanides
z. chromates
A. acetic acid
B. formic acid
C. methylethylamine
D. diethylamine
E. thiosulfide
F. methyl mercaptan

G. cobalt
H. molybdenum
I. zinc
J. cresylic acid
K. xylenols
L. thiophenes
M. thiophenol
N. nickel carbonyl
O. tetraethyl lead
P. cobalt carbonyl
Q. catalyst fines
R. coke fines
S. formaldehyde
T. aromatic amines
U. copper

1.3
HAPs FROM SYNTHETIC ORGANIC CHEMICAL MANUFACTURING INDUSTRIES

The Synthetic Organic Chemical Manufacturing Industry (SOCMI), as a source category, emits a larger volume of a variety of HAPs compared to other source categories (see Table 1.3.1). In addition, individual SOCMI sources tend to be located close to the population. As such, components of SOCMI sources have been subject to various federal, state, and local air pollution control rules. However, the existing rules do not comprehensively regulate emissions for all organic HAPs emitted from all emissions points at both new and existing plants.

By describing hazardous organic national emission standards for air pollutants (NESHAP), or the HON, this section describes the emission points common to all SOCMI manufacturing processes and the maximum achievable control technology (MACT) required for reducing these emissions.

Hazardous Organic NESHAP

The HON is one of the most comprehensive rules issued by the EPA. It covers more processes and pollutants than previous EPA air toxic programs (40 CFR Part 63). For example, one major portion of the rule applies to sources that produce any of the 396 SOCMI products (see Table 1.3.2) that use any of the 112 organic HAPs (see Table 1.3.3) either in a product or as an intermediate or reactant. An additional 37 HAPs are regulated under another part of the HON (40 CFR Part 63). The HON lists 189 HAPs regulated under the air toxic program.

The focus of this rule is the SOCMI. For purposes of the MACT standard, a SOCMI manufacturing plant is viewed as an assortment of equipment—process vents, storage tanks, transfer racks, and wastewater streams—all of which emit HAPs. The HON requires such plants to monitor and repair leaks to eliminate fugitive emissions and requires controls to reduce toxics coming from discrete emission points to minuscule concentrations. Table 1.3.4 summarizes the impacts of these emission sources.

PROCESS VENTS

A process vent is a gas stream that is continuously discharged during the unit operation from an air oxidation unit, reactor process unit, or distillation operation within a SOCMI chemical process. Process vents include gas streams discharged directly to the atmosphere after diversion through a product recovery device. The rule applies only to the process vents associated with continuous (nonbatch) processes and emitting vent streams containing more than 0.005 wt % HAP. The process vent provisions do not apply to vents from control devices installed to comply with wastewater provisions. Process vents exclude relief valve discharges and other fugitive leaks but include vents from product accumulation vessels.

Halogenated streams that use a combustion device to comply with 98% or 20 parts per million by volume (ppmv) HAP emissions must vent the emissions from the combustion device to an acid gas scrubber before venting to the atmosphere.

STORAGE VESSELS

A storage vessel is a tank or vessel storing the feed or product of a SOCMI chemical manufacturing process when the liquid is on the list of HAPs (see Table 1.3.3). The storage vessel provisions require that one of the following control systems is applied to storage vessels:

- An internal floating roof with proper seals and fittings
- An external floating roof with proper seals and fittings
- An external floating roof converted to an internal floating roof with proper seals and fittings
- A closed-vent system with 95% efficient control

TABLE 1.3.1   EMISSION POTENTIAL ACCORDING TO BASIC MANUFACTURING CATEGORY

Emission Potential[a]	Category	% Total Industry Emissions (U.S.)
1	Chemical synthesis	64
2	Fermentation	19
3	Extraction[b]	7
4	Formulation	5
5	Other[c]	5

Notes:
[a]Decreasing order.
[b]Listed as botanicals.
[c]Includes research and development, animal sources, and biological products.

TABLE 1.3.2 SOCML

Chemical Name[a]		
Acenaphthene	Chloroacetophenone (2-)	Dihydroxybenzoic acid (Resorcylic acid)
Acetal	Chloroaniline (p-)	Diisodecyl phthalate
Acetaldehyde	Chlorobenzene	Diisooctyl phthalate
Acetaldol	Chlorodifluoroethane	Dimethylbenzidine (3,3'-)
Acetamide	Chlorodifluoromethane	Dimethyl ether
Acetanilide	Chloroform	Dimethylformamide (N,N-)
Acetic acid	Chloronaphthalene	Dimethylhydrazine (1,1-)
Acetic anhydride	Chloronitrobenzene (1,3-)	Dimethyl phthalate
Acetoacetanilide	Chloronitrobenzene (o-)	Dimethyl sulfate
Acetone	Chloronitrobenzene (p-)	Dimethyl terephthalate
Acetone cyanohydrin	Chlorophenol (m-)	Dimethylamine
Acetonitrile	Chlorophenol (o-)	Dimethylaminoethanol (2-)
Acetophenone	Chlorophenol (p-)	Dimethylaniline (N,N)
Acrolein	Chloroprene	Dinitrobenzenes (NOS)
Acrylamide	Chlorotoluene (m-)	Dinitrophenol (2,4-)
Acrylic acid	Chlorotoluene (o-)	Dinitrotoluene (2,4-)
Acrylonitrile	Chlorotoluene (p-)	Dioxane
Adiponitrile	Chlorotrifluorourethane	Dioxolane (1,3-)
Alizarin	Chrysene	Diphenyl methane
Alkyl anthraquinones	Cresol and cresylic acid (m-)	Diphenyl oxide
Allyl alcohol	Cresol and cresylic acid (o-)	Diphenyl thiourea
Allyl chloride	Cresol and cresylic acid (p-)	Diphenylamine
Allyl cyanide	Cresols and cresylic acids (mixed)	Dipropylene glycol
Aminophenol sulfonic acid	Crotonaldehyde	Di(2-methoxyethyl)phthalate
Aminophenol (p-)	Cumene	Di-o-tolyguanidine
Aniline	Cumene hydroperoxide	Dodecyl benzene (branched)
Aniline hydrochloride	Cyanoacetic acid	Dodecyl phenol (branched)
Anisidine (o-)	Cyanoformamide	Dodecylaniline
Anthracene	Cyclohexane	Dodecylbenzene (n-)
Anthraquinone	Cyclohexanol	Dodecylphenol
Azobenzene	Cyclohexanone	Epichlorohydrin
Benzaldehyde	Cyclohexylamine	Ethane
Benzene	Cyclooctadienes	Ethanolamine
Benzenedisulfonic acid	Decahydronaphthalene	Ethyl acrylate
Benzenesulfonic acid	Diacetoxy-2-Butene (1,4-)	Ethylbenzene
Benzil	Dialyl phthalate	Ethyl chloride
Benzilic acid	Diaminophenol hydrochloride	Ethyl chloroacetate
Benzoic acid	Dibromomethane	Ethylamine
Benzoin	Dibutoxyethyl phthalate	Ethylaniline (n-)
Benzonitrile	Dichloroaniline (inbred isomers)	Ethylaniline (o-)
Benzophenone	Dichlorobenzene (p-)	Ethylcellulose
Benzotrichloride	Dichlorobenzene (m-)	Ethylcyanoacetate
Benzoyl chloride	Dichlorobenzene (o-)	Ethylene carbonate
Benzyl acetate	Dichlorobenzidine (3,5-)	Ethylene dibromide
Benzyl alcohol	Dichlorodifluoromethane	Ethylene glycol
Benzyl benzoate	Dichloroethane (1,2-) (Ethylene dichloride) (EDC)	Ethylene glycol diacetate
Benzyl chloride	Dichloroethyl ether	Ethylene glycol dibutyl ether
Benzyl dichloride	Dichloroethylene (1,2-)	Ethylene glycol diethyl ether (1,2-diethoxyethane)
Biphenyl	Dichlorophenol (2,4-)	Ethylene glycol dimethyl ether
Bisphenol A	Dichloropropene (1,3-)	Ethylene glycol monoacetate
Bis(Chloromethyl)Ether	Dichlorotetrafluoroethane	Ethylene glycol monobutyl ether acetate
Bromobenzene	Dichloro-1-butene (3,4-)	Ethylene glycol monobutyl ether
Bromoform	Dichloro-2-butene (1,4-)	Ethylene glycol monoethyl ether acetate
Bromonaphthalene	Diethanolamine	Ethylene glycol monoethyl ether
Butadiene (1,3-)	Diethyl phthalate	Ethylene glycol monohexyl ether
Butanediol (1,4-)	Diethyl sulfate	Ethylene glycol monomethyl ether acetate
Butyl acrylate (n-)	Diethylamine	Ethylene glycol monomethyl ether
Butylbenzyl phthalate	Diethylaniline (2,6-)	Ethylene glycol monooctyl ether
Butylene glycol (1,3-)	Diethylene glycol	Ethylene glycol monophenyl ether
Butyrolacetone	Diethylene glycol dibutyl ether	Ethylene glycol monopropyl ether
Caprolactam	Diethylene glycol diethyl ether	Ethylene oxide
Carbaryl	Diethylene glycol dimethyl ether	Ethylenediamine
Carbazole	Diethylene glycol monobutyl ether acetate	Ethylenediamine tetracetic acid
Carbon disulfide	Diethylene glycol monobutyl ether	Ethylenimine (Aziridine)
Carbon tetrabromide	Diethylene glycol monoethyl ether acetate	Ethylhexyl acrytate (2-isomer)
Carbon tetrachloride	Diethylene glycol monoethyl ether	Fluoranthene
Carbon tetrafluoride	Diethylene glycol monohexyl ether	Formaldehyde
Chloral	Diethylene glycol monomethyl ether acetate	Formamide
Chloroacetic acid	Diethylene glycol monomethyl ether	Formic acid

Continued on next page

TABLE 1.3.2 *Continued*

Chemical Name[a]		
Fumaric acid	Naphthylamine sulfonic acid (2,1-)	Styrene
Glutaraldehyde	Naphthylamine (1-)	Succinic acid
Glyceraldehyde	Naphthylamine (2-)	Succinonitrile
Glycerol	Nitroaniline (m-)	Sulfanilic acid
Glycerol tri(polyoxypropylene)ether	Nitroaniline (o-)	Sulfolane
Glycine	Nitroanisole (o-)	Tartaric acid
Glyoxal	Nitroanisole (p-)	Terephthalic acid
Hexachlorobenzene	Nitrobenzene	Tetrabromophthalic anhydride
Hexachlorobutadiene	Nitronaphthalene (1-)	Tetrachlorobenzene (1,2,4,5-)
Hexachloroethane	Nitrophenol (p-)	Tetrachloroethane (1,1,2,2-)
Hexadiene (1,4-)	Nitrophenol (o-)	Tetrachlorophthalic anhydride
Hexamethylenetetramine	Nitropropane (2-)	Tetraethyl lead
Hexane	Nitrotoluene (all isomers)	Tetraethylene glycol
Hexanetriol (1,2,6-)	Nitrotoluene (o-)	Tetraethylenepentamine
Hydroquinone	Nitrotoluene (m-)	Tetrahydrofuran
Hydroxyadipaldehyde	Nitrotoluene (p-)	Tetrahydronapthalene
Iminodiethanol (2,2-)	Nitroxylene	Tetrahydrophthalic anhydride
Isobutyl acrylate	Nonylbenzene (branched)	Tetramethylenediamine
Isobutylene	Nonylphenol	Tetramethylethylenediamine
Isophorone	N-Vinyl-2-Pyrrolidine	Tetramethyllead
Isophorone nitrile	Octene-1	Thiocarbanilide
Isophthalic acid	Octylphenol	Toluene
Isopropylphenol	Paraformaldehyde	Toluene 2,4 diamine
Lead phthalate	Paraldehyde	Toluene 2,4 diisocyanate
Linear alkylbenzene	Pentachlorophenol	Toluene diisocyanates (mixture)
Maleic anhydride	Pentaerythritol	Toluene sulfonic acids
Maleic hydrazide	Peracetic acid	Toluenesulfonyl chloride
Malic acid	Perchloroethylene	Toluidine (o-)
Metanilic acid	Perchloromethyl mercaptan	Trichloroaniline (2,4,6-)
Methacrylic acid	Phenanthrene	Trichlorobenzene (1,2,3-)
Methanol	Phenetidine (p-)	Trichlorobenzene (1,2,4-)
Methionine	Phenol	Trichloroethane (TCA) (1,1,1-)
Methyl acetate	Phenolphthalein	TCA (1,1,2-)
Methyl acrylate	Phenolsulfonic acids (all isomers)	Trichloroethylene (TCE)
Methyl bromide	Phenyl anthranilic acid (all isomers)	Trichlorofluoromethane
Methyl chloride	Phenylenediamine (p-)	Trichlorophenol (2,4,5-)
Methyl ethyl ketone	Phloroglucinol	Trichlorotrifluoroethane (1,2,2-1,1,2)
Methyl formate	Phosgene	Triethanolamine
Methyl hydrazine	Phthalic acid	Triethylamine
Methyl isobutyl carbinol	Phthalic anhydride	Triethylene glycol
Methyl isocyanate	Phthalimide	Triethylene glycol dimethyl ether
Methyl mercaptan	Phthalonitrile	Triethylene glycol monoethyl ether
Methyl methacrylate	Picoline (b-)	Triethylene glycol monomethyl ether
Methyl phenyl carbinol	Piperazine	Trimethylamine
Methyl tert-butyl ether	Polyethylene glycol	Trimethylcyclohexanol
Methylamine	Polypropylene glycol	Trimethylcyclohexanone
Methylaniline (n-)	Propiolactone (beta-)	Trimethylcyclohexylamine
Methylcyclohexane	Propionaldehyde	Trimethylolpropane
Methylcyclohexanol	Propionic acid	Trimethylpentane (2,2,4-)
Methylcyclohexanone	Propylene carbonate	Tripropylene glycol
Methylene chloride	Propylene dichloride	Vinyl acetate
Methylene dianiline (4,4′-isomer)	Propylene glycol	Vinyl chloride
Methylene diphenyl diisocyanate (4,4′-) (MDI)	Propylene glycol monomethyl ether	Vinyl toluene
Methylionones (a-)	Propylene oxide	Vinylcyclohexane (4-)
Methylpentynol	Pyrene	Vinylidene chloride
Methylstyrene (a-)	Pyridine	Vinyl(N)-pyrrolidone (2-)
Naphthalene	p-tert-Butyl toluene	Xanthates
Naphthalene sulfonic acid (a-)	Quinone	Xylene sulfonic acid
Naphthalene sulfonic acid (b-)	Resorcinol	Xylenes (NOS)
Naphthol (a-)	Salicylic acid	Xylene (m-)
Naphthol (b-)	Sodium methoxide	Xylene (o-)
Naphtholsulfonic acid (1-)	Sodium phenate	Xylene (p-)
Naphthylamine sulfonic acid (1,4-)	Stilbene	Xylenol

Source: Code of Federal Regulations, Title 40, Part 63.104, *Federal Register 57*, (31 December 1992).

[a]Isomer means all structural arrangements for the same number of atoms of each element and does not mean salts, esters, or derivatives.

TABLE 1.3.3 ORGANIC HAPs

Chemical Name[a,b]		
Acetaldehyde	Dimethylformamide	Methylene chloride (Dichloromethane)
Acetamide	1,1-Dimethylhydrazine	Methylene diphenyl diisocyanate (MDI)
Acetonitrile	Dimethyl phthalate	4,4′-Methylenedianiline
Acetophenone	Dimethyl sulfate	Naphthalene
Acrolein	2,4-Dinitrophenol	Nitrobenzene
Acrylamide	2,4-Dinitrotoluene	4-Nitrophenol
Acrylic acid	1,4-Dioxane (1,4-Diethyleneoxide)	2-Nitropropane
Acrylonitrile	1,2-Diphenylhydrazine	Phenol
Allyl chloride	Epichlorohydrin (1-Chloro-2,3-	p-Phenylenediamine
Aniline	epoxypropane)	Phosgene
o-Anisidine	Ethyl acrylate	Phthalic anhydride
Benzene	Ethylbenzene	Polycyclic organic matter[d]
Benzotrichloride	Ethyl chloride (Chloroethane)	Propiolactone (beta-isomer)
Benzyl chloride	Ethylene dibromide (Dibromoethane)	Propionaldehyde
Biphenyl	Ethylene dichloride (1,2-Dichloroethane)	Propylene dichloride (1,2-
Bis(chloromethyl)ether	Ethylene glycol	Dichloropropane)
Bromoform	Ethylene oxide	Propylene oxide
1,3-Butadiene	Ethylidene dichloride (1,1-	Quinone
Caprolactam	Dichloroethane)	Styrene
Carbon disulfide	Formaldehyde	1,1,2,2-Tetrachloroethane
Carbon tetrachloride	Glycol ethers[c]	Tetrachloroethylene (Perchloroethylene)
Chloroacetic acid	Hexachlorobenzene	Toluene
2-Chloroacetophenone	Hexachlorobutadiene	2,4-Toluene diamine
Chlorobenzene	Hexachloroethane	2,4-Toluene diisocyanate
Chloroform	Hexane	o-Toluidine
Chloroprene	Hydroquinone	1,2,4-Trichlorobenzene
Cresols and cresylic acids (mixed)	Isophorone	1,1,2-TCA
o-Cresol and o-cresylic acid	Maleic anhydride	TCB
m-Cresol and m-cresylic acid	Methanol	2,4,5-Trichlorophenol
p-Cresol and p-cresylic acid	Methyl bromide (Bromomethane)	Triethylamine
Cumene	Methyl chloride (Chloromethane)	2,2,4-Trimethylpentane
1,4-Dichlorobenzene(p-)	Methyl chloroform (1,1,1-	Vinyl acetate
3,3′-Dichlorobenzidine	Trichloroethane)	Vinyl chloride
Dichloroethyl ether (Bis(2-	Methyl ethyl ketone (2-Butanone)	Vinylidene chloride (1,1-
chloroethyl)ether)	Methyl hydrazine	Dichloroethylene)
1,3-Dichloropropene	Methyl isobutyl ketone (Hexone)	Xylenes (isomers and mixtures)
Diethanolamine	Methyl isocyanate	o-Xylene
N,N-Dimethylaniline	Methyl methacrylate	m-Xylene
Diethyl sulfate	Methyl tert-butyl ether	p-Xylene
3,3′-Dimethylbenzidine		

Source: 40 CFR Part 63.104.

Notes: [a]For all listings containing the word "Compounds" and for glycol ethers, the following applies: Unless otherwise specified, these listings include any unique chemical substance that contains the named chemical (i.e., antimony, arsenic) as part of that chemical's infrastructure.

[b]Isomer means all structural arrangements for the same number of atoms of each pigment and does not mean salts, esters, or derivatives.

[c]Includes mono- and di-ethers of ethylene glycol, diethylene glycol, and triethylene glycol R-$(OCH_2CH_2)_n$-OR where n = 1, 2, or 3; R = alkyl or aryl groups; and R′ = R, H, or groups which, when removed, yield glycol ethers with the structure: R-$(OCH_2CH_2)_n$-OH Polymers are excluded from the glycol category.

[d]Includes organic compounds with more than one benzene ring, and which have a boiling point greater than or equal to 100°C.

TABLE 1.3.4   NATIONAL PRIMARY AIR POLLUTION IMPACTS IN THE FIFTH YEAR[a]

| Emission Points | Baseline Emissions (Mg/yr) | | Emission Reductions | | | |
| | | | (Mg/yr) | | (Percent) | |
	HAP	VOC[t]	HAP	VOC[b]	HAP	VOC[b]
Equipment leaks	66,000	84,000	53,000	68,000	80	81
Process vents	317,000	551,000	292,000	460,000	92	83
Storage vessels	15,200	15,200	5,560	5,560	37	37
Wastewaster collection and treatment operations	198,000	728,000	124,000	452,000	63	62
Transfer loading operations	900	900	500	500	56	56
Total	597,000	1,380,000	475,000	986,000	80	71

Source: Code of Federal Regulations, Title 40, part 63; Clean Air Act Amendments, amended 1990, Section 112.

[a]These numbers represent estimated values for the fifth year. Existing emission points contribute 84% of the total. Emission points associated with chemical manufacturing process equipment built in the first 5 yr of the standard contribute 16% of the total.

[b]The VOC estimates consist of the sum of the HAP estimates and the nonHAP VOC estimates.

TRANSFER OPERATIONS

Transfer operations are the loading of liquid products on the list of HAPs from a transfer rack within the SOCMI chemical manufacturing process into a tank truck or railcar. The transfer rack includes the total loading arms, pumps, meters, shutoff valves, relief valves, and other piping and valves necessary to load trucks or railcars.

The proposed transfer provisions control transfer racks to achieve a 98% organic HAP reduction or an outlet concentration of 20 ppmv. Combustion devices or product recovery devices can be used. Again, halogenated streams that use combustion devices to comply with the 98% or 20 ppmv emission reduction must vent the emissions from the combustion device to an acid scrubber before venting to the atmosphere.

WASTEWATER

The wastewater to which the proposed standard applies is any organic HAP-containing water or process fluid discharged into an individual drain system. This wastewater includes process wastewater, maintenance-turnaround wastewater, and routine and routine-maintenance wastewater. Examples of process wastewater streams include those from process equipment, product or feed tank drawdown, cooling water blowdown, steam trap condensate, reflux, and fluid drained into and material recovered from waste management units. Examples of maintenance-turnaround wastewater streams are those generated by the descaling of heat exchanger tubing bundles, cleaning of distillation column traps, and draining of pumps into individual drain system. A HAP-containing wastewater stream is a wastewater stream that has a HAP concentration of 5 parts per million by weight (ppmw) or greater and a flow rate of 0.02 liters per minute (lpm) or greater.

The proposed process water provisions include equipment and work practice provisions for the transport and handling of wastewater streams between the point of generation and the wastewater treatment processes. These provisions include the use of covers, enclosures, and closed-vent systems to route organic HAP vapors from the transport and handling equipment. The provisions also require the reduction of volatile organic HAP (VOHAP) concentrations in wastewater streams.

SOLID PROCESSING

The product of synthetic organic processes can be in solid, liquid, or gas form. Emissions of solid particulates are also of concern. One reason is that particulate emissions occur with drying, packaging, and formulation operations. Additionally, these emissions can be in the respirable size range. Within this range, a significant fraction of the particulates can be inhaled directly into the lungs, thereby enhancing the likelihood of being absorbed into the body and damaging lung tissues.

Toxic Pollutants

Table 1.3.3 shows that halogenated aliphatics are the largest class of priority toxics. These chemicals can cause damage to the central nervous system and liver. Phenols are carcinogenic in mice; their toxicity increases with the

TABLE 1.3.5 HEALTH EFFECTS OF SELECTED HAPS

Pollutant	Major Health Effects
Acryronitrile (CH$_2$ CH C N)	Dermatitis; haematological changes; headaches; irritation of eyes, nose, and throat; lung cancer
Benzene (C$_6$H$_6$)	Leukemia; neurotoxic symptoms; bone marrow injury including anaemia, and chromosome aberrations
Carbon disulfide (CS$_2$)	Neurologic and psychiatric symptoms, including irritability and anger; gastrointestinal troubles; sexual interferences
1,2 Dichloroethane (C$_2$H$_2$Cl$_2$)	Damage to lungs, liver, and kidneys; heart rhythm disturbances; effects on central nervous systems, including dizziness; animal mutagen and carcinogen
Formaldehyde (HC HO)	Chromosome aberrations; irritation of eyes, nose, and throat; dermatitis; respiratory tract infections in children
Methylene chloride (CH$_2$Cl$_2$)	Nervous system disturbances
Polychlorinated bi-phenyls (PCB) (coplanar)	Spontaneous abortions; congenital birth defects; bioaccumulation in food chains
Polychlorinated dibenzo-dioxins and furans	Birth defects; skin disorders; liver damage; suppression of the immune system
Polycyclic organic matter (POM) [including benzo(a)pyrene (BaP)]	Respiratory tract and lung cancers; skin cancers
Styrene (C$_6$H$_5$ CH CH$_2$)	Central nervous system depression; respiratory tract irritations; chromosome aberrations; cancers in the lymphatic and haematopoietic tissues
Tetrachloroethylene (C$_2$Cl$_4$)	Kidney and genital cancers; lymphosarcoma; lung, cervical, and skin cancers; liver dysfunction; effects on central nervous system
Toluene (C$_6$H$_5$ CH$_3$)	Dysfunction of the central nervous system; eye irritation
TCE (C$_2$HCl$_3$)	Impairment of psychomotoric functions; skin and eye irritation; injury to liver and kidneys; urinary tract tumors and lymphomas
Vinyl chloride (CH$_2$ CHCl)	Painful vasospastic disorders of the hands; dizziness and loss of consciousness; increased risk of malformations, particularly of the central nervous systems; severe liver disease; liver cancer; cancers of the brain and central nervous system; malignancies of the lymphatic and haematopoietic system

Source: OECD.

degree of chlorination of phenolic molecules. Maleic anhydride and phthalic anhydride are irritants to the skin, eyes, and mucous membranes. Methanol vapor is irritating to the eyes, nose, and throat; this vapor explodes if ignited in an enclosed area.

Table 1.3.5 lists the health effects of selected HAPs. Because of the large number of HAPs, enumerating the potential health effects of the category as a whole is not possible. However, material safety data sheets (MSDS) for the HAPs are available from chemical suppliers on request,

and handbooks such as the *Hazardous chemical data book* (Weiss 1980) provide additional information.

—David H.F. Liu

References

Code of Federal Regulations. Title 40, Part 63. *Federal Register* 57, (31 December 1992).

Weiss, G., ed. 1980. *Hazardous chemicals data book.* Park Ridge, N.J.: Noyes Data Corp.

1.4
ATMOSPHERIC CHEMISTRY

Pollutants enter the atmosphere primarily from natural sources and human activity. This pollution is called *primary pollution,* in contrast to *secondary pollution,* which is caused by chemical changes in substances in the atmosphere. Sulfur dioxides, nitric oxides, and hydrocarbons are major primary gaseous pollutants, while ozone is a secondary pollutant, the result of atmospheric photochemistry between nitric oxide and hydrocarbons.

Pollutants do not remain unchanged in the atmosphere after release from a source. Physical changes occur, especially through dynamic phenomena, such as movement and scattering in space, turbulent diffusion, and changes in the concentration by dilution.

Changes also result from the chemistry of the atmosphere. These changes are often simple, rapid chemical reactions, such as oxidation and changes in temperature to condense some gases and vapors to yield mist and droplets. After a long residence of some gaseous pollutants in the atmosphere, these gases convert into solid, finely dispersed substances. Solar conditions cause chemical reactions in the atmosphere among various pollutants and their supporting media. Figure 1.4.1 shows simplified schemes of the main chemical changes of pollutants in the atmosphere.

Basic Chemical Processes

A basic chemical process in the atmosphere is the oxidation of substances by atmospheric oxygen. Thus, sulfur dioxide (SO_2) is oxidized to sulfur trioxide (SO_3), and nitric oxide to nitrogen dioxide. Similarly, many organic substances are oxidized, for example, aldehydes to organic acids and unsaturated hydrocarbons. While pollutant clouds are transported and dispersed to varying degrees, they also age. Pollutant cloud aging is a complex combination of homogeneous and heterogeneous reactions and physical processes (such as nucleation, coagulation, and the Brownian motion). Chemically unlike species can make contact and further branch the complex pattern (see Figure 1.4.1). Table 1.4.1 summarizes the major removal reactions and sinks. Most of these reactions are not understood in detail.

Sulfur oxides, in particular SO_2, have been studied with respect to atmospheric chemistry. However, an understanding of the chemistry of SO_2 in the atmosphere is still far from complete. Most evidence suggests that the eventual fate of atmospheric SO_2 is oxidation to sulfate. One problem that complicates understanding atmospheric SO_2 processes is that reaction paths can be homogeneous and heterogeneous. Two processes convert SO_2 to sulfate: catalytical and photochemical.

CATALYTIC OXIDATION OF SO_2

In clear air, SO_2 is slowly oxidized to SO_3 by homogeneous reactions. However, studies show that the rate of SO_2 oxidation in a power plant plume can be 10 to 100 times the clear-air photooxidation rate (Gartrell, Thomas, and Carpenter 1963). Such a rapid rate of reaction is similar to that of oxidation in solution in the presence of a catalyst.

SO_2 dissolves readily in water droplets and can be oxidized by dissolved oxygen in the presence of metal salts, such as iron and manganese. The overall reaction can be expressed as:

$$2\ SO_2 + 2\ H_2O + O_2 \xrightarrow{\text{catalyst}} 2\ H_2SO_4 \qquad 1.4(1)$$

Catalysts for the reaction include sulfates and chlorides of manganese and iron which usually exists in air as sus-

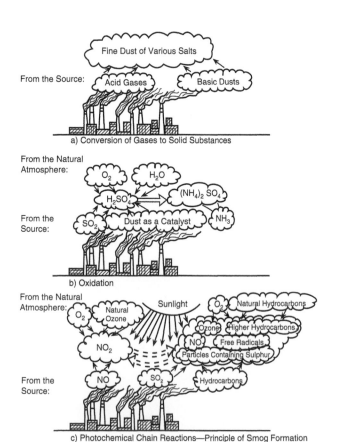

a) Conversion of Gases to Solid Substances

b) Oxidation

c) Photochemical Chain Reactions—Principle of Smog Formation

FIG. 1.4.1 Examples of chemical reactions in the atmosphere.

pended particles. At high humidities, these particles act as condensation nuclei or undergo hydration to become solution droplets. The oxidation then proceeds by absorption of both SO_2 and O_2 by the liquid aerosols with subsequent chemical reactions in the liquid phase. The oxidation slows considerably when the droplets become highly acidic because of the decreased solubility of SO_2. However if sufficient ammonia is present, the oxidation process is not impeded by the accumulation of H_2SO_4. Measurements of particulate composition in urban air often show large concentrations of ammonium sulfate.

PHOTOCHEMICAL REACTIONS

In the presence of air, SO_2 is slowly oxidized to SO_3 when exposed to solar radiation. If water is present, the SO_2 rapidly converts to sulfuric acid. Since no radiation wavelengths shorter than 2900 Å reach the earth's surface and the dissociation of SO_2 to SO and O is possible only for wavelengths below 2180 Å, the primary photochemical processes in the lower atmosphere following absorption by SO_2 involve activated SO_2 molecules and not direct dissociation. Thus, the conversion of SO_2 to SO_3 in clear air is a result of a several-step reaction sequence involving excited SO_2 molecules, oxygen, and oxides of sulfur other than SO_2. In the presence of reactive hydrocarbons and nitrogen oxides, the conversion rate of SO_2 to SO_3 increases markedly. In addition, oxidation of SO_2 in systems of this type is frequently accompanied by aerosol formation.

A survey of possible reactions by Bufalini (1971) and Sidebottom et al. (1972) concludes that the most important oxidation step for the triplet state $^3SO_2$ from among those involving radiation only is:

$$^3SO_2 + O_2 \xrightarrow{hr} SO_3 + O \quad (3400 \text{ to } 4000 \text{ Å}) \quad 1.4(2)$$

Other primary substances absorbing UV radiation include sulfur and nitrogen oxides and aldehydes. UV radiation excites the molecules of these substances, which then react with atmospheric molecular oxygen to yield atomic oxygen. Analogous to SO_2 oxidation, aldehydes react as follows:

$$HCHO + O_2 \xrightarrow{hr} HCOOH + O \quad 1.4(3)$$

Atomic oxygen can also be formed by the following reactions:

$$H_2S + O_2 \longrightarrow H_2O + S + O \quad 1.4(4)$$

$$NO + O_2 \longrightarrow NO_2 + O \quad 1.4(5)$$

$$CH_4 + O_2 \longrightarrow CH_3OH + O \quad 1.4(6)$$

$$C_2H_6 + O_2 \longrightarrow C_2H_4 + H_2O + O \quad 1.4(7)$$

$$CO + O_2 \longrightarrow CO_2 + O \quad 1.4(8)$$

SO_2 and aldehydes react irreversibly, whereby the amount of atomic oxygen formed by these processes is relatively small and corresponds to the amount of SO_2 and aldehydes in the atmosphere. In the reaction of nitrogen dioxide, however, the absorption of UV radiation leads to the destruction of one bond between the nitrogen and oxygen atoms and to the formation of atomic oxygen and nitrogen oxide. Further reactions lead to the formation of atomic oxygen and nitrogen oxide as follows:

$$NO_2 \xrightarrow{hr} NO + O \quad 1.4(9)$$

$$NO_2 + O \longrightarrow NO + O_2 \quad 1.4(10)$$

$$O + O_2 \longrightarrow O_3 \quad 1.4(11)$$

The regenerated nitrogen dioxide can reenter the reaction, and this process can repeat until the nitrogen dioxide converts into nitric acid or reacts with organic substances to form nitrocompounds. Therefore, a low concentration of nitrogen dioxide in the atmosphere can lead to the formation of a considerable amount of atomic oxygen and ozone. This nitrogen dioxide is significant in the formation of oxidation smog.

Olefins with a large number of double bonds also react photochemically to form free radicals. Inorganic substances in atomic form in the atmosphere also contribute to the formation of free radicals. On reacting with oxygen, some free radicals form peroxy compounds from which new peroxides or free radicals are produced that can cause polymerization of olefins or be a source of ozone. The photochemistry is described by the thirty-six reactions for the twenty-seven species in Table 1.4.2 which includes four reactive hydrocarbon groups: olefins, paraffins, aldehydes, and aromatics.

Particulates

Atmospheric reactions are strongly affected by the number of suspended solid particles and their properties. The particles supply the surfaces on which reactions can occur thus acting as catalysts. They can also affect the absorption spectrum through the adsorption of gases (i.e., in the wavelength range of adsorbed radiation) and thus affect the intensities of radiation absorption and photochemical reactions. Moreover, solid particles can react with industrially emitted gases in common chemical reactions.

Combustion, volcanic eruptions, dust storms, and sea spray are a few processes that emit particles. Many particulates in the air are metal compounds that can catalyze secondary reactions in the air or gas phase to produce aerosols as secondary products. Physical processes such as nucleation, condensation, absorption, adsorption, and coagulation are responsible for determining the physical properties (i.e., the number concentration, size distribution, optical properties, and settling properties) of the formed aerosols. Particles below 0.1 μ, (known as Aitken nuclei), although not significant by gravity, are capable of serving as condensation nuclei for clouds and fog.

TABLE 1.4.1 SUMMARY OF SOURCES, CONCENTRATIONS, AND SCAVENGING PROCESSES OF ATMOSPHERIC TRACE GASES

Contaminant	Major Pollutant Sources	Natural Sources	Estimated Annual Emissions Tg/yr* Pollutants	Estimated Annual Emissions Tg/yr* Natural	Atmospheric Background Concentrations	Estimated Atmospheric Residence Time	Removal Reactions and Sinks	Remarks
SO_2	Fossil fuel combustion	Volcanoes, reactions of biogenic S emissions	212[1]	20[2]	About 0.1 ppb[3]	1–4 days	Oxidation to sulfate by photochemical reactions or in liquid droplets	High reaction rates in summer due to photochemical processes
H_2S and organic sulfides[4]	Chemical processes, sewage treatment	Volcanoes, biogenic processes in soil and water	3 (as sulfur)	84[5] (as sulfur)	H_2S: 0.05–0.1 ppb; COS: 0.5 ppb[6]; CS_2: 0.05 ppb[6]	H_2S: 1–2 days; COS: 1–2 yr[6]	Oxidation to SO_2 and SO_4	Atmospheric data are incomplete; COS residence time can be 20 yr.[2]
CO	Auto exhaust, general combustion	Forest fires, photochemical reactions	700[7]	2100[7]	0.1–0.2 ppm (N. Hemisphere) 0.04–0.06 ppm (S. Hemisphere)	1–3 mon	Photochemical reactions with CH_4 and OH	No long-term changes in the atmosphere have been detected.
NO, NO_2	Combustion	Biogenic processes in soil, lightning	75[8] (as NO_2)	180[9] (as NO_2)	About 0.1 ppb[10]	2–5 days	Oxidation to nitrate	Natural processes mostly estimated; background concentrations are in doubt but may be as low as 0.01 ppb.
NH_3	Waste treatment, combustion	Biogenic processes in soil	6[11]	260[9]	About 10 ppm[9]	1–7 days	Reaction with SO_2 to form $(NH_4)_2SO_4$	Atmospheric measurements are sparse.
N_2O	Small amounts from combustion	Biogenic processes in soil	3[12]	340[13]	300 ppb	20–100 yr	Photochemical in stratosphere	Some estimates place natural source at 100 Tg or less.[12]
CH_4	Combustion, natural gas leakage	Biogenic processes in soil and water	160[14]	1050[14]	1.5 ppm	8 yr[15]	Reaction with OH to form CO	Pollutant source includes 60 Tg yr^{-1} from biomass burning.
Isoprene and terpenes	None	Biogenic plant emissions	None	830[7]	0	1–2 hr	Photochemical reactions with OH and O_3	Not found in ambient atmosphere away from source regions
Total nonCH$_4$ hydrocarbons	Combustion	Biogenic processes in soil and vegetation	40[17]	2×10^4 [16]	0–1 $\mu g\ m^{-3}$ for C_2's	Hours to a few days	Photochemical reactions with NO and O_3	Concentration given for C_2's in rural atmosphere

CO$_2$	Combustion	Biological processes	22,000[16]	10^6 [13]	345 ppm (1981)	2–4 yr	Biogenic processes, photosynthesis, absorption in oceans	Forest destruction and changes in earth's biomass may add 20–30 × 10^3 Tg CO$_2$/yr to atmosphere[18]
CH$_3$Cl	Combustion	Oceanic biological processes	2[19]	4–6[16,19]	600 ppt[19,20]	1–2 yr[19]	Stratospheric reactions	Photochemical reactions in stratosphere may impact on O$_3$ layer
HCl, Cl$_2$	Combustion, Cl manufacturing	Atmospheric reactions of NaCl, volcanoes	4[21]	100–200[23]	About 0.5 ppb[21]	About 1 wk	Precipitation	Volcanoes can release 10–20 Tg Cl yr^{-1} [22]

Source: Elmer Robinson, (Pullman, Wash.: Washington State University).

Notes: *Tg/yr = 10^{12} gm/yr or 10^6 metric tn/yr

[1]Based on 1978 global fuel usage and estimated sulfur contents.

[2]Major reference is R.D. Cadle, 1980, *Rev. Geophys. Space Phys.* 18, 746–752.

[3]P.J. Maroulis, A.L. Torres, A.B. Goldberg, and A.R. Bandy, 1980, *J. Geophys. Res.* 85, 7345–7349.

[4]Includes COS, CS$_2$, (CH$_3$)$_2$S, (CH$_3$)$_2$S$_2$, CH$_3$, and SH.

[5]Adapted from D.F. Adams, S.O. Farwell, E. Robinson, and M.R. Pack, 1980, *Biogenic sulfur emissions in the SURE region.* Final report by Washington State University for Electric Power Research Institute, EPRI Report No. EA-1516.

[6]A.L. Torres, P.J. Maroulis, A.B. Goldberg, and A.R. Bandy, 1980, *J. Geophys. Res.* 85, 7357–7360.

[7]P.R. Zimmerman, R.B. Chatfield, J. Fishman, P.J. Crutzen, and P.L. Hanst, 1978, *Geophys. Res. Lett.* 5, 679–682.

[8]Based on 1978 global combustion estimates.

[9]I.E. Galbally, *Tellus* 27, 67–70.

[10]Approximate value combining values given in several references.

[11]R. Söderlund, and B.H. Svensson, 1976, The global nitrogen cycle, in *SCOPE Report 7,* Swedish National Science Research Council, Stockholm.

[12]1978 fuel usage figures apply to the following references: R.F. Weiss, and H. Craig, *Geophys. Res. Lett.* 3, 751–753; and D. Pierotti, and R.A. Rasmussen, 1976, *Geophys. Res. Lett.* 3, 265–267.

[13]E. Robinson, and R.C. Robbins, Emissions, concentrations, and fate of gaseous atmospheric pollutants, in *Air pollution control,* edited by W. Strauss, 1–93, Part 2 of New York: Wiley.

[14]J.C. Sheppard, H. Westberg, J.F. Hopper, and K. Ganesan, 1982, *J. Geophys. Res.* 87, 1305–1312.

[15]L.E. Heidt, J.P. Krasnec, R.A. Lueb, W.H. Pollock, B.E. Henry, and P.J. Crutzen, 1980, *J. Geophys. Res.* 85, 7329–7336.

[16]R.E. Graedel, 1979, *J. Geophys. Res.* 84, 273–286.

[17]Reference 13 tabulation updated to approximate 1978 emissions.

[18]G.M. Woodwell, R.H. Whittaker, W.A. Reiners, G.E. Likens, C.C. Delwiche, and D.B. Botkin, 1978, *Science* 199, 141–146.

[19]R.A. Rasmussen, L.E. Rasmussen, M.A.K. Khalil, and R.W. Dalluge, 1980, *J. Geophys. Res.* 85, 7350–7356.

[20]E. Robinson, R.A. Rasmussen, J. Krasnec, D. Pierotti, and M. Jakubovic, 1977, *Atm. Environ.* 11, 213–215.

[21]J.A. Ryan, and N.R. Mukherjee, 1975, *Rev. Geophys. Space Phys.* 13, 650–658.

[22]R.D. Cadle, 1980, *Rev. Geophys. Space Phys.* 18, 746–752.

[23]Based on estimated reaction of NaCl to form Cl$_2$.

TABLE 1.4.2   GENERALIZED CHEMICAL KINETIC MECHANISM IN PHOTOCHEMICAL BOX MODEL

1.	$NO_2 \xrightarrow{h\nu} NO + O$	19.	$OLEF + O \longrightarrow RO_2 + ALD + HO_2$
2.	$O + O_2 + M \longrightarrow O_3 + M$	20.	$OLEF + O_3 \longrightarrow RO_2 + ALD + HO_2$
3.	$O_3 + NO \longrightarrow NO_2 + O_2$	21.	$OLEF + HO \longrightarrow RO_2 + ALD$
4.	$O_3 + NO_2 \longrightarrow NO_3 + O_2$	22.	$PARAF + HO \longrightarrow RO_2$
5.	$NO_3 + NO \longrightarrow 2NO_2$	23.	$ALD \xrightarrow{h\nu} 0.5RO_2 + 1.5HO_2 + 1.0CO$
6.	$NO_3 + NO_2 + H_2O \longrightarrow 2HONO_2$	24.	$ALD + HO \longrightarrow 0.5RlO_2 + 0.5HO_2 + HO_2$
7.	$HONO \xrightarrow{h\nu} HO + NO$	25.	$RO_2 + NO \longrightarrow RO + NO_2$
8.	$HO + NO \xrightarrow{(O_3)} HO_2 + CO_2$	26.	$RO + O_2 \longrightarrow ALD + HO_2$
9.	$HO_2 + NO_2 \longrightarrow HONO + O_2$	27.	$RlO_2 + NO_2 \longrightarrow PAN$
10.	$HO_2 + NO \longrightarrow HO + NO_2$	28.	$RO + NO_2 \longrightarrow RONO_2$
11.	$HO_2 + NO_2 + M \longrightarrow HOONO_2 + M$	29.	$RO_2 + O_3 \longrightarrow RO + 2O_2$
12.	$HOONO_2 \longrightarrow HO_2 + NO_2$	30.	$RlO_2 + NO \xrightarrow{(O_3)} RO_2 + NO_2$
13.	$HO + HONO \longrightarrow NO_2 + H_2O$	31.	$PAN \longrightarrow RlO_2 + NO_2$
14.	$HO + NO_2 + M \longrightarrow HONO_2 + M$	32.	$AROM + HO \xrightarrow{(O_3)} R_2O_2 + 2ALD + CO$
15.	$HO + NO + M \longrightarrow HONO + M$	33.	$R_2O_2 + NO \longrightarrow R_2O + NO_2$
16.	$HO_2 + O_3 \longrightarrow HO + 2O_2$	34.	$R_2O + O_2 \longrightarrow ALD + HO_2 + 2CO$
17.	$HO + O_3 \longrightarrow HO_2 + O_2$	35.	$R_2O_2 + O_3 \longrightarrow R_2O + 2O_2$
18.	$HO_2 + HO_2 \longrightarrow H_2O_2 + O_2$	36.	$RlO_2 + O_3 \longrightarrow RO_2 + 2O_2$

Source: K.L. Demerjian and K.L. Schere, 1979, *Proceedings, Ozone/Oxidants: Interactions with Total Environment II* (Pittsburgh: Air Pollution Association).

Note: M stands for any available atom or molecule which by collision with the reaction product carries off the excess energy of the reaction and prevents the reaction product from flying apart as soon as it is formed.

Secondary effects are the results of gas-phase chemistry and photochemistry that form aerosols.

The removal of particles (aerosols and dust) from the atmosphere involves dry deposition by sedimentation, washout by rainfalls and snowfalls, and dry deposition by impact on vegetation and rough surfaces.

A volcanic eruption is a point source which has local effects (settling of particles and fumes) and global effects since the emissions can circulate in the upper atmosphere (i.e., the stratosphere) and increase the atmospheric aerosol content.

From the point of view of atmospheric protection, some of these reactions are favorable as they quickly yield products that are less harmful to humans and the biosphere. However, the products of some reactions are even more toxic than the reactants, an example being peroxylacetyl nitrate.

The atmospheric chemical reactions of solid and gaseous substances in industrial emissions are complex. A deeper analysis and description is beyond the scope of this section.

Long-Range Planning

Other long-range problems caused by atmospheric chemical reactions occur in addition to those of sulfur and nitrogen compounds. States and provinces must formulate strategies to achieve oxidant air quality standards. They must assess both the transport of oxidants from outside local areas and the estimated influx of precursors that create additional oxidants. Lamb and Novak (1984) give the principal features of a four-layer regional oxidant model (see Figure 1.4.2) designed to simulate photochemical processes over time scales of several days and space scales of 1000 km. Temporal resolution yields hourly concentrations from time steps of 30 min and spatial resolution of about 18 km. The model includes the following processes:

- Horizontal transport
- Photochemistry using thirty-five reactions of twenty-three species
- Nighttime chemistry
- Nighttime wind shear, thermal stratification, and turbulent episodes associated with nocturnal jet
- Cumulus cloud effects, including venting from the mixed layer and photochemical reactions caused by their shadows
- Mesoscale vertical motion induced by terrain and horizontal divergence
- Mesoscale eddy effects on trajectories and growth rates of urban plumes
- Terrain effects on flow and diffusion
- Subgrid-scale chemical processes due to subgrid-scale emissions

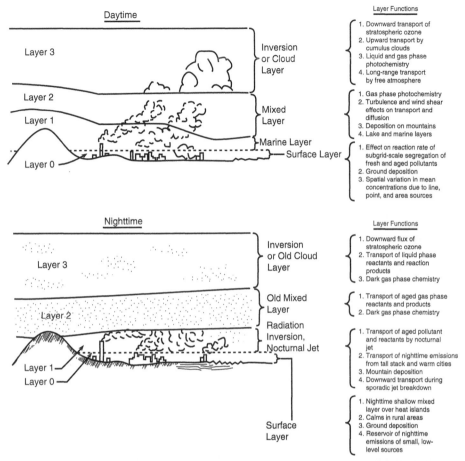

FIG. 1.4.2 Schematic diagram of the dynamic layer structure of the regional model. (Reprinted, with permission, from R.G. Lamb and J.H. Novak, 1984, *Proceedings of EPA–DECD International Conference on Long Range Transport Models for Photochemical Oxidants and Their Precursors*, EPA-600/9-84/006, Research Triangle Park, N.C.: U.S. EPA.)

- Natural sources of hydrocarbons and nitrogen oxides
- Wet and dry removal processes

The model was initially applied to the northeastern quarter of the United States. A 1980 emissions inventory gathered data on nitrogen oxides, VOCs, carbon dioxide, sulfur oxides, and total suspended particulate matter. In the model, volatile organics are considered as four reactive classes: olefins, paraffins, aldehydes, and aromatics. Applying the model requires acquiring and preparing emission and meteorological information for an area and a three- to four-month commitment of a person with knowledge of the model (Turner 1986).

—David H.F. Liu

References

Bufalini, M. 1971. The oxidation of sulfur dioxide in polluted atmospheres: A review. *Environ. Sci. Technol.* 5, no. 685.

Gartrell, F.E., F.W. Thomas, and S.B. Carpenter. 1963. Atmospheric oxidation of SO_2 in coal burning power plant plumes. *Am. Ind. Hygiene Assoc. J.* 24, no. 113.

Lamb, R.G., and J.H. Novak. 1984. *Proceedings of EPA-OECD International Conference on Long Range Transport Models for Photochemical Oxidants and Their Precursors.* EPA-600/9-84/006. Research Triangle Park, N.C.: U.S. EPA.

Sidebottom, H.W., C.D. Badcock, G.E. Jackson, J.G. Calvert, G.W. Reinhardt, and E.K. Damon. 1972. Photooxidation of sulfur dioxide. *Environ. Sci. Technol.* 6, no. 72.

Turner, D. Bruce. 1986. The transport of pollutants. Vol. VI in *Air pollution*, edited by Arthur C. Stern. Academic Press, Inc.

1.5
MACRO AIR POLLUTION EFFECTS

Macro air pollution effects refer to those consequences of air pollution exhibited on a large geographical scale, with the scale ranging from regional to global. Examples of such effects include acid rain, losses in the stratospheric ozone layer, and global warming.

Acid Rain Effects

Acid precipitation causes multiple effects on both terrestrial and aquatic ecosystems. Also, acid precipitation and dry deposition can affect materials and even human health. Demonstrated effects on terrestrial ecosystems include necrotic lesions on foliage, nutrient loss from foliar organs, reduced resistance to pathogens, accelerated erosion of the waxes on leaf surfaces, reduced rates of decomposition of leaf litter, inhibited formation of terminal buds, increased seedling mortality, and heavy metal accumulation (Cowling and Davey 1981). Soil and vegetation and crop-related effects include soil acidification, calcium removal, aluminum and manganese solubilization, tree growth reduction, reduction of crop quality and quantity, elimination of useful soil microorganisms, and selective exchange of heavy metal elements for more beneficial mono- and divalent cations (Glass, Glass, and Rennie 1979). Soil microbiological processes such as nitrogen fixation, mineralization of forest litter, and nitrification of ammonium compounds can be inhibited, the degree depending on the amount of cultivation and soil buffering capacity (Cowling and Davey 1981).

EFFECTS ON FORESTS

Field studies of the effects of acid precipitation on forests have been conducted in the United States and Europe. Reports of decreased growth and increased mortality of forest trees in areas receiving high rates of atmospheric pollutants emphasize the need to understand and quantify both the mechanisms and kinetics of changes in forest productivity. The complex chemical nature of combined pollutant exposures and the fact that these changes can involve both direct effects to vegetation and indirect and possibly beneficial effects mediated by a variety of soil processes make quantification of such effects challenging. However, evidence is growing on the severity of forest problems in central Europe due to acid precipitation. For example, in West Germany, fully 560,000 hectares of forests have been damaged (Wetstone and Foster 1983).

EFFECTS ON SOIL

Acid precipitation can affect soil chemistry, leaching, and microbiological processes. In addition, various types of soils exhibit a range of sensitivities to the effects of acid rain; for example, some soils are more sensitive than others. Factors influencing soil sensitivity to acidification include the lime capacity, soil profile buffer capacity, and water–soil reactions (Bache 1980). Wiklander (1980) reviews the sensitivity of various soils, and Peterson (1980) identifies soil orders and classifications according to their response to acid precipitation.

Two important effects of acid precipitation on soil are associated with changes in the leaching patterns of soil constituents and with the potential removal and subsequent leaching of chemical constituents in the precipitation. For example, Cronan (1981) describes the results of an investigation of the effects of regional acid precipitation on forest soils and watershed biogeochemistry in New England. Key findings include the following:

1. Acid precipitation can cause increased aluminum mobilization and leaching from soils to sensitive aquatic systems
2. Acid deposition can shift the historic carbonic acid/organic acid leaching regime in forest soils to one dominated by atmospheric H_2SO_4
3. Acid precipitation can accelerate nutrient cation leaching from forest soils and can pose a threat to the potassium resources of northeastern forested ecosystems
4. Progressive acid dissolution of soils in the laboratory is an important tool for predicting the patterns of aluminum leaching from soils exposed to acid deposition.

Soil microorganisms and microbiological processes can be altered by acid precipitation. The effects of acid precipitation include changes in bacterial numbers and activity, alterations in nutrient and mineral cycling, and changes in the decomposition of organic matter.

EFFECTS ON GROUNDWATER

As groundwater quality is becoming increasingly important, a concern is growing related to the effects of acid precipitation on quality constituents. Direct precipitation in recharge areas is of particular concern. The most pronounced effects are associated with increased acidity causing accelerated weathering and chemical reactions as the precipitation passes through soil and rock in the process of recharging an aquifer. The net effect on groundwater is reduced water quality because of increased mineralization.

EFFECTS ON SURFACE WATER

Acid precipitation causes many observable, as well as nonobservable, effects on aquatic ecosystems. Included are changes in water chemistry and aquatic faunal and floral species. One reason for changes in surface water chemistry is the release of metals from stream or lake sediments. For example, Wright and Gjessing (1976) note that concentrations of aluminum, manganese, and other heavy metals are higher in acid lakes due to enhanced mobilization of these elements in acidified areas.

Due to the extant water chemistry and sediment characteristics, some surface water is more susceptible to changes in water chemistry than others. Several surface water sensitivity studies leading to classification schemes have been conducted. For example, Hendrey et al. (1980) analyzed bedrock geology maps of the eastern United States to determine the relationship between geological material and surface water pH and alkalinity. They verified map accuracy by examining the current alkalinity and pH of water in several test states, including Maine, New Hampshire, New York, Virginia, and North Carolina. In regions predicted to be highly sensitive, the alkalinity in upstream sites was generally low, less than 200 microequivalents per liter. They pinpoint many areas of the eastern United States in which some of the surface water, especially upstream reaches, are sensitive to acidification.

Acid precipitation affects microdecomposers, algae, aquatic macrophytes, zooplankton, benthos, and fish (Hendry et al. 1976). For example, many of the 2000 lakes in the Adirondack Region of New York are experiencing acidification and declines or loss of fish populations. Baker (1981) found that, on the average, aluminum complexed with organic ligands was the dominant aluminum form in the dilute acidified Adirondack surface water studied. In laboratory bioassays, speciation of aluminum had a substantial effect on aluminum and hydrogen ions, and these ions appeared to be important factors for fish survival in Adirondack surface water affected by acidification.

EFFECTS ON MATERIALS

Acid precipitation can damage manmade materials such as buildings, metals, paints, and statuary (Glass, Glass, and Rennie 1980). For example, Kucera (1976) has reported data on the corrosion rates of unprotected carbon steel, zinc and galvanized steel, nickel and nickel-plated steel, copper, aluminum, and antirust painted steel due to sulfur dioxide and acid precipitation in Sweden. Corrosion rates are higher in polluted urban atmospheres than in rural atmospheres because of the high concentrations of airborne sulfur pollutants in urbanized areas. Economic damage is significant in galvanized, nickel-plated, and painted steel and painted wood.

EFFECTS ON HEALTH

Acid precipitation affects water supplies which in turn affects their users. Taylor and Symons (1984) report the results of the first study concerning the impact of acid precipitation on drinking water; the results report health effects in humans as measured by U.S. EPA maximum contaminant levels. The study sampled surface water and groundwater supplies in the New England states, but it also included other sites in the northeast and the Appalachians. No adverse effects on human health were demonstrated, although the highly corrosive nature of New England water may be at least partly attributable to acidic deposition in poorly buffered watersheds and aquifers.

Losses in Stratospheric Ozone Layer

The stratospheric ozone layer occurs from 12 to 50 km above the earth; the actual ozone concentration in the layer is in the order of ppmv (Francis 1994). Ozone can be both formed and destroyed by reactions with NO_x; of recent concern is the enhanced destruction of stratospheric ozone by chlorofluorocarbons (CFCs) and other manmade oxidizing air pollutants. The natural ozone layer fulfills several functions related to absorbing a significant fraction of the ultraviolet (uv) component of sunlight and terrestrial infrared radiation, and it also emits infrared radiation.

Several potential deleterious effects result from decreasing the stratospheric ozone concentration. Of major concern is increased skin cancer in humans resulting from greater UV radiation reaching the earth's surface. Additional potential concerns include the effects on some marine or aquatic organisms, damage to some crops, and alterations in the climate (Francis 1994). While environmental engineers are uncertain about all seasonal and geographic characteristics of the natural ozone layer and quantifying these effects, the effects are recognized via precursor pollutant control measures included in the 1990 CAAAs.

Precursor pollutants that reduce stratospheric ozone concentrations via atmospheric reactions include CFCs and nitrous oxide. Principal CFCs include methylchloroform and carbon tetrachloride; these CFCs are emitted to the atmosphere as a result of their use as aerosol propellants, refrigerants, foam-blowing agents, and solvents. Example reactions for one CFC (CFC-12) and ozone follow (Francis 1994):

$$CCl_2F_2 + UV \longrightarrow Cl\cdot + CClF_2 \qquad 1.5(1)$$

$$Cl\cdot + O_3 \longrightarrow ClO + O_2 \qquad 1.5(2)$$

Cl· denotes atomic chlorine. ClO is also chemically reactive and combines with atomic oxygen as follows:

$$ClO + O\cdot \longrightarrow Cl\cdot + O_2 \qquad 1.5(3)$$

Because of the preceding cycling of Cl·, environmental engineers estimate that a single Cl atom can destroy an average of 100,000 ozone molecules (Francis 1994).

Global Warming

The potential effects of global climate change can be considered in terms of ecological systems, sea-level rise, water resources, agriculture, electric demand, air quality, and health effects (Smith and Tirpak 1988). Since climate influences the location and composition of plants and animals in the natural environment, changes in climate have numerous consequences on ecological systems. One consequence includes shifts in forests in geographic range and composition; for example, the current southern boundary of hemlock and sugar maple in the eastern United States could move northward by about 400 mi (Smith and Tirpak 1988). An example of compositional change is that the mixed boreal and northern hardwood forest in northern Minnesota could become all northern hardwood.

ECOLOGICAL IMPACT

Biodiversity is also impacted by climate change, with *diversity* defined as the variety of species in ecosystems and the variability within each species. Examples of impact include an increased extinction of many species; growth or losses in freshwater fish populations depending on geographic location; and mixed effects on migratory birds, with some arctic-nesting herbivores benefiting and continental nesters and shorebirds suffering (Smith and Tirpak 1988).

Sea-level rises can cause increased losses of coastal wetlands, inundation of coastal lowlands, increased erosion of beaches, and increased salinity in estuaries. Coastal wetlands currently total 13,145 sq mi; with a 1-m rise in sea level, 26 to 66% of these wetlands can be lost, with the majority occurring in states bordering the Gulf of Mexico (Smith and Tirpak 1988).

IMPACT ON WATER RESOURCES

The main consequences of climatic changes to inland waters include the following (da Cunha 1988): (a) changes in the global amount of water resources and in the spatial and temporal distribution of these resources; (b) changes in soil moisture; (c) changes in extreme phenomena related to water resources, i.e., floods and droughts; (d) changes in water quality; (e) changes in sedimentation processes; and (f) changes in water demand.

The consequences of climatic change on water quality include possible changes in the precipitation regime and the occurrence of acid rain. Direct consequences of climatic changes on water quality occur. For example, temperature increases can decrease levels of dissolved oxygen in the water. Second, the biochemical oxygen demand (BOD) also increases with temperature. These two effects can decrease the dissolved oxygen concentration in a surface water system. Also, climatic changes can have indirect consequences on water quality since a decrease of river discharges, particularly during the dry season, can increase the concentration of pollutants in water bodies.

Climatic changes influence not only water availability but also water demand. For example, water demand for irrigation is largely affected by climatic change which conditions evapotranspiration. Water demands for domestic or industrial use are also affected by climatic change, for example, as a result of temperature increases that influence water consumption for cooling systems, bathing, washing, and gardening.

The simplest way to view the implications of global climate change on water resources is to consider the relationship between increasing atmospheric CO_2 and the hydrologic cycle; this relationship is shown in Figure 1.5.1. The following comments relate to the implications of Figure 1.5.1 (Waggoner and Revelle 1990):

1. When the atmosphere reaches a new and warmer equilibrium, more precipitation balances faster evaporation. This faster hydrologic cycle is predicted to raise the global averages of the up-and-down arrows of precipitation and evaporation in Figure 1.5.1 by 7 to 15%. However, the predicted change in precipitation is not a uniform increase. The net of precipitation and evaporation, that is, soil moisture, is predicted to fall in some places. Also, precipitation is predicted to increase in some seasons and decrease in others. Although global climatic models disagree on where precipitation will decrease and although they do not simulate the present seasonal change in precipitation correctly, disregarding the warning of a drier climate is unwise.

2. To be most usable for water resource considerations, frequency distributions of precipitation (and flood and drought projections) are needed. Models to develop such information are in their infancy.

3. A dimensionless elasticity for runoff and precipitation is:

$$[\text{Percentage change in runoff/percentage change in precipitation}] \quad 1.5(4)$$

If the elasticity for runoff and precipitation is greater than 1, the percentage change in runoff is greater than the percentage change in precipitation that causes it. Therefore, a general conclusion about the transformation of climate change into runoff change can be stated. Over diverse climates, the elasticities of percentage change in runoff to percentage change in precipitation and evaporation are 1 to 4. The percentage change in runoff is greater than the percentage change in the forcing factor.

Based on this brief review of water resource issues related to global climate change, the following summary comments can be made:

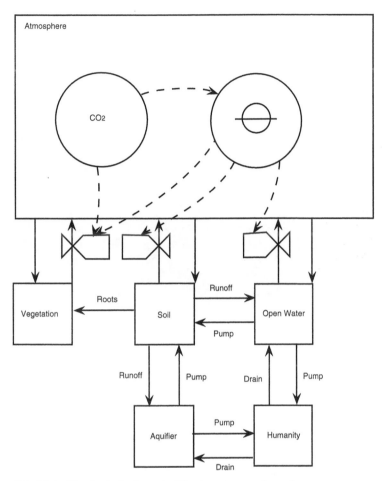

FIG. 1.5.1 The hydrologic cycle. The large rectangle at the top represents the water in the atmosphere. The rectangles beneath represent the water in vegetation, in the soil reached by roots, in the aquifers below, and in open bodies of water. The rectangle for humanity represents the water in people and pipes. Arrows represent fluxes. Valves are placed on three fluxes to show their control by CO_2. CO_2 directly affects transpiration from foliage by enlarging it through faster photosynthesis and by narrowing leaf pores. Indirectly, CO_2 warms the air temperature, speeding the evaporation from vegetation, soil, and open water. Faster evaporation decreases the water that runs off to bodies of water or into aquifers, increases pumping to irrigate soil, and even raises humanity's demands. The arrows from air down to vegetation, soil, and open water are the flux and addition to them by precipitation. Water resources are the pools signified by the rectangles for open water and aquifers. (Reprinted, with permission, from P.E. Waggoner and R.R. Revelle, 1990, Summary, Chap. 19 in *Climate change and U.S. water resources,* edited by P.E. Waggoner, New York: John Wiley and Sons.)

1. Global climate change caused by the greenhouse effect appears to be a reality although scientific opinion differs as to the rate of change.
2. Numerous inland and coastal water resource management issues are impacted when temperature and precipitation patterns change.
3. The global perspective presented in most studies is not specific enough to address the water resource implications in regional and local areas.
4. Global climate change might be controlled by effective programs to reduce the atmospheric emissions of greenhouse gases.
5. Global climate change must rise higher on national and international political agendas before effective societal measures can be implemented to control or manage the water resources implications.

IMPACT ON AGRICULTURE

Agricultural productivity in the United States is based on the temperate climate and rich soils. Global warming exhibits direct and indirect geographical effects on agricul-

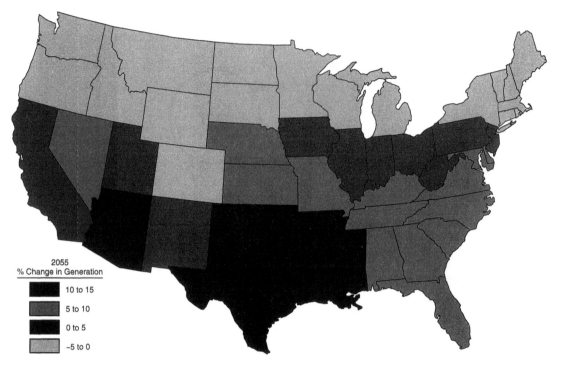

FIG. 1.5.2 Changes in electric generation by state induced by climate change to 2055. (Reprinted from J.B. Smith and D.A. Tirpak, 1988, *The potential effects of global climate change on the United States,* Washington, D.C.: U.S. EPA [October].)

tural productivity. Direct effects occur through changes in the length of the growing season, the frequency of heat waves, and altered patterns of rainfall; while indirect effects result from changes in topsoil management practices. Dryland yields of corn, wheat, and soybeans could decrease in many regions as a result of higher temperatures which shorten a crop's life cycle (Smith and Tirpak 1988). In contrast, in northern areas such as Minnesota, dryland yields of corn and soybeans could double as warmer temperatures extend the frost-free growing season. Crop acreage could decrease by 5 to 25% in Appalachia, the southeast, and the southern Great Plains areas; conversely, acreage could increase by 5 to 17% in the northern Great Lakes states, the northern Great Plains, and the Pacific northwest areas (Smith and Tirpak 1988). Irrigation would probably increase in many areas because irrigated yields are more stable than dry-land yields under conditions of increased heat stress and reduced precipitation.

Climate-sensitive electric end uses include space heating and cooling and, to a lesser degree, water heating and refrigeration. Summer cooling electric demands would increase, while winter heating demands would decrease. As a result of climate change, annual electric demands are expected to increase by 4 to 6% by the year 2055. Figure 1.5.2 summarizes regional demand changes. Additional power plants will be required to meet peak demands.

IMPACT ON AIR QUALITY

Global climate change also has implications for ambient air quality. Increased emissions of SO_x, NO_x, and CO are associated with power plants meeting increased electric demands. Plume rises from stacks would decrease due to higher ambient air temperatures and reduced buoyancy effects; these decreases would be manifested in higher ground-level concentrations of air pollutants located closer to their stacks. Air pollutant dispersion is also affected by weather variables such as windspeed and direction, temperature, precipitation patterns, cloud cover, atmospheric water vapor, and global circulation patterns (Smith and Tirpak 1988). Ozone pollution problems in many urban areas would worsen due to higher reaction rates at higher temperatures and lengthened summer seasons.

IMPACT ON HUMAN HEALTH

Human health effects are manifested by changes in morbidity and increases in mortality, particularly for the elderly during hotter and extended summer periods. Geographical patterns in relation to health effects are also expected.

—Larry W. Canter

References

Bache, B.W. 1980. The sensitivity of soils to acidification. In *Effects of acid rain precipitation on terrestrial ecosystems.* 569–572. New York: Plenum Press.

Baker, J.P. 1981. Aluminum toxicity to fish as related to acid precipitation and Adirondack surface water quality. Ph.D. diss. Cornell University, Ithaca, New York (January).

Cowling, E.B., and C.B. Davey. 1981. Acid precipitation: Basic principles and ecological consequences. *Pulp and Paper,* vol. 55, no. 8 (August): 182–185.

Cronan, C.S. 1981. Effects of acid precipitation on cation transport in New Hampshire forest soils. DOE/EV/04498-1. Washington, D.C.: U.S. Department of Energy (July).

da Cunha, L.V. 1988. Climatic changes and water resources development. *Symposium on Climate and Geo-Sciences: A Challenge for Science and Society in the 21st Century,* May 22–27, 1988, Louvain-la-Neuve, Belgium.

Francis, B.M. 1994. *Toxic substances in the environment.* 42–47. New York: John Wiley and Sons, Inc.

Glass, N.R., G.E. Glass, and P.J. Rennie. 1979. Effects of acid precipitation. *Environmental Science and Technology,* vol. 13, no. 11 (November): 1350–1352.

———. 1980. Effects of acid precipitation in North America. *Environmental International,* vol. 4, no. 5–6: 443–452.

Hendrey, G.R., et al. 1976. Acid precipitation: Some hydrobiological changes. *Ambio,* vol. 5, no. 5–6: 224–228.

Hendrey, G.R., et al. 1980. *Geological and hydrochemical sensitivity of the eastern United States to acid precipitation.* EPA-600/3-80-024. Washington, D.C.: U.S. EPA (January).

Kucera, V. 1976. Effects of sulfur dioxide and acid precipitation on metals and anti-rust painted steel. *Ambio,* vol. 5, no. 5–6: 248–254.

Petersen, L. 1980. Sensitivity of different soils to acid precipitation. In *Effects of acid rain precipitation on terrestrial ecosystems.* 573–577. New York: Plenum Press.

Smith, J.B., and D.A. Tirpak. 1988. *The potential effects of global climate change on the United States.* Washington, D.C.: U.S. EPA (October): 8–33.

Taylor, F.B., and G.E. Symons. 1984. Effects of acid rain on water supplies in the northeast. *Journal of the American Water Works Association,* vol. 76, no. 3 (March): 34–41.

Waggoner, P.E., and R.R. Revelle. 1990. Summary. Chap. 19 in *Climate change and U.S. water resources,* edited by P.E. Waggoner, 447–477. New York: John Wiley and Sons, Inc.

Wetstone, G.S., and S.A. Foster. 1983. Acid precipitation: What is it doing to our forests. *Environment,* vol. 25, no. 4 (May): 10–12, 38–40.

Wiklander, L. 1980. The sensitivity of soils to acid precipitation. In *Effects of acid rain precipitation on terrestrial ecosystems,* 553–567. New York: Plenum Press.

Wright, R.F., and E.T. Gjessing. 1976. Acid precipitation: changes in the chemical composition on lakes. *Ambio,* vol. 5, no. 5–6: 219–224.

1.6
METEOROLOGY

Clinging to the surface of the earth is a thin mantle of air known as the atmosphere (Figure 1.6.1). Calling the atmosphere thin may be confusing; however, 99% of the atmosphere mass lies within just 30 km (19 mi) of the earth's surface, and 90% of the atmosphere's mass lies within just 15 km (9 mi) of the surface.

The atmosphere is often classified in terms of temperature. Starting at the earth's surface and moving upward, temperature generally decreases with increasing altitude. This region, termed the *troposphere,* is of most interest to meteorologists because it is where weather and air pollution problems occur.

The *tropopause* is the boundary between the troposphere and the *stratosphere.* Below the tropopause, atmospheric processes are governed by turbulent mixing of air; but above it, they are not. In the stratosphere, temperature increases with height because of the high ozone concentration. Ozone absorbs radiation from the sun, resulting in an increase in stratospheric temperature.

The meteorological elements that have the most direct and significant effects on the distribution of air pollutants in the atmosphere are wind speed, wind direction, solar radiation, atmospheric stability, and precipitation.

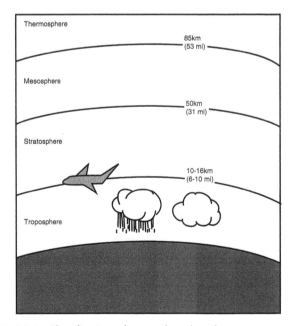

FIG. 1.6.1 Classification of atmosphere based on temperature. (This figure is not drawn to scale.)

Wind

The effects of wind on the distribution of air pollutants in the atmosphere involve understanding the scales of air motion, wind rose, and turbulence.

SCALES OF AIR MOTION

Wind is the motion of air relative to earth's surface. On the macroscale, the movement originates in the unequal distribution of atmospheric temperature and pressure over the earth's surface and is influenced by the earth's rotation. The direction of wind flow is characteristically from high pressure to low, but the Coriolis force deflects the air current out of these expected patterns (see Figure 1.6.2). These phenomena occur on scales of thousands of kilometers and are exemplified by the semipermanent high- and low-pressure areas over oceans and continents.

On the mesoscale and microscale, topographical features critically influence wind flow. Surface variations have an obvious effect on wind velocity and the direction of air flow. Monsoons, sea and land breezes, mountain–valley winds, coastal fogs, windward precipitation systems, and urban heat islands are all examples of the influence of regional and local topography on atmospheric conditions. Mesoscale phenomena occur over hundreds of kilometers; microscale phenomena, over areas less than 10 kilometers.

For an area, the total effect of these circulations establishes the hourly, daily, and seasonal variation in wind speed and direction. The frequency distribution of wind direction indicates the areas toward which pollutants are most frequently transported.

WIND ROSE

Wind speed determines the travel time of a pollutant from its source to a receptor and accounts for the amount of pollutant diffusion in the windward direction. Therefore, the concentration of pollutant at any receptor is inversely proportional to the wind speed. Wind direction determines

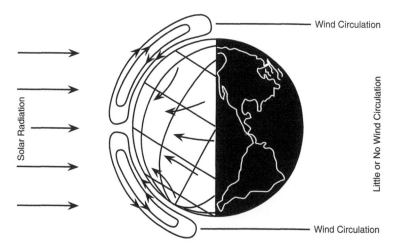

A. If the earth did not turn, air would circulate in a fixed pattern.

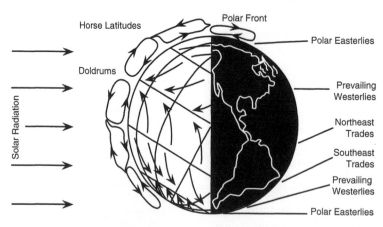

B. The earth turns, creating variable wind patterns.

FIG. 1.6.2 Global wind patterns. (Reprinted, with permission, from the American Lung Association.)

in what direction a pollutant travels and what receptor is affected at a given time. Wind direction is normally defined by a *wind rose*, a graphic display of the distribution of wind direction at a location during a defined period. The characteristic patterns can be presented in either tabular or graphic forms.

Wind speed is usually measured by an anemometer, which consists of three or four hemispherical cups arranged around a vertical axis. The faster the rotation of the cups, the higher the speed of the wind. A wind vane indicates wind direction. Although wind is three-dimensional in its movement, generally only the horizontal component is denoted because the vertical component is much smaller.

A wind rose is a set of wind statistics that describes the frequency, direction, force, and speed (see Figure 1.6.3). In this plot, the average wind direction is shown as one of the sixteen compass points, each separated by 22.5° measured from true north. The length of the bar for a direction indicates the percent of time the wind came from that direction. Since the direction is constantly changing, the time percentage for a compass point includes those times for wind direction at 11.25° on either side of the point. The percentage of time for a velocity is shown by the thickness of the direction bar. Figure 1.6.3 shows that the average wind direction from the southwest direction is 19% of the time and 7% of the time the southwesterly wind velocity is 16–30 mph.

Figure 1.6.4 shows the particulate fallout around an emission source and a wind rose based on the same time period.

The wind rose is imprecise in describing a point in a study region because the data are collected at one location in the region and not at each location. The data are often a seasonal or yearly average and therefore not accurate in

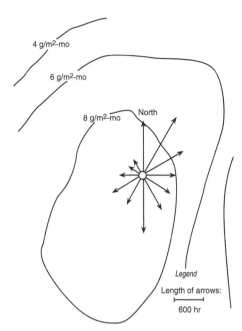

FIG. 1.6.4 Wind rose and corresponding particulate fallout pattern.

describing any point in time in an ideal representation of atmospheric diffusion. A final limitation of wind rose is that the wind is only measured in the horizontal plane and is assumed identical at any height above the earth's surface. (Note: wind speed generally increases with height in lower levels due to the decrease of the frictional drag effect of the underlying ground surface features.)

TURBULENCE

In general, atmospheric pollutants are dispersed by two mechanisms: wind speed and atmospheric turbulence. Atmospheric turbulence usually includes those wind flow fluctuations that have a frequency of more than 2 cycles/hr. The more important fluctuations have frequencies in the 1-to-0.01-cycles/sec range. Turbulent fluctuations occur randomly in both vertical and horizontal directions. This air motion provides the most effective mechanism to disperse or dilute a cloud or plume of pollutants. Figure 1.6.5 shows the dispersive effect of fluctuations in horizontal wind direction.

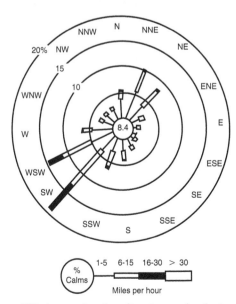

FIG. 1.6.3 Wind rose showing direction and velocity frequencies.

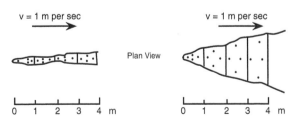

FIG. 1.6.5 Effect of wind direction variability or pollutant concentration from constant source. (Continuous emission of 4 units per sec.)

Turbulence is induced in air flow in two ways: by thermal current from heating below (*thermal turbulence*) and by disturbances or eddies from the passage of air over irregular, rough ground surfaces (*mechanical turbulence*). These small eddies feed large ones.

Generally turbulent motion and, in turn, the dispersive ability of the atmosphere, are enhanced during solar heating over rough terrain. Conversely, turbulence is suppressed during clear nights over smooth terrain.

Lapse Rates and Stability

Lapse rates, stability, and inversions also affect the dispersion of pollutants in the atmosphere.

LAPSE RATES

In the stratosphere, the temperature of the ambient air usually decreases with an increase in altitude. This rate of temperature change is called the *elapse rate*. Environmental engineers can determine this rate for a place and time by sending up a balloon equipped with a thermometer. The balloon moves through the air, not with it, and measures the temperature gradient of ambient air, called the *ambient lapse rate,* the *environmental lapse rate,* or the *prevailing lapse rate.*

Using the ideal-gas law and the law of conservation of energy, environmental engineers have established a mathematical ratio for expressing temperature change against altitude under adiabatic conditions (Petterssen 1968). This rate of decrease is termed the *adiabatic elapse rate,* which is independent of the prevailing atmospheric temperature.

Dry air, expanding adiabatically, cools at 9.8°C per km (or 5.4°F per 1000 ft), which is the dry adiabatic lapse rate (Smith 1973). In a wet as in a dry adiabatic process, a parcel of air rises and cools adiabatically, but a second factor affects its temperature. Latent heat is released as water vapor condenses within the saturated parcel of rising air. Temperature changes in the air are then due to the liberation of latent heat as well as the expansion of air. The wet adiabatic lapse rate (6°C/km) is thus less than the dry adiabatic lapse rate. Since a rising parcel of effluent gases is seldom completely saturated or completely dry, the adiabatic lapse rate generally falls somewhere between these two extremes.

STABILITY

Ambient and adiabatic lapse rates are a measure of atmospheric stability. Figure 1.6.6 shows these stability conditions. The atmosphere is *unstable* as long as a parcel of air moving upward cools at a slower rate than the surrounding air and is accelerated upward by buoyancy force.

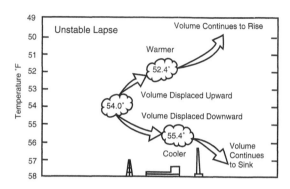

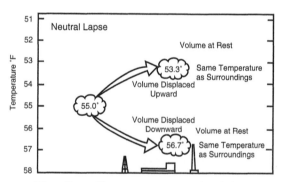

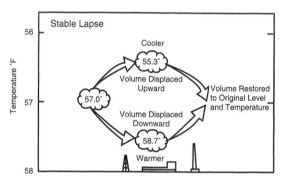

FIG. 1.6.6   The effect of lapse rate on vertical stability.

Moving downward, the parcel cools slower and is accelerated downward. Under these conditions, vertical air motions and turbulence are enhanced.

Conversely, when a rising parcel of air is cooler than the surrounding air, the parcel settles back to its original elevation. Downward movement produces a warmer parcel, which rises to its original elevation. Under these conditions, vertical movement is dampened out by adiabatic cooling or warming, and the atmosphere is *stable.*

Figure 1.6.7 shows that the boundary line between stability and instability is the dry adiabatic lapse line. When the ambient lapse rate exceeds the adiabatic lapse rate, the ambient lapse rate is termed *superadiabatic,* and the atmosphere is highly unstable. When the two lapse rates are equal, the atmosphere is *neutral.* When the ambient lapse rate is less than the dry adiabatic lapse rate, the ambient lapse rate is termed *subadiabatic,* and the atmosphere is

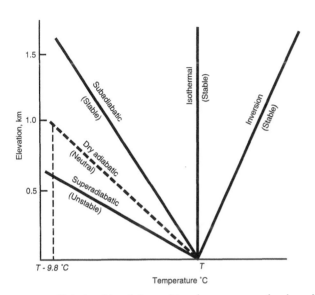

FIG. 1.6.7 Relationship of the ambient lapse rates to the dry adiabatic rate.

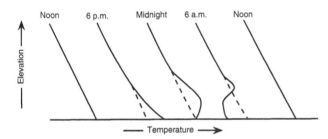

FIG. 1.6.8 Typical ambient lapse rates during a sunny day and clear night.

stable (Figure 1.6.8). If the air temperature is constant throughout a layer of atmosphere, the ambient lapse rate is zero, the atmosphere is described as *isothermal,* and the atmosphere is stable (Battan 1979).

When the temperature of ambient air increases (rather than decreases) with altitude, the lapse rate is negative, or inverted, from the normal state. A negative lapse rate occurs under conditions referred to as an *inversion,* a state in which warmer air blankets colder air. Thermal or temperature inversions represent a high degree of atmospheric stability (Battan 1979). An inversion is an extreme subadiabatic condition, thus almost no vertical air movement occurs.

INVERSIONS

Three types of inversions develop in the atmosphere: radiational (surface), subsidence (aloft), and frontal (aloft).

Radiational Inversions

A radiational inversion occurs at low levels, seldom above a few hundred feet, and dissipates quickly. This type of inversion occurs during periods of clear weather and light to calm winds and is caused by rapid cooling of the ground by radiation. The inversion develops at dusk and continues until the surface warms again the following day. Initially, only the air close to the surface cools, but after several hours, the top of the inversion can extend to 500 ft (see Figure 1.6.9). Pollution emitted during the night is caught under this "inversion lid."

Subsidence Inversions

A subsidence inversion is important in pollution control because it can affect large areas for several days. A subsidence inversion is associated with either a stagnant high-pressure cell or a flow aloft of cold air from an ocean over land surrounded by mountains (Cooper and Alley 1986). Figure 1.6.9 shows the inversion mechanism.

A significant condition is the subsidence inversion that develops with a stagnating high-pressure system (generally associated with fair weather). Under these conditions, the pressure gradient becomes progressively weaker so that winds become light. These light winds greatly reduce the horizontal transport and dispersion of pollutants. At the same time, the subsidence inversion aloft continuously descends, acting as a barrier to the vertical dispersion of the pollutants. These conditions can persist for several days, and the resulting accumulation of pollutants can cause serious health hazards.

Fog almost always accompanies serious air pollution episodes. These tiny droplets of water are detrimental in two ways: (1) fogs makes the conversion of SO_3 to H_2SO_4 possible, and (2) fogs sits in valleys and prevents the sun from warming the valley floor to break inversions, thus prolonging air pollution episodes.

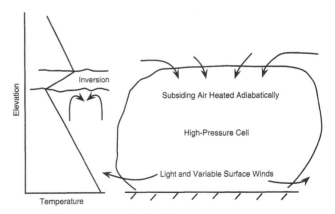

FIG. 1.6.9 Formation of subsidence inversion.

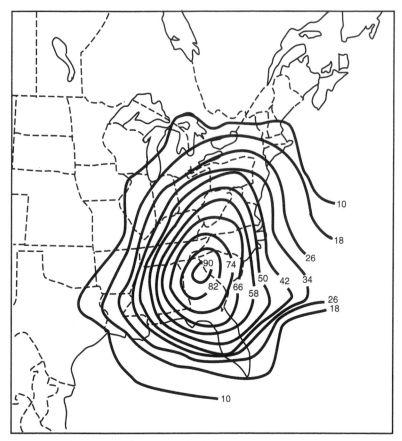

FIG. 1.6.10 Frequency of stagnating high-pressure cells over the eastern United States. The contours give the number of periods of four or more successive days between 1936–1965.

Figure 1.6.10 shows the frequency of stagnation periods of high-pressure cells over the eastern United States.

Frontal Inversions

A frontal inversion usually occurs at high altitudes and results when a warm air mass overruns a cold air mass below. This type of inversion is not important from a pollution control standpoint.

Precipitation

Precipitation serves an effective cleansing process of pollutants in the atmosphere as follows:

The washing out or scavenging of large particles by falling raindrops or snowflakes (washout)
The accumulation of small particles in the formation of raindrops or snowflakes in clouds (rainout)

The removal of gaseous pollutants by dissolution or absorption

The efficiencies of these processes depend on complex relationships between the properties of the pollutants and the characteristics of precipitation. The most effective and prevalent process is the washout of large particles in the lower layer of the atmosphere where most pollutants are released.

Topography

The topographic features of a region include both natural (hills, bridges, roads, canals, oceans, rivers, lakes, and foliages) and manmade (cities, bridges, roads, and canals) elements in a region. The prime significance of topography is its effects on meteorological elements, particularly the local or small-scale circulations that develop. These circu-

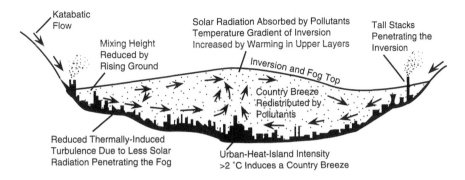

Katabatic Flow

Mixing Height Reduced by Rising Ground

Solar Radiation Absorbed by Pollutants
Temperature Gradient of Inversion Increased by Warming in Upper Layers

Tall Stacks Penetrating the Inversion

Inversion and Fog Top

Country Breeze Redistributed by Pollutants

Reduced Thermally-Induced Turbulence Due to Less Solar Radiation Penetrating the Fog

Urban-Heat-Island Intensity >2 °C Induces a Country Breeze

FIG. 1.6.11 Meteorology–pollution relationships during a smog in a valley location. (Reprinted, with permission, from D.M. Elsom, 1992, *Atmospheric pollution,* 2d ed., Oxford, U.K.: Blackwell Publishers.)

lations contribute either favorably or unfavorably to the transport and dispersion of the pollutants.

LAND–SEA BREEZE

In the daytime, land heats rapidly, which heats the air above it. The water temperature remains relatively constant. The air over the heated land surface rises producing low pressure compared with the pressure over water. The resulting pressure gradient produces a surface flow off the water toward land. This circulation can extend to a considerable distance inland. Initially, the flow is onto the land, but as the breeze develops, the Coriolis force gradually shifts the direction so that the flow is more parallel to the land mass. After sunset and several hours of cooling by radiation, the land mass is cooler than the water temperature. Then, the reverse flow pattern develops, resulting in a wind off the land. During a stagnating high-pressure system when the transport and dispersion of pollutants are reduced, this short-period, afternoon increase in airflow can prevent the critical accumulation of pollutants.

MOUNTAIN–VALLEY WINDS

In the valley region, particularly in winter, intensive surface inversions develop from air cooled by the radiationally cooled valley wall surfaces. Populated and industrialized bottom valley areas are subject to a critical accumulation of pollutants during this period.

Areas on the windward side of mountain ranges expect added precipitation because of the forced rising, expan-

sion, and cooling of the moving air mass with the resultant release of moisture. The precipitation increases the removal of pollutants.

URBAN-HEAT-ISLAND EFFECT

The increased surface roughness created by buildings throughout a city enhances the turbulence of airflow over the city, thus improving the dispersion of the pollutants emitted. However, at the same time, the city's buildings and asphalt streets act as a heat reservoir for the radiation received during the day. This heat plus the added heat from nighttime heating during cool months creates a temperature and pressure differential between the city and surrounding rural area so that a local circulation inward to the city develops. This circulation concentrates the pollutants in the city.

Figure 1.6.11 shows the combined effects of the urban-heat-island effect and katabatic winds.

—David H.F. Liu

References

Battan, L.J. 1979. *Fundamentals of meteorology.* Englewood Cliffs, N.J.: Prentice-Hall.

Cooper, C.D., and F.C. Alley. 1986. *Air pollution control—A design approach.* Prospect Heights, Ill.: Waveland Press, Inc.

Petterssen, S. 1968. *Air pollution.* 2d ed. New York: McGraw-Hill.

Smith, M.E., ed. 1973. *Recommended guide for the prediction of the dispersion of airborne effluents.* New York: ASME.

1.7
METEOROLOGIC APPLICATIONS IN AIR POLLUTION CONTROL

This section gives examples of meteorologic applications in air pollution control problems.

Air Pollution Surveys

Air pollution surveys are unique in their development and conduct. A common goal is to obtain a representative sample from an unconfined volume of air in the vicinity of one or more emission sources.

Depending on the objectives of an air pollution survey, a mobile or fixed sampler can be used. Other than the obvious considerations such as accessibility and the relationship to interfering pollutant sources, the principal factors in site selection are meteorology and topography. The controlling factor for site selection is wind movement. With some knowledge of the predominant wind direction, the environmental engineer can predict the path of pollution from the emission source to the point of ground-level impact and determine the most suitable location for an air monitoring site. The most convenient method for performing this analysis is to use the wind rose described in Section 1.6.

Besides wind direction and wind speed, other meteorological data necessary for sample correlations are temperature, cloud cover, and lapse rate where possible. The environmental engineer uses local temperatures to estimate the contribution of home heating to the total pollutant emission rates.

The simplest case is one where one wind direction predominates over a uniform topography for an isolated plant emitting a single pollutant that remains unchanged in the atmosphere. Two monitors are used: one monitors the effects of the source and the other is placed upwind to provide background concentrations. Where wind directions vary and other emission sources are operating nearby, the environmental engineer requires additional samples to identify the concentrations attributable to the source.

Environmental engineers often use a variation of the wind rose, called a *pollution rose*, to determine the source of a pollutant. Instead of plotting all winds on a radial graph, they use only those days when the concentration of a pollutant is above a minimum. Figure 1.7.1 is a plot of pollution roses. Only winds carrying SO_3 levels greater than 250 $\mu g/m$ are plotted. The fingers of the roses point to plant three. Pollution roses can be plotted for other pollutants and are useful for pinpointing sources of atmospheric contamination.

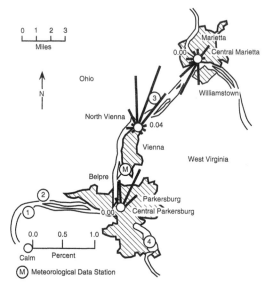

FIG. 1.7.1 Pollution roses, with SO_2 concentrations greater than 250 $\mu g/m$. The major suspected sources are the four chemical plants, but the data indicates that Plant Three is the primary culprit. (Reprinted, with permission, from P.A. Vesilind, 1983, *Environmental pollution and control*, Ann Arbor Science Publishers.)

Selection of Plant Site

In selecting a plant site, planners should consider the air pollution climatology of the area. They should prepare seasonal wind roses to estimate pollution dispersion patterns. Wind roses based on average winds excluding frontal weather systems are especially helpful. Planners must consider the frequency of stagnant weather periods and the effects of topography and local wind systems, such as land–sea breezes and mountain–valley winds, with respect to dispersion patterns and nearby residential and industrial areas.

The location of the plant within an area can depend on local wind speed and directions data. For example, residential areas may lie downwind of a proposed plant, in line with the prevailing wind direction. Considering a more suitable site would reduce the air pollution impact of the plant. Figure 1.7.2 illustrates this point.

Data on temperature, humidity, wind speed and direction, and precipitation are generally available through official weather agencies. Other potential sources of information are local airports, military installations, public utilities, and colleges and universities. The National Climatological Data Center, Asheville, North Carolina, is a major source of information. The center also contracts to prepare specific weather summaries and frequency.

Allowable Emission Rates

In plant planning, planners should consider local, state, and federal air pollution authorities, which can shutdown or curtail plant emission activities during times of air pollution emergency. Plants must have standby plans ready for reducing the emission of air contaminants into the atmosphere.

A plant must control emission rates to ensure that problems do not occur even during poor dispersion conditions. This control requires full knowledge of the frequency of poor dispersion weather. In addition, weather conditions should be considered when plant start ups are scheduled or major repairs that may produce more emissions are undertaken.

Stack Design

Section 1.4 presents stack design procedures and lists the weather parameters required. The stack height design must consider the average height at stack elevation, average temperature, average mixing conditions (stability), and average lapse rate. The stack height design must also consider the average height and frequencies of inversions. For emission sources such as generating stations, the ideal stack height should exceed the most frequent inversion height. Also, planners should consider not only the averages of

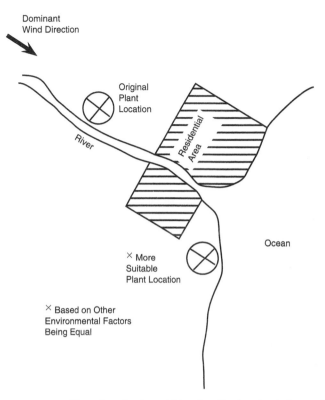

FIG. 1.7.2 Plant site selection within a localized region influenced by meteorological conditions.

temperature, wind speed, and stability, but also the frequency with which worst-case combinations of these parameters occur.

—David H.F. Liu

1.8
ATMOSPHERIC DISPERSION MODELING

The Gaussian Model

The goal of air quality dispersion modeling is to estimate a pollutant's concentration at a point downwind of one or more emission sources. Since the early 1970s, the U.S. EPA has developed several computer models based on the Gaussian (or normal) distribution function curve. The models were developed from the research of Turner (1964; 1970); Pasquill (1974; 1967), Gifford (1968; 1975), and others. The Gaussian-based model is effective for representing the plume diffusion for a range of atmospheric conditions. The technique applies the standard deviations of the Gaussian distribution in two directions to represent the characteristics of the plume downwind of its origin. The

plume's shape, and hence the standard deviations, varies according to different meteorological conditions. The following equation gives the ordinate value of the Gaussian distribution:

$$y = [1/(\sqrt{2\pi}\sigma)]\{\exp[(-1/2)(x - \bar{x}/\sigma)^2]\} \qquad 1.8(1)$$

which is depicted as a bell-shaped curve as shown in Figure 1.8.1.

The coordinate system used in models dealing with the Gaussian equation defines the x axis as downwind of the source, the y axis as horizontal (lateral) to the x axis, and the z axis as the vertical direction. The Gaussian lateral distribution can be restated as follows:

$$\chi(y) = [1/(\sqrt{2\pi}\sigma_y)]\{\exp[(-1/2)(y/\sigma_y)^2]\} \qquad 1.8(2)$$

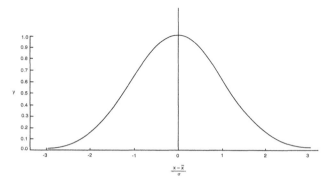

FIG. 1.8.1 The Gaussian distribution (or normal) curve. (Reprinted from D.B. Turner, 1970, *Workbook of atmospheric dispersion estimates (Revised),* Office of Air Programs Pub. No. AP-26, Research Triangle Park, N.C.: U.S. EPA.)

Shown in Figure 1.8.2.

A second, similar Gaussian distribution describes the distribution of the plume in the vertical, or z, direction. The distribution of the plume around the centerline in both the y and z directions can be represented when the two single distributions in each of the two coordinate directions are multiplied to give a double Gaussian distribution. Projecting this distribution downwind through x gives the volume of space that contains the plume as shown in Figure 1.8.3.

Shifting the centerline upward a distance H corrects the equation for emissions at the effective stack height (stack height plus plume rise above stack) as follows:

$$\chi(x,y,z;H) = [Q/(2\pi\sigma_y\sigma_z u)]\{\exp[(-1/2)(y/\sigma_y)^2]\}$$
$$\{\exp[(-1/2)(z - H/\sigma_z)^2]\} \quad 1.8(3)$$

where:

$\chi(x,y,z; H)$ = the downwind concentration at a point x,y,z, $\mu g/m^3$

Q = the emission rate of pollutants, g/s

σ_y, σ_z = the plume standard deviations, m

u = the mean vertical wind speed across the plume height, m/s

y = the lateral distance, m

z = the vertical distance, m

H = the effective stack height, m

As the plume propagates downwind, at some point the lowest edge of the plume strikes the ground. At that point, the portion of the plume impacting the ground is reflected upward since no absorption or deposition is assumed to occur on the ground (conservation of matter). This reflection causes the concentration of the plume to be greater in that area downwind and near the ground from the impact site. Functionally this effect can be mimicked, within the model with a virtual point source created identical to the original, emitting from a mirror image below the stack base as shown in Figure 1.8.4. Adding another term to the equation can account for this reflection of the pollutants as follows:

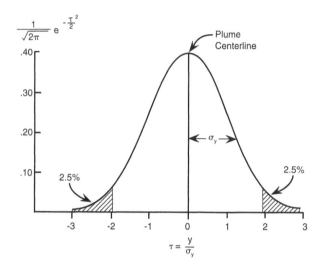

FIG. 1.8.2 Properties of the Gaussian distribution. Adapted, with permission, from H.A. Panofsky and J.A. Dutton, 1984. *Atmospheric Turbulence: Models and Methods for Engineering Applications,* John Wiley & Sons, New York.

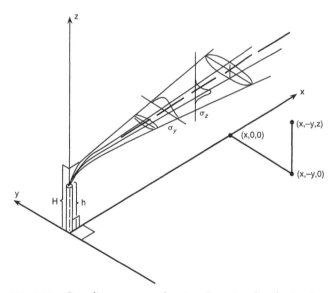

FIG. 1.8.3 Coordinate system showing Gaussian distribution in the horizontal and vertical. (Reprinted from D.B. Turner, 1970, *Workbook of atmospheric dispersion estimates (Revised),* Office of Air Programs Pub. No. AP-26, Research Triangle Park, N.C.: U.S. EPA.)

$$\chi(x,y,z; H) = [Q/(2\pi\sigma_y\sigma_z u)]\{\exp[(-1/2)(y/\sigma_y)^2]\}$$
$$\{\exp[(-1/2)(z - H/\sigma_z)^2] + \exp[(-1/2)(z + H/\sigma_z)^2]\} \quad 1.8(4)$$

When the plume reaches equilibrium (total mixing) in the layer, several more iterations of the last two terms can be added to the equation to represent the reflection of the plume at the mixing layer and the ground. Generally, no more than four additional terms are needed to approximate total mixing in the layer.

For the concentrations at ground level, z can be set equal to zero, and Equation 1.8(4) reduces as follows:

$$\chi(x,y,H) = [Q/(\pi\sigma_y\sigma_z u)]\{\exp[(-1/2)(y/\sigma_y)^2]\}$$
$$\{\exp[(-1/2)(H/\sigma_z)^2]\} \quad 1.8(5)$$

In addition, the plume centerline gives the maximum values. Therefore, setting y equal to zero gives the following equation:

$$\chi(x,H) = [Q/(\pi\sigma_y\sigma_z u)]\{\exp[(-1/2)(H/\sigma_z)^2]\} \quad 1.8(6)$$

which can estimate the concentration of pollutants for a distance x.

Finally, if the emission source is located at ground level with no effective plume rise, the equation can be reduced to its minimum as follows:

$$\chi(x) = Q/(\pi\sigma_y\sigma_z u) \quad 1.8(7)$$

A number of assumptions are typically used for Gaussian modeling. First, the analysis assumes a steady-state system (i.e., a source continuously emits at a constant strength; the wind speed, direction, and diffusion charac-teristics of the plume remain steady; and no chemical trans-formations take place in the plume). Second, diffusion in the x direction is ignored although transport in this direc-tion is accounted for by wind speed. Third, the plume is reflected up at the ground rather than being deposited, ac-cording to the rules of conservation of matter (i.e., none of the pollutant is removed from the plume as it moves downwind). Fourth, the model applies to an ideal aerosol or an inert gas. Particles greater than 20 μm in diameter tend to settle out of the atmosphere at an appreciable rate. More sophisticated EPA models consider this deposition, as well as the decay or scavenging of gases. Finally, the calculations are only valid for wind speeds greater than or equal to 1 m per sec.

Application of the Gaussian models is limited to no more than 50 km due to extrapolation of the dispersion coefficients. Other factors that influence the Gaussian dis-persion equations are ground roughness, thermal charac-teristics, and meteorological conditions. Plume dispersion

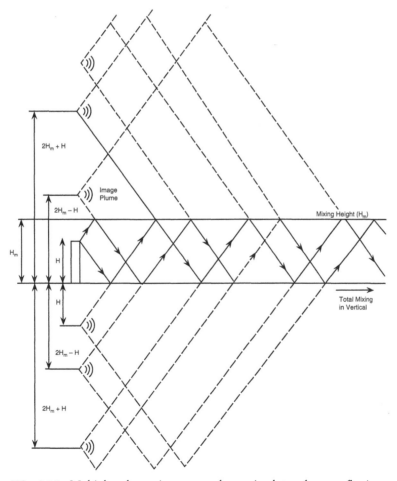

FIG. 1.8.4 Multiple plume images used to simulate plume reflections. (Reprinted from U.S. Environmental Protection Agency (EPA), 1987, *Industrial source complex (ISC) dispersion model user's guide—Second edition (revised)*, Vol. 1, EPA-450/4-88-002a, Research Triangle Park, N.C.: Office of Air Quality Planning and Standards, [December].)

tends to increase as each of these factors increases, with the models most sensitive to atmospheric stability.

PLUME CHARACTERISTICS

Boilers or industrial furnaces that have a well-defined stack and emissions resulting from products of combustion are the most common sources of pollutants modeled. The hot plume emitted from the stack rises until it has expanded and cooled sufficiently to be in volumetric and thermal equilibrium with the surrounding atmosphere. The height at which the plume stabilizes is referred to as the effective plume height (H) and is defined as:

$$H = h_s + \Delta H \qquad\qquad 1.8(8)$$

where (h_s) is the physical stack height portion and (ΔH) is the plume rise portion as shown in Figure 1.8.5. The plume rise is the increase in height induced by both the momentum and buoyancy effects of the plume. The momentum component of the rise is the physical speed at which the effluent is ejected from the stack, while the buoyancy component is due to the thermal characteristics of the plume in relation to ambient air. In the modeling, pollutants are assumed either to emit from a point directly above the stack at the effective plume height or to gradually rise over some distance downwind of the emission point until the effective plume height is reached.

The standard plume rise formula used in most EPA air dispersion models is based on a review of empirical data performed by Briggs (1969). (The next section includes a more detailed description.) The Briggs formula takes into account both momentum and buoyancy factors as well as meteorological conditions.

DISPERSION COEFFICIENTS

Sigma y (σ_y) and sigma z (σ_z) are the standard deviations of the Gaussian distribution functions. Since they are used in describing a plume as it disperses, they increase with time and distance traveled. The rates of growth for σ_y and σ_z depend upon meteorological conditions. Sigma y is generally larger than σ_z, since no stratification obstacles are in the y (horizontal) direction. Sigma y and σ_z are strongly influenced by heat convection and mechanical turbulence, and, as these forces become more pronounced, the sigmas increase more quickly.

As a standard deviation, σ_y and σ_z characterize the broadness or sharpness of the normal distribution of pollutants within the plume. As both sigmas increase, the concentration value of a pollutant at the plume centerline decreases. However, the total amount of the pollutant in the plume remains the same; it is merely spread out over a wider range and thus, the concentration changes. Approximately two thirds of the plume is found between plus or minus one sigma, while 95% of the plume is found between plus or minus 2 sigma as shown in Figure 1.8.2. The plume edge is considered to be that concentration that is one-tenth the concentration of the centerline.

Sigma y and σ_z are generally determined from equations derived using empirical data obtained by Briggs (1969) and McElroy and Pooler (1968) and research performed by Bowne (1974). Figures 1.8.6 and 1.8.7 show plots of these curves. These plots were developed from the observed dispersion of a tracer gas over open, level terrain.

STABILITY CLASSES

In the late 1960s, Pasquill developed a method for classifying atmospheric conditions which was later modified by

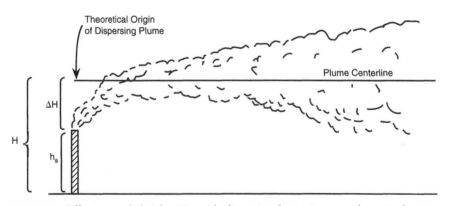

FIG. 1.8.5 Effective stack height (H), with dispersion beginning at a theoretical point above the stack. (Adapted with permission from American Society for Mechanical Engineers [ASME] Air Pollution Control Div., 1973, *Recommended guide for the prediction of the dispersion of airborne effluents*, 2d ed., New York: ASME.)

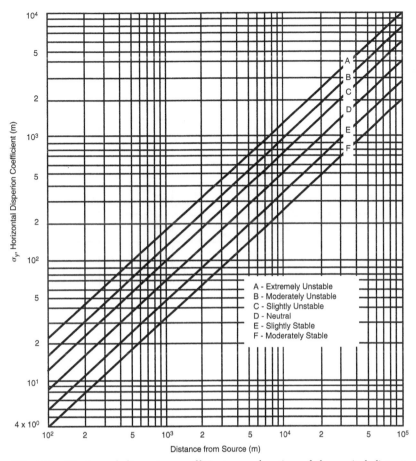

FIG. 1.8.6 Horizontal dispersion coefficient as a function of downwind distance from the source. (Reprinted from D.B. Turner, 1970, *Workbook of atmospheric dispersion estimates (Revised)*, Office of Air Programs Pub. No. AP-26, Research Triangle Park, N.C.: U.S. EPA.)

Gifford (1975), resulting in six stability classes, labeled A through F. The method was based on the amount of incoming solar radiation, cloud cover, and surface wind speed as shown in Table 1.8.1.

Stability greatly affects plume behavior as demonstrated by the dispersion curves discussed above. Classes E and F indicate stable air in which stratification strongly dampens mechanical turbulence, typically with strong winds in a constant wind direction. These conditions can produce a fanning plume that does not rise much and retains a narrow shape in the vertical dimension for a long distance downwind as shown in part (a) of Figure 1.8.8.

A situation where a plume in a stable layer is brought quickly to the surface by turbulence in a less stable layer is termed *fumigation* and is shown in part (b) of Figure 1.8.8. This can occur as the result of heat convection in the morning.

Class D stability is neutral, with moderate winds and even mixing properties. These conditions produce a coning plume as shown in part (c) of Figure 1.8.8. Classes A, B, and C represent unstable conditions which indicate various levels of extensive mixing. These conditions can produce a looping plume as shown in part (d). If the effective stack height exceeds the mixing height, the plume is assumed to remain above it, and no ground-level concentrations are calculated. This effect is known as lofting and is shown in part (e).

Rough terrain or heat islands from cities increase the amount of turbulence and change the classification of ambient conditions, usually upward one stability class.

In general, a plume under stability class A conditions affects areas immediately near the emission source with high concentrations. Class F stability causes the plume to reach ground level further away, with a lower concentration (unless terrain is a factor).

WIND SPEED

The wind speed is the mean wind speed over the vertical distribution of a plume. However, usually the only wind

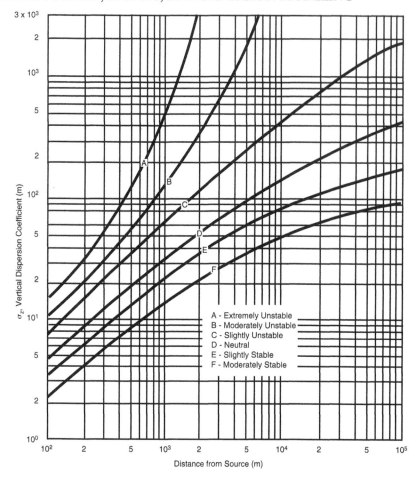

FIG. 1.8.7 Vertical dispersion coefficient as a function of downwind distance from the source. (Reprinted from Turner, 1970.)

TABLE 1.8.1   KEY TO PASQUILL STABILITY CATEGORIES

Surface Wind Speed (at 10 m) (m/sec)	Day			Night	
	Incoming Solar Radiation			Thinly Overcast or $\geq \frac{4}{8}$ Low Cloud	Clear or $\leq \frac{3}{8}$ Cloud
	Strong	Moderate	Slight		
<2	A	A–B	B		
2–3	A–B	B	C	E	F
3–5	B	B–C	C	D	E
5–6	C	C–D	D	D	D
>6	C	D	D	D	D

Source: D.B. Turner, 1970, *Workbook of atmospheric dispersion estimates* (Revised), Office of Air Programs Pub. No. AP-26 (Research Triangle Park, N.C.: U.S. EPA).

Note: The neutral class *D* should be assumed for overcast conditions during day or night.

speed available is that monitored at ground-level meteorological stations. These stations record ambient atmospheric characteristics, usually at the 10-m level, and typically with lower wind speeds than those affecting the plume. These lower speeds are due to the friction caused by the surface as shown in Figure 1.8.9. Therefore, the wind speed power law must be used to convert near-surface wind speed data into a wind speed representative of

the conditions at the effective plume height. The wind speed power law equation is as follows:

$$u_2 = u_1 * (z_2/z_1)^p \qquad 1.8(9)$$

where u_1 and z_1 correspond to the wind speed and vertical height of the wind station, while u_2 and z_2 pertain to the characteristics at the upper elevation. This formula is

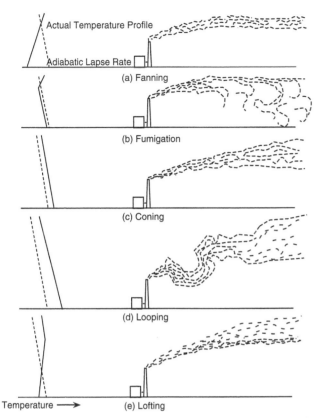

FIG. 1.8.8 Effect of temperature profile on plume rise and diffusion. (Reprinted from G.A. Briggs, 1969, *Plume rise*, U.S. Atomic Energy Commission Critical Review Series T10-25075. Clearinghouse for Federal Scientific and Technical Information.)

empirical, with the exponent derived from observed data. The exponent (p) varies with the type of ambient weather conditions, generally increasing with stability and surface roughness (Irwin 1979). It can range from 0.1 for calm conditions to 0.4 for turbulent weather conditions. Table 1.8.2 shows exponents for various types of surface characteristics. Table 1.8.3 shows selected values for both urban and rural modes as used for the six stability categories in the industrial source complex (ISC3) models.

Plume Rise and Stack Height Considerations

In dispersion modeling, the plume height is critical to the basic equation for determining downwind concentrations at receptors. Several factors affect the initial dispersion of the plume emitted from a stack, including the plume rise, the presence of buildings or other features disturbing the wind stream flow, and the physical stack height. This section discusses these factors.

PLUME RISE

Various attempts have been made to estimate the plume rise from stationary sources. Two types of equations have resulted: theoretical and empirical. Theoretical models are generally derived from the laws of buoyancy and momentum. They are often adjusted for empirical data. Empirical models are developed from large amounts of observed data such as tracer studies, wind tunnel experi-

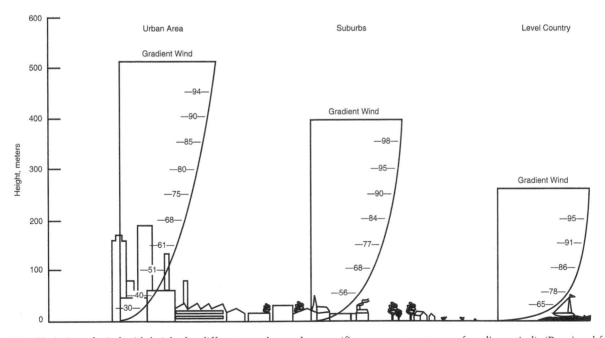

FIG. 1.8.9 Variation of wind with height for different roughness elements (figures are percentages of gradient wind). (Reprinted from D.B. Turner, 1970, *Workbook of atmospheric dispersion estimates (Revised)*, Office of Air Programs Pub. No. AP-26, Research Triangle Park, N.C.: U.S. EPA and based on A.G. Davenport, 1963, the relationship of wind structure to wind loading. Presented at Int. Conf. on the Wind Effects on Buildings and Structures, Nat. Physical Laboratory, Teddington, Middlesex, England, 26-28 June.)

TABLE 1.8.2 EXPONENTS FOR POWER LAW VELOCITY PROFILE EQUATION

Surface Configuration	Stability	p
Smooth open country	Unstable	0.11
	Neutral	0.14
	Moderate stability	0.20
	Large stability	0.33
Nonurban—varying roughness and terrain	Daytime—unstable and neutral	0.1–0.3
	Nighttime—stable including inversion	0.2–0.8
Urban (Liverpool)	Unstable, $\Delta\theta^a < 0$	0.20
	Neutral, $\Delta\theta = 0$	0.21
	Stable, $0 \le \Delta\theta < 0.75$	$0.21 + 0.33\ \Delta\theta$
Flat open country, $z_G^b = 274$ m	Neutral	0.16
Woodland forest, $z_G = 396$ m	Neutral	0.28
Urban area, $z_G = 518$ m	Neutral	0.40

Source: Gordon H. Strom, 1976, Transport and diffusion of stack effluents, Vol. 1, in *Air pollution,* 3d ed., edited by Arthur C. Stern, p. 412 (New York: Academic Press).

Notes: [a]$\Delta\theta$ = Potential temperature difference between 162- and 9-m elevations.
   [b]z_G = Height of planetary boundary layer to gradient wind.

TABLE 1.8.3 WIND PROFILE EXPONENTS IN ISC3 DISPERSION MODELS

Stability Category	Rural Exponent	Urban Exponent
A	0.07	0.15
B	0.07	0.15
C	0.10	0.20
D	0.15	0.25
E	0.35	0.30
F	0.55	0.30

Source: U.S. Environmental Protection Agency (EPA), 1987, *Industrial source complex (ISC) dispersion model user's guide—Second edition (Revised),* Vol. 1, EPA-450/4-88-002a (Research Triangle Park, N.C.: Office of Air Quality Planning and Standards [December]).

ments, and photographic evidence. Most plume rise equations apply to uniform or smoothly varying atmospheric conditions. An important consideration is that while they often predict the plume rise reasonably well under similar conditions, they can give wrong answers for other conditions.

Momentum and Buoyancy Factors

The plume rises mainly due to two factors: 1) the velocity of the exhaust gas, which imparts momentum to the plume, and 2) the temperature of the exhaust gas, which gives the plume buoyancy in ambient air. The momentum flux comes from mechanical fans in duct systems and the natural draft that occurs in the stack. If the gas in the plume is less dense than the surrounding atmosphere, this condition also causes the plume to rise and adds to its upward momentum. The plume rise also depends upon the growth of the plume, which is caused by turbulence, whether pre-existing in the atmosphere or induced by interaction with the plume. One method of increasing the upward momentum of the plume is to constrict the exit diameter of the stack.

The buoyancy flux is an effect of the plume's increased temperature. For most large combustion sources (such as power plants), the buoyancy flux dominates the momentum flux. Buoyant plumes contribute to both the vertical and horizontal velocity of the plume, in addition to that caused by ambient turbulent levels. This condition is caused by entrainment of the surrounding air into the expanding plume in relation to its surroundings. Buoyancy can also affect both σ_y and σ_z because of buoyancy-induced atmospheric turbulence.

Meteorological Factors

Some meteorological factors affecting plume rise include wind speed, air temperature, atmospheric stability, turbulence, and wind shear at the stack height. As these conditions vary, the rise of the plume is enhanced or reduced. The final effective plume height is reached only at some point downwind of the stack; this condition is known as gradual plume rise. Many dispersion models (especially where terrain is an important consideration) incorporate special algorithms to analyze ambient concentrations in the region before the final rise is attained. However, models usually apply the final effective stack height at the point above the source to simplify calculations.

An early equation to predict plume rise was developed by Holland (1953) for the Atomic Energy Commission. This equation accounted for momentum and buoyancy by having a separate term for each. A major problem with the equation was that it did not account for meteorological effects. At the time of his work, Holland recognized that the formula was appropriate only for class D stability conditions.

BRIGGS PLUME RISE

Briggs (1969) developed a plume rise formula that is applicable to all stability cases. The Briggs plume rise formula is now the standard used by the EPA in its dispersion models, but it is not as straightforward as Holland's approach. For the Briggs equation, determining the buoyancy flux, F_b, is usually necessary using the following equation:

$$F_b = gv_sd_s^2[(\Delta T)/4T_s] \qquad 1.8(10)$$

where:

g = the gravitational constant (9.8 m/s^2)
v_s = the stack gas exit velocity, m/s
d_s = the diameter of the stack, m
T_s = the stack gas temperature, deg. K
ΔT = the difference between T_s and T_a (the ambient temperature), deg. K

When the ambient temperature is less than the exhaust gas temperature, it must be determined whether momentum or buoyancy dominates. Briggs determines a crossover temperature difference $(\Delta T)_c$ for $F_b \geq 55$, and one for $F_b < 55$. If ΔT exceeds $(\Delta T)_c$, then a buoyant plume rise algorithm is used; if less, then a momentum plume rise equation is employed. The actual algorithms developed by Briggs to calculate the plume rise further depend upon other factors, such as the atmospheric stability, whether the plume has reached the distance to its final rise (i.e., gradual rise), and building downwash effects (U.S. EPA 1992d).

DOWNWASH

All large structures distort the atmosphere and interfere with wind flow to some extent. These atmospheric distortions usually take the form of a wake, which consists of a pocket of slower, more turbulent air. If a plume is emitted near a wake, it is usually pulled down because of the lower pressure in the wake region. This effect is termed *downwash*. When downwash occurs, the plume is brought down to the ground near the emission source more quickly.

A wake that causes downwash usually occurs as the result of one of three physical conditions: 1) the stack, referred to as stack-tip downwash, 2) local topography, or 3) nearby large structures or building downwash. Figure 1.8.10 shows examples of each of these conditions.

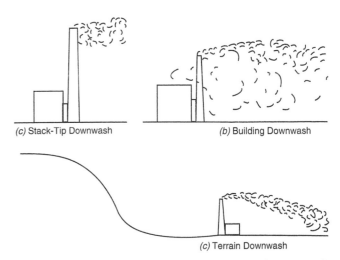

(c) Stack-Tip Downwash   *(b) Building Downwash*

(c) Terrain Downwash

FIG. 1.8.10 Physical conditions that cause downwash. Reprinted from G.A. Briggs, 1969. *Plume rise,* U.S. Atomic Energy Commission Critical Review Series TID-25075, Clearinghouse for Federal Scientific and Technical Information.

Stack-Tip Downwash

Stack-tip downwash occurs when the ambient wind speed is high enough relative to the exit velocity of the plume so that some or all of the plume is pulled into the wake directly downwind of the stack, as shown in Figure 1.8.11. This downwash has two effects on plume rise. First, the pollutants drawn into the stack wake leave the stack region at a lower height than that of the stack and with a lower upward velocity. Second, the downwash increases the plume cross section, which decreases the concentration.

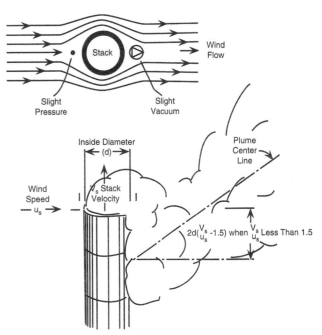

FIG. 1.8.11 Stack-tip downwash. (Reprinted, with permission, from Trinity Consultants, Inc., 1989, *Atmospheric diffusion notes,* Issue no. 13, Dallas, Tex. [June].)

To avoid stack-tip downwash, environmental engineers should consider the ratio of emission velocity (v_s) to wind speed velocity at the stack height (u_s) in the stack design. If $v_s < 1.5\ u_s$, then the physical stack height should be adjusted by the following equation:

$$h_{std} = 2d_s[(v_s/u_s) - 1.5] \qquad 1.8(11)$$

If $v_s/u_s > 1.5$, then stack-tip downwash is avoided since the exhaust gas is emitted from the stack at sufficient velocity to clear the downwash area on the downwind side of the stack. If $v_s/u_s < 1.0$, downwash will probably occur, possibly seriously. For intermediate values of v_s/u_s, downwash may occur depending on the ambient conditions at the time.

Downwash Caused by Topography

Downwash can also be caused by local topography. Large hills or mountains can change the normal wind patterns of an area. If the stack is located closely downwind of a hill above stack height, the air flowing off the hill can cause the plume to impact closer to the stack than normal as shown in part (c) of Figure 1.8.10. Modeling of these situations often employs physical models in wind tunnels. A recently developed model, the complex terrain dispersion model plus algorithms for unstable situations (CTDM-PLUS), employs a critical hill height calculation that determines if the plume impacts the hill or follows the uninterrupted laminar flow around it (U.S. EPA 1989).

Building Downwash

Large structures surrounding the stack also affect ambient wind conditions. The boundaries of the wake region resulting from surrounding structures are not sharply defined. They depend on the three-dimensional characteristics of the structure and are time dependent. The extent of distortion depends extensively upon building structure geometry and wind direction. Generally, a single cylindrical structure (e.g., a free-standing silo) has little influence on the wind flow compared to a rectangular structure.

Part (b) in Figure 1.8.10 shows the building downwash that occurs when the plume is drawn into a wake from a nearby structure. Two zones exist within the downwash area of a structure. The first zone, which extends approximately three building heights downwind, is the cavity region where plume entrapment can occur. The second zone, which extends from the cavity region to about ten times the lesser dimension of the height or projected width, is the wake region where turbulent eddies exist as a result of structure disturbance to the wind flow. Figure 1.8.12 shows an example of these zones. Generally, the cavity region concentration is higher than the wake region concentration due to plume entrapment. Bittle and Borowsky (1985) examined the impact of pollutants in cavity zones

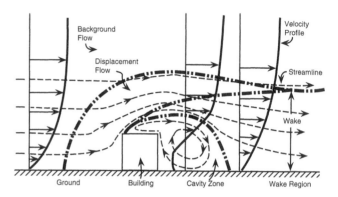

FIG. 1.8.12 Building structure downwash with cavity zone and wake effect. (Reprinted, with permission, from American Society of Mechanical Engineers [ASME].)

and found several calculations apply depending on the building and stack geometry. Beyond these zones, wind flow is unaffected by the structure.

The first downwash calculations were developed as the result of studies in a wind tunnel by Snyder and Lawson (1976) and Huber (1977). However, these studies were limited to a specific stability, structure shape, and orientation to the wind. Additional work by Hosker (1984), Schulman and Hanna (1986), and Schulman and Scire (1980) refined these calculations. Figure 1.8.13 shows the areas where the Huber–Snyder and Schulman–Scire downwash calculations apply. The following equation determines whether the Huber–Snyder or Schulman–Scire downwash calculations apply:

$$h_s = H + 0.5L \qquad 1.8(12)$$

where:

h_s = the physical stack height
H = the structure height
L = the lesser dimension of the height or projected width

The adjustments are made to the dispersion parameters.

To avoid building downwash, the EPA has developed a general method for designing the minimum stack height needed to prevent emissions from being entrained into any wake created by the surrounding buildings. In this way, emissions from a stack do not result in an excessive concentration of the pollutant close to the stack. This approach is called the good engineering practice (GEP) of stack height design (U.S. EPA 1985) and retroactively covers all stacks built since December 31, 1970. The following equation determines the GEP stack height:

$$H_{GEP} = H + 1.5(L) \qquad 1.8(13)$$

where:

H_{GEP} = the GEP stack height
H = the maximum height of an adjacent or nearby structure
L = the lesser dimension of the height or projected width of an adjacent or nearby structure

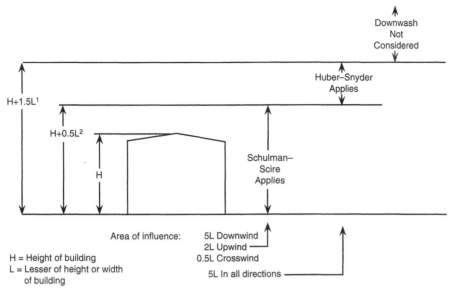

¹Compare H+1.5L to plume elevation (stack height above grade + momentum plume rise 2L
downwind-stack-tip downwash)

²Compare H+.5L to physical stack height above grade

FIG. 1.8.13 Selection of downwash algorithms in the ISCST model. (Reprinted, with permission, from Trinity Consultants, Inc., 1989, *Atmospheric diffusion notes*, Issue no. 13, Dallas, Tex. [June].)

The projected width of a structure is the exposed area perpendicular to the wind as shown in Figure 1.8.14. Environmental engineers should check all structures within 5L of the stack for their possible effect on the plume. The range of influence for a given structure is defined by Tickvart (1988) as 2L upwind of the structure, 5L downwind, and 0.5L on the sides parallel to the wind flow.

TALL STACKS

If concentration impacts are excessive, constructing a tall stack is one approach to reduce them. A tall stack dilutes ambient ground-level concentrations near the emission source. However, this approach does not reduce emission levels or total pollution loadings in a region; it merely provides greater initial dispersion at the source. The EPA regulates stack height under its tall stacks policy to encourage better control technology application.

Tall stacks describes stacks that are greater than the GEP stack height. For stacks which are taller than GEP guidelines, dispersion modeling is conducted as if the emission source has a GEP stack height. When the stack height is at the GEP level, downwash is not likely to occur, and modeling can proceed without further changes. If the proposed or existing stack is less than the GEP stack height, surrounding structures must be investigated as possible downwash sources for modifying dispersion parameters in the air dispersion model. Under GEP guidelines, however, a source that has a GEP stack height less than 65 m can

raise it to that level (and still be considered GEP).

A consideration for building a tall stack is the cost of construction; investments for tall stacks usually start at $1 million, and costs of $4–5 million are not uncommon (Vatavuk 1990). Stacks of this size typically consist of an outer shell and a liner. The outer shell is usually constructed of concrete, while the liner is usually steel or acid-resistant brick. The choice of liner depends on whether the exhaust gas is above the acid dew point (steel above it, brick below). Given these costs, building a stack taller than GEP is rare; however, constructing a stack below GEP and conducting the additional downwash modeling required may be worthwhile. In some cases, if the environmental engineer is involved early in the design phase of the project, designing the height and shape of buildings and nearby structures to lessen the need for a taller stack is possible.

Computer Programs for Dispersion Modeling

Air quality dispersion models are useful tools for determining potential concentration impacts from proposed as well as existing sources. The models can be categorized into four general classes: Gaussian, numerical, statistical (empirical), and physical. The first three models are computer based, with numerical and Gaussian models dominating the field. This section focuses on Gaussian-based models since they are the most widely applied. This wide application is almost entirely due to their ease of applica-

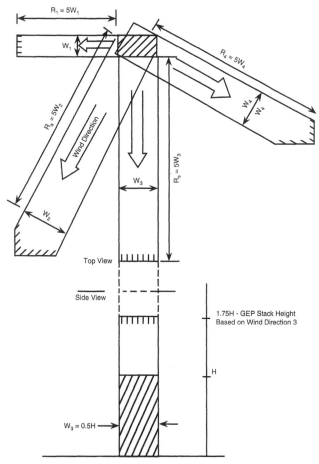

FIG. 1.8.14 GEP determination of projected structure width and associated region of adverse influence for a structure in four different wind directions. (Reprinted from U.S. Environmental Protection Agency (EPA), 1985, *Guideline for determination of good engineering practice stack height [Technical support document for the stack height regulations] [Revised]*, EPA-450/4-80-023R, Research Triangle Park, N.C.: U.S. EPA, Office of Air Quality Planning and Standards [June].)

tion and the conservative estimates they provide, despite any of their shortcomings in precisely describing a plume's diffusion in the atmosphere.

Gaussian-based models generally require three types of input data: source emission data, receptor data, and meteorological data, though the latter two can be assumed in some cases. Source emission data provide the characteristics of the pollutant released to the atmosphere. Receptor data provide the location where a predicted concentration is desired. Meteorological data provide the conditions for the model to determine how the emissions are transported from the source to the receptor.

GUIDELINE MODELS

To promote consistency in applying models, the U.S. EPA has developed a document entitled *Guideline on Air Quality Models* (1978; 1986; 1987; 1993a; 1996). In this document, the EPA summarizes the performance of models in several comparative analyses and suggests the best applications of the model. As a result of the EPA's evaluations, those that perform well for a general set of conditions are classified as *Appendix A* models, and these models are listed in Table 1.8.4. Of the models listed, only the urban airshed model (UAM) and the emission and dispersion model (EDMS) are not Gaussian-based.

Models not classified as Appendix A but recognized as having potential application for a specific case can be used pending EPA approval. These models are designated as Appendix B models and are listed in Table 1.8.5. However, a performance demonstration may be required for an Appendix B model to demonstrate its suitability over a standard Appendix A model. This section addresses Appendix A models.

All Appendix A and B models and user's documentation are available from the U.S. Department of Commerce's National Technical Information Service (NTIS), in Springfield, Virginia 22161. In addition, model codes and selected abridged user's guides are available from the U.S. EPA's Support Center for Regulatory Air Models Bulletin Board System (SCRAM BBS), which can be accessed on the Internet (at http://ttnwww.rtpnc.epa.gov).

MODEL OPTIONS

Due to the range and combinations of physical conditions, dispersion models are often simplified and designed for specialized applications in limited situations. The EPA has designated Appendix A models for use under such specific applications and may require the selection of predetermined options for regulatory application. Some examples of specialized functions include simple screening versus refined modeling, terrain features, surrounding land use, pollutant averaging period, number and type of sources to be modeled, and additional influences to the release. These functions are discussed next.

SCREENING AND REFINED MODELS

Dispersion models have two levels of sophistication. The first, referred to as *screening modeling*, is a preliminary approach designed to simplify a source's emissions and provide conservative plume concentration impact estimates. The model user compares the results of screening modeling to the NAAQS, prevention of significant deterioration (PSD) increments, and/or ambient significance levels to determine if a second level of analysis, *refined modeling*, is required for a better estimate of the predicted concentrations. The purpose of screening is to identify if the potential for exceeding applicable air quality threshold levels exists, and thus the need for refined analysis. Screening eliminates the time and expense of refined modeling if the predicted concentrations do not approach the applicable levels.

TABLE 1.8.4   U.S. EPA PREFERRED AIR QUALITY DISPERSION MODELS
The following is a list of U.S. EPA approved Appendix A guideline models and their intended application. The user is referred to the *Guideline on Air Quality Models* (U.S. EPA 1978; 1986; 1987; 1993) and the appropriate user's guide (see the following references) to select and apply the appropriate model.

Terrain	Mode	Model	Reference
Screening			
Simple	Both	SCREEN3	U.S. EPA 1988; 1992a
Simple	Both	ISC3	Bowers, Bjorklund, and Cheney 1979; U.S. EPA 1987; 1992b; 1995
Simple	Both	TSCREEN	U.S. EPA 1990b
Simple	Urban	RAM	Turner and Novak 1978; Catalano, Turner, and Novak 1987
Complex	Rural	COMPLEXI	Chico and Catalano 1986; Source code.
Complex	Urban	SHORTZ	Bjorklund and Bowers 1982
Complex	Rural	RTDM3.2	Paine and Egan 1987
Complex	Rural	VALLEY	Burt 1977
Complex	Both	CTSCREEN	U.S. EPA 1989; Perry, Burns, and Cinnorelli 1990
Line	Both	BLP	Schulman and Scire 1980
Refined			
Simple	Urban	RAM	Turner and Novak 1978; Catalano, Turner, and Novak 1987
Simple	Both	ISC3	Bowers, Bjorklund, and Cheney 1979; U.S. EPA 1987; 1992b; 1995
Simple		EDMS	Segal 1991; Segal and Hamilton 1988; Segal 1988
Simple	Urban	CDM2.0	Irwin, Chico, and Catalano 1985
Complex	Both	CTDMPLUS	Paine et al. 1987; Perry et al. 1989; U.S. EPA 1990a
Line	Both	BLP	Schulman and Scire 1980
Line	Both	CALINE3	Benson 1979
Ozone	Urban	UAM-V	U.S. EPA 1990a
Coastal		OCD	DiCristofaro and Hanna 1989

References for Table 1.8.4

Benson, P.E. 1979. CALINE3—*A versatile dispersion model for predicting air pollutant levels near highways and arterial streets, Interim Report.* FHWA/CA/TL-79/23. Washington, D.C.: Federal Highway Administration.

Bjorklund, J.R., and J.F. Bowers. 1982. *User's instructions for the SHORTZ and LONGZ computer programs.* EPA 903/9-82-004a and b. Philadelphia: U.S. EPA Region 3.

Bowers, J.F., J.R. Bjorklund, and C.S. Cheney. 1979. *Industrial source complex (ISC) dispersion model user's guide.* EPA-450/4-79-030 and 031. Research Triangle Park, N.C.: U.S. EPA, Office of Air Quality Planning and Standards.

Burt, E.W. 1977. VALLEY model user's guide. EPA-450/2-77-018. Research Triangle Park, N.C.: U.S. EPA.

Catalano, J.A., D.B. Turner, and J.H. Novak. 1987. *User's guide for RAM.* 2d ed. EPA-600/3-87-046. Research Triangle Park, N.C.: U.S. EPA, Office of Air Quality Planning and Standards (October).

DiCristofaro, D.C., and S.R. Hanna. 1989. *OCD: The offshore and coastal dispersion model, version 4.* Vols. I & II. Westford, MA.: Sigma Research Corporation.

Irwin, J.S., T. Chico, and J. Catalano. 1985. *CDM 2.0 (climatological dispersion model) user's guide.* Research Triangle Park, N.C.: U.S. EPA.

Paine, R.J., and B.A. Egan. 1987. *User's guide to the rough terrain diffusion model (RTDM) (Rev. 3.20).*

Paine, R.J., D.G. Strimaitis, M.G. Dennis, M.T. Mills, and E.M. Insley. 1987. *User's guide to the complex terrain dispersion model.* Vol. 1. EPA-600/8-87-058a. Research Triangle Park, N.C.: U.S. Environmental Protection Agency.

Perry, S.G., D.J. Burns, and A.J. Cimorelli. 1990. *User's guide to CDTMPLUS: Volume 2. The screening mode (CTSCREEN).* EPA-600/8-90-087. Research Triangle Park, N.C.: U.S. EPA.

Perry, S.G., D.J. Burns, L.H. Adams, R.J. Paine, M.G. Dennis, M.T. Mills, D.G. Strimaitis, R.J. Yamartino, and E.M. Insley. 1989. *User's guide to the complex terrain dispersion model plus algorithms for unstable situations (CTDMPLUS).* Volume 1: Model Descriptions and User Instructions. EPA-600/8-89-041. Research Triangle Park, N.C.: U.S. EPA.

Segal, H.M. 1988. *A microcomputer pollution model for civilian airports and Air Force bases—Model application and background.* FAA-EE-88-5. Washington, D.C.: Federal Aviation Administration.

Segal, H.M. 1991. *EDMA-Microcomputer pollution model for civilian airports and Air Force bases: User's guide.* FAA-EE-91-3. Washington, D.C.: Federal Aviation Administration.

Segal, H.M., and P.L. Hamilton. 1988. *A microcomputer pollution model for civilian airports and Air Force bases—Model description.* FAA-EE-88-4. Washington, D.C.: Federal Aviation Administration.

Schulman, L.L., and J.S. Scire. 1980. *Buoyant line and point source (BLP) dispersion model user's guide.* Doc. P-7304B. Concord, Mass.: Environmental Research & Technology, Inc.

Turner, D.B., and J.H. Novak. 1978. *User's guide for RAM.* Vols. 1 & 2. EPA-600/8-78-016a and b. Research Triangle Park, N.C.: U.S. EPA.

U.S. Environmental Protection Agency (EPA). 1988. *Screening procedures for estimating the air quality impact of stationary sources, draft for public comment.* EPA-450/4-88-010. Research Triangle Park, N.C.: U.S. EPA, Office of Air Quality Planning and Standards (August).

———. 1989. *User's guide to the complex terrain dispersion model plus algorithms for unstable situations (CTDMPLUS).* Vol. 1. EPA-600/8-89-041. Research Triangle Park, N.C.: U.S. EPA, Atmospheric Research and Exposure Assessment Laboratory (March).

———. 1990a. *User's guide to the urban airshed model.* Vols. 1–8. EPA 450/4-90-007a-c, d(R), e-g, EPA-454/B-93-004. Research Triangle Park, N.C.: U.S. EPA.

———. 1990b. *User's guide to TSCREEN: A model for screening toxic air pollutant concentrations.* EPA-450/4-91-013. Research Triangle Park, N.C.: U.S. EPA, Office of Air Quality Planning and Standards.

———. 1992a. *The SCREEN2 model user's guide.* EPA-450/4-92-006. Research Triangle Park, N.C.: U.S. EPA.

———. 1992b. *User's guide for the industrial source complex (ISC2) dispersion models.* Vols. 1–3. EPA-450/4-92-008a, b, and c. Research Triangle Park, N.C.: U.S. EPA, Office of Air Quality Planning and Standards (March).

———. 1995. *User's guide for the industrial source complex (ISC3) dispersion models.* Vols 1 & 2. EPA-454/B-95-003a & b. Research Triangle Park, N.C.: U.S. EPA.

TABLE 1.8.5 ALTERNATIVE MODELS
The following is a list of U.S. EPA selected Appendix B models to the
preferred guideline models. The user is referred to the *Guideline on Air
Quality Models Appendix B* (U.S. EPA 1993), the appropriate user's guide
(see the following references), and contact with the U.S. EPA Regional
Modeler to select and apply the model.

Model	Reference
AVACTA-II	Zannetti, Carboni, and Lewis 1985
DEGADIS 2.1	U.S. EPA 1989
ERT Visibility	ENSR Consulting and Engineering 1990
HGSYSTEM	Post 1994a; Post 1994b
HOTMAC/RAPTAD	Mellor and Yamada 1974; 1982; Yamada and Bunker 1988
LONGZ	Bjorklund and Bowers 1982
MESOPUFF-II	Scire et al. 1984; U.S. EPA 1992b
MTDDIS	Wang and Waldron 1980
PANACHE	Transoft Group 1994
PLUVUE-II	U.S. EPA 1992a
PAL-DS	Petersen 1978; Rao and Snodgrass 1982
PPSP	Brower 1982; Weil and Brower 1982
RPM-IV	U.S. EPA 1993
SCSTER	Malik and Baldwin 1980
SHORTZ	Bjorklund and Bowers 1982
SDM	PEI Associates 1988
SLAB	Ermak 1990
Simple Line Source Model	Chock 1980
WYNDvalley	Harrison 1992

References for Table 1.8.5

Bjorklund, J.R., and J.F. Bowers. 1982. *User's instructions for the SHORTZ and LONGZ computer programs.* EPA 903/9-82-004a and b. Philadelphia: U.S. EPA, Region 3.

Brower, R. 1982. *The Maryland power plant siting program (PPSP) air quality model user's guide.* PPSP-MP-38. Baltimore: Maryland Department of Natural Resources.

Chock, D.P. 1980. *User's guide for the simple line-source model for vehicle exhaust dispersion near a road.* Warren, Mich.: Environmental Science Department, General Motors Research Laboratories.

ENSR Consulting and Engineering. 1990. *ERT visibility model: Version 4; Technical description and user's guide.* M2020-003. Acton, Mass.: ENSR Consulting and Engineering.

Ermak, D.L. 1990. *User's manual for SLAB: An atmospheric dispersion model for denser-than-air releases (UCRL-MA-105607).* Lawrence Livermore National Laboratory.

Harrison, H. 1992. *A user's guide to WYNDvalley 3.11, An Eulerian-grid air quality dispersion model with versatile boundaries, sources, and winds.* Mercer Island, Wash.: WYNDsoft, Inc.

Malik, M.H., and B. Baldwin. 1980. *Program documentation for multi-source (SCSTER) model.* EN7408SS. Atlanta: Southern Company Services, Inc.

Mellor, G.L., and T. Yamada. 1974. A hierarchy of turbulence closure models for planetary boundary layers. *Journal of Atmospheric Sciences* 31:1791–1806.

Mellor, G.L., and T. Yamada. 1982. Development of a turbulence closure model for geophysical fludi problems. *Rev. Geophys. Space Phys.* 20:851–875.

PEI Associates. 1988. *User's guide to SDM—A shoreline dispersion model.* EPA-450/4-88-017. Research Triangle Park, N.C.: U.S. EPA.

Post, L. (ed.). 1994a. *HGSYSTEM 3.0 technical reference manual.* Chester, United Kingdom: Shell Research Limited, Thornton Research Centre.

Post, L. 1994b. *HGSYSTEM 3.0 user's manual.* Chester, United Kingdom: Shell Research Limited, Thornton Research Centre.

Rao, K.S., and H.F. Snodgrass. 1982. *PAL-DS model: The PAL model including deposition and sedimentation.* EPA-600/8-82-023. Research Triangle Park, N.C.: U.S. EPA, Office of Research and Development.

Scire, J.S., F.W. Lurmann, A. Bass, and S.R. Hanna. 1984. *User's guide to the Mesopuff II model and related processor programs.* EPA-600/8-84-013. Research Triangle Park, N.C.: U.S. EPA.

Transoft Group. 1994. *User's guide to fluidyn-PANACHE, a three-dimensional deterministic simulation of pollutants dispersion model for complex terrain.* Cary, N.C.: Transoft Group.

U.S. Environmental Protection Agency (EPA). 1989. *User's guide for the DEGADIS 2.1—Dense gas dispersion model.* EPA-450/4-89-019. Research Triangle Park, N.C.: U.S. EPA.

———. 1992a. *User's manual for the plume visibility model, PLUVUE II (Revised).* EPA-454/B-92-008. Research Triangle Park, N.C.: U.S. EPA.

———. 1992b. *A modeling protocol for applying MESOPUFF II to long range transport problems.* EPA-454/R-92-021. Research Triangle Park, N.C.: U.S. EPA.

———. 1993. *Reactive Plume Model IV (RPM-IV) User's Guide.* EPA-454/B-93-012. Research Triangle Park, N.C.: U.S. EPA, (ESRL).

Wang, I.T., and T.L. Waldron. 1980. *User's guide to MTDDIS mesoscale transport, diffusion, and deposition model for industrial sources.* EMSC6062.1UR(R2). Newbury Park, Calif.: Combustion Engineering.

Weil, J.C., and R.P. Brower. 1982. *The Maryland PPSP dispersion model for tall stacks.* PPSP-MP-36. Baltimore, Md.: Maryland Department of Natural Resources.

Yamada, T., and S. Bunker, 1988. Development of a nested grid, second moment turbulence closure model and application to the 1982 ASCOT Brush Creek data simulation. *Journal of Applied Meteorology* 27:562–578.

Zannetti, P., G. Carboni, and R. Lewis. 1985. *AVACTA-II user's guide (Release 3).* AV-OM-85/520. Monrovia, Calif.: AeroVironment, Inc.

A refined model provides a more detailed analysis of the parameters and thus gives a more accurate estimate of the pollutant concentration at receptors. However, a refined model demands more specific input data. The specific data can include topography, better receptor grid resolution, downwash or other plume adjustments, and pollutant decay or deposition.

SIMPLE AND COMPLEX TERRAIN

Air quality dispersion models can also be divided into three categories based on their application to terrain features (Wilson 1993) which address the relationship between receptor elevations and the top of the stack. Early models were developed without any terrain consideration (i.e., they were flat). In simple terrain models, receptors are located below the stack top, while in complex terrain models, receptors are located at or above the stack top.

More recently, model developers have focused attention on elevations which are located between the stack top and the final height of the plume rise. These elevations are classified as intermediate terrain (see Figure 1.8.15), and calculations within the intermediate region can be evaluated by either simple or complex models. The model predicting the highest concentration for a receptor is conservatively selected for that point. Thus, intermediate terrain is an overlapping region for model predictions.

A model, CTDMPLUS, is included in the EPA's Appendix A to address this concern. The model can predict concentrations at the stack top or above for stability conditions and has been approved for intermediate and complex terrain applications (U.S. EPA 1989).

URBAN AND RURAL CLASSIFICATION

EPA recommends models for either urban or rural applications, and most models contain an option for selecting urban or rural dispersion coefficients. These coefficients affect the plume's spread in the y and z directions and are determined by the characteristics of the area surrounding the site. In heavily developed areas such as cities, the urban heat island and structures affect the surrounding atmosphere, increasing turbulence and air temperature. In the country, the foliage and undeveloped land reduce long-wave radiation and generate less turbulence. The amount of turbulence generated by each of these locations affects how the plume is dispersed in the atmosphere.

Irwin (1978) recommends two methods for categorizing the surrounding area as urban or rural. One technique relies on a methodology developed by Auer (1978), which characterizes the land use within a 3-km radial area. The second technique is based on a population density threshold (750 people/sq km) within a 3-km area. The Auer method is typically the preferred approach; it would identify a large industrial plant with storage yards (i.e., a steel mill) as an industrial (urban) site instead of an unpopulated rural site. Table 1.8.6 lists the Auer land use categories.

AVERAGING PERIODS

While several averaging periods are available for pollutants, the models are divided into two groups. The first group, referred to as short-term models, handles averaging periods from an hour to a year using hourly meteorological conditions. Note that most refined short term models calculate concentrations for block averaging periods rather than for running average periods within a day. Screening models fall into this category and predict concentrations for one averaging period only.

Long-term models use meteorological conditions ranging from a month or season to one or more years. The climatological data used in these models are generated from hourly data into joint frequency–distribution tables of wind speed, wind direction, and Pasquill–Gifford stability categories. These data are referred to as stability array (STAR) data sets.

SINGLE AND MULTIPLE SOURCES

To simplify calculations, some models predict concentrations for only one source at a time, while others can calculate concentrations for a complex combination of sources. Screening models are usually limited to one source. As a compromise to these two extremes, some models colocate numerous sources at the same point and add individual source concentrations together. This approach is usually applied for conservative screening estimates.

TYPE OF RELEASE

Four types of releases can be simulated: point, area, line, and volume sources. Some models can be limited to the

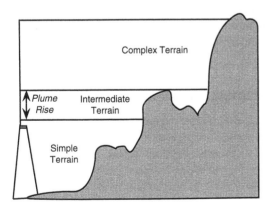

FIG. 1.8.15 Terrain categories. (Reprinted, with permission, from Trinity Consultants, Inc., 1993, *Air issues review,* Issue no. 5, Dallas, Tex. [September].)

TABLE 1.8.6 IDENTIFICATION AND CLASSIFICATION OF LAND USE TYPES

Type	Use and Structures	Vegetation
I1	Heavy industrial Major chemical, steel, and fabrication industries generally 3–5 story buildings, flat roofs	Grass and tree growth extremely rare; <5% vegetation
I2	Light–moderate industrial Rail yards, truck depots, warehouse, industrial parks, minor fabrications; generally 1–3 story buildings, flat roofs	Very limited grass, trees almost totally absent; <5% vegetation
C1	Commercial Office and apartment buildings, hotels; >10 story heights, flat roofs	Limited grass and trees; <15% vegetation
R1	Common residential Single-family dwelling with normal easements; generally one story, pitched roof structures; frequent driveways	Abundant grass lawns and light–moderately wooded; >70% vegetation
R2	Compact residential Single-, some multiple-, family dwellings with close spacing; generally <2 story height, pitched roof structures; garages (via alley), no driveways	Limited lawn sizes and shade trees; <30% vegetation
R3	Compact residential Old multifamily dwellings with close (<2m) lateral separation; generally 2 story height, flat roof structures; garages (via alley) and ash pits, no driveways	Limited lawn sizes, old established shade trees; <35% vegetation
R4	Estate residential Expansive family dwelling on multiacre tracts	Abundant grass lawns and light wooded; >80% vegetation
A1	Metropolitan natural Major municipal, state, or federal parks, golf courses, cemeteries, campuses; occasional single-story structures	Nearly total grass and light wooded; >95% vegetation
A2	Agricultural rural	Local crops (e.g., corn, soybean); 95% vegetation
A3	Undeveloped Uncultivated, wasteland	Mostly wild grasses and weeds, lightly wooded; >90% vegetation
A4	Undeveloped rural	Heavy wooded; 95% vegetation
A5	Water surfaces Rivers, lakes	

Source: A.H. Auer, 1978, Correlation of land use and cover with meteorological anomalies, *Journal of Applied Meteorology* 17.

types of releases they handle. For most applications, line sources can be treated as area or volume sources, as well as in individual line models (see the subsection on mobile and line source modeling). Point sources are the most common modeling application and are defined by height, temperature, diameter, and velocity parameters.

In some cases, small or insignificant point sources may be grouped into either point or area sources to simplify the calculation. Area sources are only defined by a release height and the length of a side (which represents a square area). Models calculate the emissions from area sources using a virtual point source located some distance upwind so that the plume dispersion is equal to the width of the box at the center of the area source (see Figure 1.8.16). The models typically ignore concentrations calculated upwind and over the area source.

Volume sources represent continuous releases over a length, such as a conveyor or monitor. They are defined in the models by a release height, length of a side, and initial dispersions in the y and z directions (σ_y and σ_z). Simulation of these sources is similar to the virtual point source used for area sources.

ADDITIONAL PLUME INFLUENCES

Some models handle various factors that affect the plume. These options include stack-tip downwash, buoyancy-induced dispersion, gradual plume rise, and building or topography downwash factors as described in previous sections. Other factors include pollutant decay (half-life) and deposition. Pollutant decay considers chemical changes or scavenging of the pollutant in the atmosphere, typically through the use of a half-life duration factor.

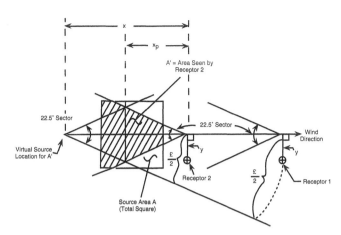

FIG. 1.8.16 Schematic of the virtual point source as projected from an area source. (Reprinted from TRW Systems Group, 1969, *Air quality display model,* Washington, D.C.: National Air Pollution Control Administration, DHEW, U.S. Public Health Service.)

Deposition accounts for the settling of the pollutant from the plume. This deposition can either occur through dry (gravitational) settling or wet (rainwash) removal. Dry deposition modeling usually requires particle size and settling velocities for the pollutant emitted; wet deposition also requires meteorological data about precipitation frequency and characteristics.

METEOROLOGY

As previously mentioned, the type of meteorology used in a model depends on the pollutant averaging period as well as the level of modeling and site characteristics. If the modeling is performed for a state implementation plan (SIP) or a PSD permit, five years of meteorological data are typically required.

Most screening models use a preselected set of meteorological conditions that provide the worst-case plume transport and ground-level concentrations. The basic data used by models include wind speed, wind direction, stability, temperature, and mixing height.

A minimum of one year of site-specific data is preferred for refined and complex terrain models. However, these data often either are not available or require a year to collect. In most cases (i.e., noncomplex terrain), the National Weather Service (NWS) observations from a nearby station can be substituted in refined modeling applications. Five years of meteorological data from a representative station provide the best, reasonable representation of climatology at the station.

For refined, simple terrain modeling, representative data from both a NWS surface and upper air station are the required minimum. The model user can modify the data

for input using standard EPA meteorological preprocessor programs. For complex terrain modeling, site-specific data are critical to represent the conditions within the local topographic regime. In most cases, site-specific data are incomplete for all the required parameters and must usually be supplemented by NWS preprocessed data. To retain consistency, the EPA has a protocol to follow when data are substituted for missing observations.

Both short- and long-term models also require mixing height data to define the upper limit of the area where effluent mixing occurs (the ground being the lower limit). Holzworth (1972) developed a set of figures and tables for seasonal and annual mixing heights, which are typically used in long-term modeling. For short-term applications, model users can interpolate hourly mixing height values (using the EPA's RAMMET preprocessor) based on twice-a-day upper air data collected by radiosonde measurements at numerous sites throughout the country and available through the National Climatic Center (NCC).

OTHER MODELS

In addition to the Gaussian-based models in Appendix A, computer programs have also been created for modeling other specific cases or pollutants. These other models include photochemical reactions or Eulerian or LaGrangian equations. Some, such as the chemical mass balance (CMB7) or fugitive dust model (FDM), are available for addressing specific pollutants, while others are implemented on a case-by-case basis for unique source or topographic conditions. The environmental engineer is directed to the appropriate U.S. EPA Region office to consult on selection and approval of the appropriate model for these cases.

The urban airshed model (UAM) is an Appendix A model applicable to most regions of the United States (USEPA 1990). The reactive plume model (RPM-IV) (USEPA 1993) accounts for ozone and other photochemically reactive pollutants over a smaller area than the urban airshed model.

Toxic or hazardous gas releases can be simulated with several models incorporated into Appendix A. These models are the toxic modeling system short-term (TOXST) (U.S. EPA 1992b) and toxic modeling system long-term (TOXLT) (U.S. EPA 1992a) models which are based on the ISC3 algorithms as well as the dense gas dispersion model (DEGADIS2.1) (U.S. EPA 1989a).

Mobile and Line Source Modeling

Mobile sources are difficult to model because the source is moving and may not be continuously emitting pollutants at a constant emission rate. Nonetheless, environmental engineers often model these sources using line source techniques that assign emissions at fixed points along a line. The U.S. EPA has developed a program called MOBILE5b to calculate emissions from mobile sources based on EPA tests of emissions from a variety of vehicle classes and types. In certain cases, environmental engineers can use the ISC3 models to estimate concentrations at receptors, but the EPA has identified the California line (CALINE3), CAL3QHC or CAL3QHCR, and the buoyant line and point source (BLP) models as the most appropriate for calculating emission concentrations from line sources.

CALINE3 MODEL

The CALINE3 model (Benson 1979) was designed to estimate highway traffic concentrations of nonreactive pollutants for the state of California. The model is a steady-state, Gaussian-based model capable of predicting 1-hr to 24-hr average concentrations from urban or rural roadway emission sources located at grade, in cut sections, in fill sections, and over bridges as shown in Figure 1.8.17. The program assumes all roadways have uncomplicated topography (i.e., simple terrain). Any wind direction and roadway orientation can be modeled. Primary pollutants can be modeled, including particulate matter, for which

the model uses deposition and settling velocity factors. However, unlike the other models, CALINE3 does not account for any plume rise calculations.

CAL3QHC MODEL

While the CAL3QHC model (U.S. EPA 1992c) is not officially part of the Appendix A models, it is a recommended model for CO emissions in the *Guidelines on Air Quality Models* (U.S. EPA 1993). The model predicts carbon monoxide concentrations at signalized intersections. In effect, the model determines the increase in emissions and their resultant concentrations during queuing periods at stoplights. A version of the model, referred to as CAL3QHCR, has recently been created to process up to a year of meteorology, vehicle emissions, traffic volume, and signalization data.

BLP MODEL

The BLP model (Schulman and Scire 1980) is a Gaussian-based, plume dispersion model that deals with plume rise and downwash effects from stationary line sources. It was originally designed to simulate emissions from an aluminum reduction plant. The model can predict short-term concentrations in rural areas for buoyant, elevated line sources. However, it does not treat deposition or settling of particles.

—Roger K. Raufer
Curtis P. Wagner

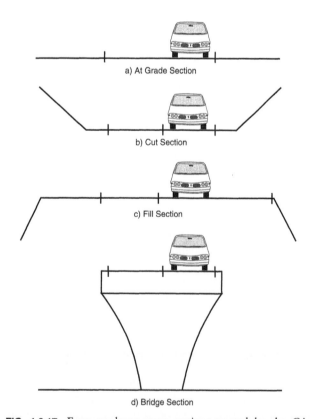

FIG. 1.8.17 Four roadway cross sections treated by the CALINE3 model.

References

Auer, A.H. 1978. Correlation of land use and cover with meteorological anomalies. *Journal of Applied Meteorology* 17:636–643.

Benson, P.E. 1979. CALINE3—A versatile dispersion model for predicting air pollutant levels near highways and arterial streets, Interim report. FHWA/CA/TL-79/23, Washington, D.C.: Federal Highway Administration.

Bittle, C.R., and A.R. Borowsky. 1985. A review of dispersion modeling methods for assessing toxic releases. Presented at the 78th Annual Meeting of the Air Pollution Control Association, June, 1985. 85–25B.2.

Bowne, N.E. 1974. Diffusion rates. *Journal of the Air Pollution Control Association* 24:832–835.

Briggs, G.A. 1969. *Plume rise.* U.S. Atomic Energy Commission Critical Review Series TID-25075. Clearinghouse for Federal Scientific and Technical Information.

Davenport, A.G. 1963. The relationship of wind structure to wind loading. Presented at Int. Conf. on the Wind Effects on Buildings and Structures, Nat. Physical Laboratory, Teddington, Middlesex, England, 26–28 June. Cited in Turner 1970.

Gifford, F.A. 1968. Meteorology and atomic energy—1968. Chap. 3 in *U.S. atomic energy report.* TID-24190. Oak Ridge, Tenn.

———. 1975. *Lectures on air pollution and environmental impact analyses.* Chap. 2. Boston, Mass.: American Meterological Society.

Holland, J.Z. 1953. *A meteorological survey of the Oak Ridge Area.* ORO-99, 540. Washington, D.C.: Atomic Energy Commission.

Holzworth, G.C. 1972. *Mixing heights, wind speeds, and potential for urban air pollution throughout the contiguous United States.* Office of Air Programs Pub. no. AP-101. Research Triangle Park, N.C.: U.S. EPA.

Hosker, R.P. 1984. Flow and diffusion near obstacles. In *Atmospheric science and power production,* edited by D. Randerson, DOE/TIC-27601. Washington, D.C.: U.S. Department of Energy.

Huber, A.H. 1977. Incorporating building/terrain wake effects on stack effluents. Preprint vol. *AMS-APCA Joint Conference on Applications of Air Pollution Meteorology, Salt Lake City, UT, Nov. 29–Dec. 2, 1977.*

Irwin, J.S. 1978. *Proposed criteria for urban versus rural dispersion co-efficients (Draft staff report).* Research Triangle Park, N.C.: U.S. EPA, Meteorology and Assessment Division.

———. 1979. A theoretical variation of the wind profile law exponent as a function of surface roughness and stability. *Atmospheric Environment* 13:191–194.

McElroy, J. and F. Pooler. 1968. St. Louis dispersion study. Vol. 2, *Analysis.* AP-53. Arlington, Va.: National Air Pollution Control Administration, U.S. Department of Health, Education, and Welfare.

Pasquill, F. 1974. *Atmospheric diffusion.* 2d ed. London: Van Nostrand.

———. 1976. *Atmospheric dispersion parameters in Gaussian-plume modeling, Part 2: Possible requirements for a change in the Turner workbook values.* EPA-600/4-76-030b. Research Triangle Park, N.C.: U.S. EPA, Office of Research and Development.

Schulman, L.L., and S.R. Hanna. 1986. Evaluation of downwash modifications to the industrial source complex model. *Journal of the Air Pollution Control Association* 24:258–264.

Schulman, L.L., and J.S. Scire. 1980. *Buoyant line and point source (BLP) dispersion model user's guide.* Doc. P-7304B. Concord, Mass.: Environmental Research & Technology, Inc.

Snyder, W.H., and R.E. Lawson. 1976. Determination of a necessary height for a stack close to a building—A wind tunnel study. *Atmospheric Environment* 10:683–691.

Stern, A.C., H.C. Wohlers, R.W. Boubel, and W.P. Lowry. 1973. *Fundamentals of air pollution.* New York: Academic Press.

Tickvart, J.A. 1988. *Memorandum on stack-structure relationships.* Research Triangle Park, N.C.: U.S. EPA, Office of Air Quality Planning and Standards (11 May).

Turner, D.B. 1964. A diffusion model for an urban area. *Journal of Applied Meteorology* 3:83–91.

———. 1970. *Workbook of atmospheric dispersion estimates (Revised).* Office of Air Programs Publication No. AP-26. Research Triangle Park, N.C.: U.S. EPA.

U.S. Environmental Protection Agency (EPA). 1978. *Guideline on air quality models.* EPA-450/2-78-027. Research Triangle Park, N.C.:

U.S. EPA, Office of Air Quality Planning and Standards (April).

———. 1985. *Guideline for determination of good engineering practice stack height (Technical support document for the stack height regulations) (Revised).* EPA-450/4-80-023R. Research Triangle Park, N.C.: U.S. EPA, Office of Air Quality Planning and Standards (June).

———. 1986. *Guideline on air quality models (Revised).* EPA-450/2-78-027R. Research Triangle Park, N.C.: U.S. EPA, Office of Air Quality Planning and Standards (July).

———. 1987. *Supplement A to the guideline on air quality models (Revised).* EPA-450/2-78-027R. Research Triangle Park, N.C.: U.S. EPA, Office of Air Quality Planning and Standards (July).

———. 1989a. *User's guide for the DEGADIS 2.1—Dense gas dispersion model.* EPA-450/4-89-019. Research Triangle Park, N.C.: U.S. EPA.

———. 1989b. *User's guide to the complex terrain dispersion model plus algorithms for unstable situations (CTDMPLUS).* Vol. 1, EPA-600/8-89-041. Research Triangle Park, N.C.: U.S. EPA Atmospheric Research and Exposure Assessment Laboratory (March).

———. 1990. *User's guide to the urban airshed model.* Vols. 1-8. EPA-450/4-90-007a to g, EPA-454/B-93-004. Research Triangle Park, N.C.: U.S. EPA.

———. 1992a. *Toxic modeling system long-term (TOXLT) user's guide.* EPA-450/4-92-003. Research Triangle Park, N.C.: U.S. EPA.

———. 1992b. *Toxic modeling system short-term (TOXST) user's guide.* EPA-450/4-92-002. Research Triangle Park, N.C.: U.S. EPA.

———. 1992c. *User's guide for CAL3QHC version 2: A modeling methodology for predicting pollutant concentrations near roadway intersections.* EPA-454/R-92-006. Research Triangle Park, N.C.: U.S. EPA.

———. 1992d. *User's guide for the industrial source complex (ISC2) dispersion models.* Vols. 1-3. EPA-450/4-92-008a, b, and c. Research Triangle Park, N.C.: U.S. EPA, Office of Air Quality Planning and Standards (March).

———. 1993a. Supplement B to the guideline on air quality models (Revised). *Federal Register* 58, no. 137. Washington, D.C. (20 July).

———. 1993b. *Reactive plume model IV (RPM-IV) user's guide.* EPA-454/B-93-012. Research Triangle Park, N.C.: U.S. EPA (ESRL).

———. 1995. Supplement C to the guideline on air quality models (Revised). *Federal Register 60,* no. 153. Washington, D.C. (August 9).

———. 1996. Guideline on air quality models (direct final rule). *Federal Register 61,* no. 156. Washington, D.C. (August 12).

Vatavuk, W.M. 1990. *Estimating costs of air pollution control.* 82. Chelsea, Mich.: Lewis Publishers.

Wilson, Robert B. 1993. Review of development and application of CRSTER and MPTER models. *Atmospheric Environment* 27B:41–57.

2

Air Quality

R.A. Herrick | Béla G. Lipták | David H.F. Liu | R.G. Smith

2.1
EMISSION MEASUREMENTS

Data from emission measurements benefit a company in many ways. Companies use these data for permits and compliance audits, for determining the effectiveness of control equipment, for the design of pollution control strategies, and for implementation of waste minimization and pollution prevention programs. The work involves sampling and testing procedures and physical and chemical measurements.

Planning an Emissions Testing Program

The first step in planning an emissions testing program is to define its objectives. Essentially, the testing objectives dictate the accuracy of the data needed, which dictate the following four conditions of a testing program:

Stream—This testing condition specifies whether sampling is performed directly at the point where the pollutant is generated (e.g., valves, pumps, or compressors) or on ambient air.

Frequency—This testing condition specifies whether samples are taken periodically or continuously.

Method—This testing condition specifies whether standardized reference methods are used to analyze for a particular compound or if a customized method must be developed.

Location—This testing condition specifies whether samples are taken from the field to the laboratory or are tested directly in the field. For stack sampling, environmental engineers must refine arrangements to ob-

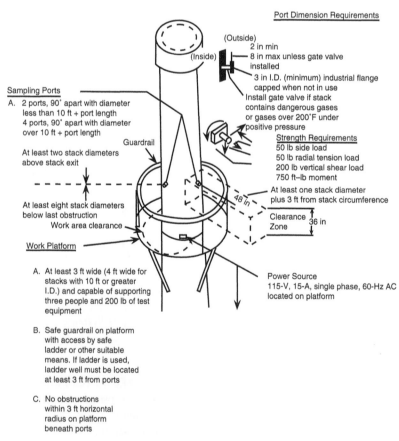

FIG. 2.1.1 Typical sampling point provisions. (Reprinted from U.S. Environmental Protection Agency [EPA].)

tain representative samples for analysis. Figure 2.1.1 shows the requirements for stack sampling.

Table 2.1.1 provides recommendations for the four preceding primary testing conditions based on the end uses of the data.

Analyzing Air Emissions

The technologies for sampling air are many and varied. The choice of the air sampling and analysis method depends on the source. For example, emissions from stacks can be measured directly at release points. If the amounts and concentrations of contaminants are needed for a large area, then area or remote methods are preferable to source testing. To detect fugitive emissions from equipment, such as valves, flanges, pumps, or motors, screening or bagging tests are used, in which the device is enclosed to capture leaks.

STACK SAMPLING

The U.S. EPA has published standard sampling routines for stack gases. These reference methods give representative concentration data for compounds. These test methods are validated by laboratory and field studies, and the

TABLE 2.1.1   TEST PLANNING MATRIX

Primary Objective	Stream		Frequency		Method		Location	
	Source	Ambient	Interval	Continuous	Reference	Custom	Lab	Field
Test compliance with regulations	√	=	=	=	√√	X	√	=
Improve overall emission inventory	=	=	=	=	=	=	=	=
Identity and characterize emission sources	√	X	=	=	=	=	=	=
Conduct performance testing of process controls	√	X	√	=	=	=	=	=
Detect intermittent or transient emissions	=	=	X	√	=	=	X	√
Gather personnel exposure data	X	√	=	=	=	=	=	=
Provide early warning of leaks	=	√	X	√	=	=	XX	√√
Develop defensible data	=	=	=	=	√√	X	=	=
Track effluent quality	X	√	=	=	√	X	=	=
Characterize water and wastes	√	X	√	X	√	=	√	=
Support waste minimization and pollution prevention	√	X	√	X	=	=	=	=

Source: Graham E. Harris, Michael R. Fuchs, and Larry J. Holcombe, 1992, A guide to environmental testing, *Chem. Eng.* (November): 98–108.

Notes: The construction of an environmental testing program begins with a list of the objectives that must be met. When that list is complete, the table indicates the proper elements to employ. Depending upon the end use of the data, testing can be performed directly in the field with portable instrumentation.

Key:
√√ Required or highly preferred
√ Preferred or likely to be used
= Neutral—Either is acceptable
X Discouraged or unlikely to be used
XX Not permitted or not available

TABLE 2.1.2 SELECTED EPA REFERENCE METHODS

Method 1—Sample and velocity traverses for stationary sources

Method 2—Determination of stack gas velocity and volumetric flow rate (type S pitot tube)

Method 2A—Direct measurement of gas volume through pipes and small ducts

Method 2B—Determination of exhaust gas volume flow rate from gasoline vapor incinerators

Method 3—Gas analysis for carbon dioxide, oxygen, excess air, and dry molecular weight

Method 3A—Determination of oxygen and carbon dioxide concentrations in emissions from stationary sources (instrumental analyzer procedure)

Method 4—Determination of moisture content in stack gases

Method 5—Determination of particulate emissions from stationary sources

Method 5A—Determination of particulate emissions from the asphalt processing and asphalt roofing industry

Method 5B—Determination of nonsulfuric acid particulate matter from stationary sources

Method 5D—Determination of particulate matter emissions from positive pressure fabric filters

Method 5E—Determination of particulate emissions from the wool fiberglass insulation manufacturing industry

Method 5F—Determination of nonsulfate particulate matter from stationary sources

Method 6—Determination of sulfur dioxide emissions from stationary sources

Method 6A—Determination of sulfur dioxide, moisture, and carbon dioxide emissions from fossil fuel combustion sources

Method 6B—Determination of sulfur dioxide and carbon dioxide daily average emissions from fossil fuel combustion sources

Method 6C—Determination of sulfur dioxide emissions from stationary sources (instrumental analyzer procedure)

Method 7—Determination of nitrogen oxide emissions from stationary sources

Method 7A—Determination of nitrogen oxide emissions from stationary sources

Method 7B—Determination of nitrogen oxide emissions from stationary sources (UV spectrophotometry)

Method 7C—Determination of nitrogen oxide emissions from stationary sources

Method 7D—Determination of nitrogen oxide emissions from stationary sources

Method 7E—Determination of nitrogen oxide emissions from stationary sources (instrumental analyzer procedure)

Method 8—Determination of sulfuric acid mist and sulfur dioxide emissions from stationary sources

Method 9—Visual determination of the opacity of emissions from stationary sources

Method 10—Determination of carbon monoxide emissions from stationary sources

Method 10A—Determination of carbon monoxide emissions in certifying continuous emission monitoring systems at petroleum refineries

Method 11—Determination of hydrogen sulfide content of fuel gas streams in petroleum refineries

Method 12—Determination of inorganic lead emissions from stationary sources

Method 13A—Determination of total fluoride emissions from stationary sources (SPADNS zirconium lake method)

Method 13B—Determination of total fluoride emissions from stationary sources (specific ion electrode method)

Method 14—Determination of fluoride emissions from potroom roof monitors of primary aluminum plants

Method 15—Determination of hydrogen sulfide, carbonyl sulfide, and carbon disulfide emissions from stationary

Method 15A—Determination of total reduced sulfur emissions from sulfur recovery plants in petroleum refineries

Method 16—Semicontinuous determination of sulfur emissions from stationary sources

Method 16A—Determination of total reduced sulfur emissions from stationary sources (impinger technique)

Method 16B—Determination of total reduced sulfur emissions from stationary sources

Method 17—Determination of particulate emissions from stationary sources (instack filtration method)

Method 18—Measurement of gaseous organic compound emissions by GC

Method 19—Determination of sulfur dioxide removal efficiency and particulate matter, sulfur dioxide, and nitrogen oxide emission rates

Method 20—Determination of nitrogen oxides, sulfur dioxide, and oxygen emissions from stationary gas turbines

Method 21—Determination of VOC leaks

Method 22—Visual determination of fugitive emissions from material sources and smoke emissions from flares

Method 24—Determination of volatile matter content, water content, density, volume solids, and weight solids of surface coating

Continued on next page

Method 24A—Determination of volatile matter content and density of printing inks and related coatings

Method 25—Determination of total gaseous nonmethane organic emissions as carbon

Method 25A—Determination of total gaseous organic concentration using a flame ionization analyzer

Method 25B—Determination of total gaseous organic concentration using a nondispersive infrared analyzer

Method 27—Determination of vapor tightness of gasoline delivery tank using pressure-vacuum test

Appendix B—Performance Specifications

Performance Specification 1—Performance specifications and specification test procedures for transmissometer systems for continuous measurement of the opacity of stack emissions

Performance Specification 2—Specifications and test procedures for SO_2 and NO_x continuous emission monitoring systems in stationary sources

Performance Specification 3—Specifications and test procedures for O_2 and CO_2 continuous emission monitoring systems in stationary sources

Source: Code of Federal Regulations, Title 40, part 60, App. A.

data obtained from them have predictable accuracies and reproducibilities. Table 2.1.2 lists the standard sampling routines for stack gases (see also, *Code of Federal Regulations*, Title 40, Part 60 App. A).

The sampling train described in reference methods 5 and 8 together with operating procedures are discussed in Section 3.1. The following sampling method numbers refer to the methods identified in EPA guidance documents and reports (U.S. EPA 1989; 1990; Harris et al. 1984):

Semivolatile Organics (Method 0010)

The sampling train in reference method 5 must be modified to include an adsorbent trap (see Figures 2.1.2 and 2.1.3) when organic compounds with a boiling point between 100 and 300°C are present in the gas. The adsorbent module is inserted between the filter and the first impinger. Gas chromatography–mass spectrometry (GC–MS) is typically used to analyze for organic compounds.

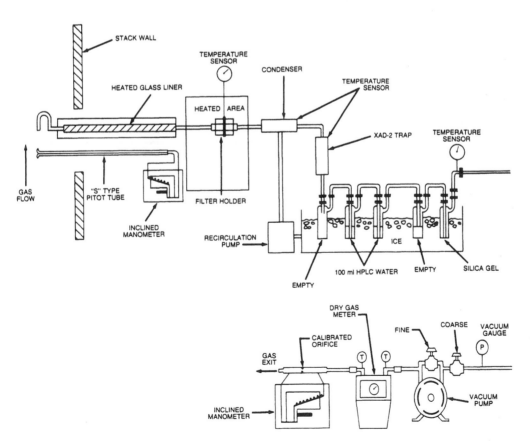

FIG. 2.1.2 Sampling train.

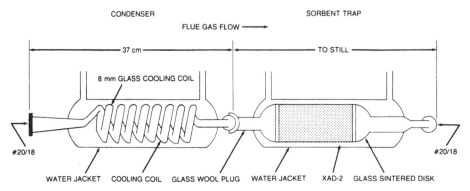

FIG. 2.1.3 Condenser and absorbent trap.

Volatile Organics (Method 0030)

The volatile organics sampling train (VOST) is used for organics that boil below 100°C. The VOST system involves drawing a stack gas sample through two adsorbent tubes in series. The first tube contains Tenax resins, and the second contains Tenax and activated carbon. The pollutants adsorbed on these tubes are then desorbed. GC–MS is a typical method for low-boiling-point organics.

HCl and Cl$_2$ (Method 0050)

This method is used to collect HCl and Cl$_2$ in stack gases. It collects emission samples isokinetically and can be combined with reference method 5 for particulate determination.

Trace Metals (Method 0012)

The metals sampling train is used to determine the total chromium, cadmium, arsenic, nickel, manganese, beryllium, copper, zinc, lead, selenium, phosphorous, thallium, silver, antimony, barium, and mercury in incinerator stack emissions. The stack gas is withdrawn isokinetically from the source with the particulate emissions collected in a probe and on heated filters and the gaseous emissions collected in a series of chilled impingers. These impingers contain a solution of dilute nitric acid combined with hydrogen peroxide in two impingers and an acidic potassium permanganate solution in two impingers. Analyzing the metals in particulates involves using either inductive coupled argon plasma (ICAP) spectroscopy or a graphite furnace atomic adsorption (GFAA) spectrometer.

Formaldehyde (Method 0011)

This method is used to collect formaldehyde in stack emissions.

Other Sampling Methods

Reference methods are continuously being developed to keep up with the demands of new regulations. Among the most recent introductions are the aldehydes–ketones sampling train and the hexavalent chromium sampling train (U.S. EPA 1990).

AIR TOXICS IN AMBIENT AIR

In addition to the reference methods, the EPA has developed methods to detect toxic and radioactive pollutants under NESHAP. Table 2.1.3 lists these test methods as well as methods for the analysis of low concentrations of organics in ambient air.

The EPA's Atmospheric Research and Exposure Assessment Laboratory (AREAL) has also developed a compendium of methods for quantifying HAPs in ambient air. Table 2.1.4 identifies the methods, and Figure 2.1.4 shows the categories of HAPs in the compendium. Two popular methods are compendium method TO-13 for semivolatiles and TO-12 for VOCs. Method TO-13 de-

TABLE 2.1.3   NESHAPS TEST METHODS

101	Particulate and gaseous mercury emissions from the air streams of chloralkali plants
101A	Particulate and gaseous mercury emissions from sewage sludge incinerators
102	Particulate and gaseous mercury emissions from the hydrogen streams of chloralkali plants
103	Beryllium screening
104	Beryllium emissions from stationary sources
105	Mercury in sewage sludges from wastewater treatment plants
106	Vinyl chloride emissions from stationary sources
107	Vinyl chloride in process wastewater, PVC resin, slurry, wet cake, and latex samples
107A	Vinyl chloride in solvents, resin–solvent solutions, PVC resin, resin slurries, wet resin, and latex
108	Particulate and gaseous arsenic emissions
111	Polonium-210 emissions from stationary sources

Source: Code of Federal Regulations, Title 40, Part 61, App. B.
Note: Analytical methods specified under NESHAPS cover toxic and radioactive pollutants not addressed by standard EPA methods.

TABLE 2.1.4 COMPENDIUM METHODS

Number	Method
TO-1	Determination of VOCs in ambient air using Tenax adsorption and GC–MS.
TO-2	Determination of VOCs in ambient air by carbon molecular sieve adsorption and GC–MS.
TO-3	Determination of VOCs in ambient air using cryogenic preconcentration techniques and GC with flame ionization and electron capture detection.
TO-4	Determination of organochlorine pesticides and PCBs in ambient air.
TO-5	Determination of aldehydes and ketones in ambient air using high-performance liquid chromatography (HPLC).
TO-6	Determination of phosgene in ambient air using HPLC.
TO-7	Determination of N-nitrosodimethylamine in ambient air using GC.
TO-8	Determination of phenol and methylphenols (cresols) in ambient air using HPLC.
TO-9	Determination of polychlorinated dibenzo-p-dioxins (PCDDs) in ambient air using high-resolution GC–high-resolution mass spectrometry.
TO-10	Determination of organochlorine pesticides in ambient air using low-volume polyurethane foam (PUF) sampling with GC–electron capture detector (GC–ECD).
TO-11	Determination of formaldehyde in ambient air using adsorbent cartridge followed by HPLC.
TO-12	Determination of nonmethane organic compounds (NMOC) in ambient air using cryogenic preconcentration and direct flame ionization detection (PDFID).
TO-13	Determination of PAHs in ambient air using high-volume sampling with GC–MS and HPLC analysis.
TO-14	Determination of VOCs in ambient air using SUMMA polished canister sampling and GC analysis.

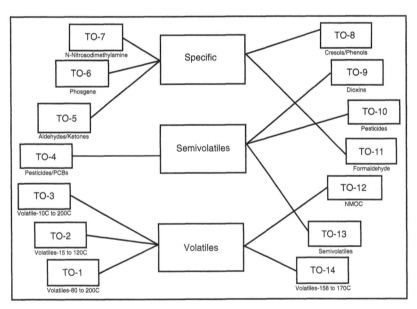

FIG. 2.1.4 Compendium of methods.

scribes a sampling and analysis procedure for semivolatiles, such as benzo(a)pyrene and polynuclear aromatic hydrocarbons (PAHs). Compendium method TO-14 involves the collection of VOHAPs in stainless canisters.

A copy of the compendium of methods can be obtained from the following:

U.S. EPA, Atmospheric Research and Exposure Assessment Laboratory (AREAL), MD-77, Research Triangle, NC 27711.

Noyes Publication, Mill Road at Grand Ave., Park Ridge, NJ 07656.

A copy of the statement of work (SOWs) for HAPs from superfund sites can be obtained from the U.S. EPA, Office of Solid Waste and Energy Response, Analytical Operation Branch, 401 M St., S.W., Washington, DC 20460.

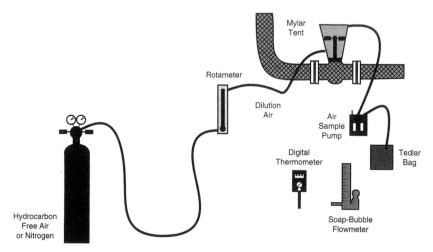

FIG. 2.1.5 Bagging test—Blow-through method.

Many of the sampling and analytic procedures recommended most likely need additional development and validation to improve accuracy and precision. A method that requires validation is not an inferior method; the method simply requires additional experimentation to define accuracy, precision, and bias. The environmental engineer begins experimental work with analyses of a known concentration of the target pollutant. Then, the engineer determines the potential interferences by repeating the tests with a gas that models the emission matrix. Validation is completed with field tests on an actual emission stream. The EPA has prepared standard protocols for validation of sampling and analysis methods (U.S. EPA 1991).

EQUIPMENT EMISSIONS

Equipment emission measurements are made by a portable hydrocarbon analyzer at potential leak points. The analyzer is a flame ionization detector (FID) that measures total hydrocarbons over a range of 1 to 10,000 ppm on a logarithmic scale. Since these measurements are basic for regulatory compliance, procedures are formalized in the EPA's reference method 21.

Environmental engineers can determine emission rates from stand-alone equipment such as valves and pumps by using bagging tests in which the source is enclosed and leaks are captured in a known value of air or inert gas.

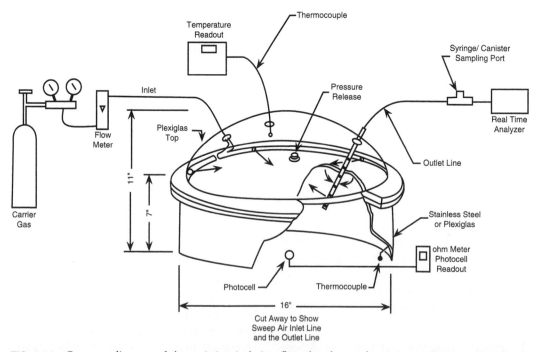

FIG. 2.1.6 Cutaway diagram of the emission isolation flux chamber and support equipment. (Reprinted, with permission, from J.C. Harris, C.E. Rechsteiner, and K.E. Thrun, 1984, *Sampling and analysis methods for hazardous waste combustion*, EPA-600/8-84/002, PB 84-155545, U.S. EPA [February].)

Two different methods are commonly used: the vacuum method and the blow-through method. Figure 2.1.5 shows the blow-through method. Multiplying the concentration of the exhaust stream by the flow rate of air through the equipment calculates the emission rates. Samples are usually analyzed with a total-hydrocarbon analyzer. Toxics are detected by gas chromatography (GC).

Monitoring Area Emissions

Measuring the concentration of air pollutants in a large area presents different problems. The potential sources of these emissions are many. For example, VOCs are emitted from hazardous waste sites; organic and inorganic gases are emitted from landfills, lagoons, storage piles, and spill sites. Area emissions can be measured either directly or indirectly. In direct measurements, ambient air is sampled at various points in the area of emission. Indirect methods sample ambient air at points upwind and downwind of the emission.

DIRECT MEASUREMENTS

Environmental engineers use an emission-flux chamber to make direct measurements of concentrations (see Figure 2.1.6). The atmospheric emissions in an area enter the chamber where they are mixed with clean dry air or nitrogen that is fed in at a fixed rate. The analyzer measures the pollutant concentration. The emission flux is the exit gas concentration multiplied by the flow rate divided by the surface area covered by the chamber.

When the area is several acres in size, measurements must be taken at various points to develop an overall emission rate. The number of measurements needed depends on the precision required and the size of the source.

Flux chambers are best suited to measure small to medium size areas (as large as a few acres) in which the pollutant concentration is fairly homogeneous. Because the flux chamber is isolated, measurements are independent from environmental influences such as wind; therefore, the measurement data are independent of the meteorological conditions at the site and are comparable from day to day and site to site.

INDIRECT METHODS

Indirect methods are best suited for the measurement of emission rates from large, heterogeneous sources. The environmental engineer measures the emission flux indirectly by collecting ambient concentrations upwind and downwind of the emission source.

The main disadvantage of indirect methods is that the analytical results rely on meteorological conditions, which can invalidate the data or preclude the collection altogether. In addition to weather patterns, buildings and hills also influence dispersion characteristics and limit sampling. The sensitivity of the analytical method is also critical since the ambient concentration of the emission can be low.

The downwind measurement determines the average concentration of contaminants in the plume. The upwind measurement monitors background readings. The volume of air passing over the monitors in a time period is recorded, and a computer model calculates an emission rate from the concentration data. The average emission rate is equal to the difference in mass measurements (downwind minus upward concentration) divided by the transit time across the source. Environmental engineers have tried various approaches for making these estimates using conventional ambient air monitoring methods (Schmidt et al. 1990).

In some cases, gaseous tracers are released at the emission source. These tracers mimic the dispersion of the emissions. When a tracer is released at a known rate, the environmental engineer can determine the emission rates of compounds by comparing the measured pollutant concentration to that of the tracer at the same location. Sections 2.4 and 2.5 describe other techniques such as continuous emission monitoring and remote sensing.

—David H.F. Liu

References

Harris, J.C., D.J. Larsen, C.E. Rechsteiner, and K.E. Thrun. 1984. *Sampling and analysis method for hazardous waste combustion.* EPA-600/8-84/002, PB 84-155545. U.S. EPA (February).

Schmidt, C. et al. 1990. *Procedures for conducting air–pathway analyses for Superfund activities, Interim final documents: Volume 2—Estimation of baseline air emission at Superfund sites.* EPA-450/1-89-002a, NTIS PB90-270588 (August).

U.S. Environmental Protection Agency (EPA). 1989. *Hazardous waste incineration measurement guidance manual.* EPA 625/6-89/021 (June).

———. 1990. *Methods manual for compliance with the BIF regulations.* EPA/530-SW-91-010, NTIS PB 90-120-006 (December).

———. 1991. *Protocol for the field validation of emission concentrations from stationary sources.* EPA 450/4-90-015 (April).

2.2
AIR QUALITY MONITORING

This section describes ambient air sampling, air quality monitoring systems, and microprocessor-based portable ambient air analyzers.

Sampling of Ambient Air

All substances in ambient air exist as either particulate matter, gases, or vapors. In general, the distinction is easily made; gases and vapors consist of substances dispersed as molecules in the atmosphere, while particulate matter consists of aggregates of molecules large enough to behave like particles. Particulate matter, or particulates, are filterable, can be precipitated, and can settle out in still air. By contrast, gases and vapors do not behave in this way and are homogeneously mixed with the air molecules.

SAMPLING METHOD SELECTION

A substance, such as carbon monoxide, exists only as a gas; an inorganic compound like iron oxide exists only as a particle. Many substances exist as either particles or vapors, however; substances that are gases can be attached by some means to particulate matter in the air.

To conduct sampling, environmental engineers must have prior knowledge of the physical state in which a substance exists or make a judgment. Devices suitable for collecting particulate matter do not usually collect gases or vapors; hence, selection of an incorrect sampling method can lead to erroneous results. Fortunately, considerable knowledge concerning the more common pollutants is available, and in most cases, selecting a suitable sampling method is not difficult.

GENERAL AIR SAMPLING PROBLEMS

Certain general observations related to sampling ambient air must be recognized. For example, the quantity of a substance contained in a volume of air is often extremely small; therefore, the sample size for the analytical method must be adequate. Even heavily polluted air is not likely to contain more than a few milligrams per cubic meter of most contaminants; and frequently, the amount present is best measured in micrograms, or even nanograms, per cubic meter.

For example, the air quality standard for particulates is $75 \ \mu g/M^3$. A cubic meter of air, or 35.3 cu ft, is a large volume for many sampling devices, and a considerable sampling period is required to draw such a quantity of air through the sampler. When atmospheric mercury analyses are made, the environmental engineer must realize that background levels are likely to be as low as several nanograms per cubic meter. In general, most substances are of concern at quite low levels in ambient air.

In addition to the problems from low concentrations of the substances being sampled, the reactivity of some substances causes other problems, resulting in changes after collection and necessitating special measures to minimize such changes.

Whenever a substance is removed from a volume of air by sampling procedures, the substance is altered, and the analysis can be less informative or even misleading. Ideally, environmental engineers should perform analyses of the unchanged atmosphere using direct-reading devices that give accurate information concerning the chemical and physical state of contaminants as well as concentration information. Such instruments exist for some substances, and many more are being developed. However, conventional air sampling methods are still used in many instances and will continue to be required for some time.

Gas and Vapor Sampling

The methods for gas and vapor sampling include collection in containers or bags, absorption, adsorption, and freeze-out sampling.

COLLECTION IN CONTAINERS OR BAGS

The simplest method of collecting a sample of air for analysis is to fill a bottle or other rigid container with it or to use a bag of a suitable material. Although sampling by this method is easy, the sample size is distinctly limited, and collecting a large enough sample for subsequent analysis may not be possible.

Bottles larger than several liters in capacity are awkward to transport; and while bags of any size are conveniently transported when empty, they can be difficult to handle when inflated. Nevertheless, collecting several samples in small bags can prove more convenient than taking more complex sampling apparatus to several sampling sites. If analyzing the contaminant is possible by GC procedures or gas-phase infrared spectroscopy, samples as small as a liter or less can be adequate and can be easily collected with bags.

Several methods exist for filling a rigid container such as a bottle. One method is to evacuate the bottle beforehand, then fill it at the sampling site by drawing air into the bottle and resealing it (see Figure 2.2.1). Alternatively, a bottle can be filled with water, which is then allowed to drain and fill with the air. A third method consists of passing a sufficient amount of air through the bottle using a pumping device until the original air is completely displaced by the air being sampled.

Plastic bags are frequently filled by use of a simple hand-operated squeeze bulb with valves on each end (see Figure 2.2.1) that are connected to a piece of tubing attached to the sampling inlet of the bag. In most cases, this procedure is satisfactory; but the environmental engineer must be careful to avoid contaminating the sampled air with the sampling bulb or losing the constituent on the walls of the sampling bulb. Problems of this kind can be avoided when the bag is placed in a rigid container, such as a box, and air is withdrawn from the box so that a negative pressure is created, resulting in air being drawn into the bag.

Selecting bag materials requires care; some bags permit losses of contaminants by diffusion through the walls, and others contribute contaminants to the air being sampled. A number of polymers have been studied, and several are suitable for many air sampling purposes. Materials suitable for use as sampling bags include Mylar, Saran, Scotchpak (a laminate of polyethylene, aluminum foil, and Teflon), and Teflon.

Even when a bag is made of inert materials, gas-phase chemical reactions are always possible, and after a period of time, the contents of the bag are not identical in composition to the air originally sampled. Thus, a reactive gas like sulfur dioxide or nitric oxide gradually oxidizes, depending on the storage temperature. Generally, analyses should be performed as soon as possible after the samples are collected. Losses by adsorption or diffusion are also greater with the passage of time and occur to some extent even when the best available bag materials are used.

The use of small bags permits the collection of samples to be analyzed for a relatively stable gas, such as carbon monoxide, at a number of locations throughout a community, thus permitting routine air quality measurements that might otherwise be inordinately expensive.

ABSORPTION

Most air sampling for gases and vapors involves absorbing the contaminant in a suitable sampling medium. Ordinarily, this medium is a liquid, but absorption can also take place in solid absorbents or on supporting materials such as filter papers impregnated with suitable absorbents. Carbon dioxide, for example, is absorbed in a granular bed of alkaline material, and sulfur dioxide is frequently measured by the absorption of reactive chemicals placed on a cloth or ceramic support. Environmental engineers also detect a number of gases by passing them through filter papers or glass tubes containing reactive chemicals. The reaction produces an immediate color change that can be evaluated by eye to measure the concentration of a substance.

Most commonly, however, gases and vapors are absorbed when they are passed through a liquid in which they are soluble or which contains reactive chemicals that combine with the substance being sampled. Many absorption vessels have been designed, ranging from simple bubblers, made with a piece of tubing inserted beneath the surface of a liquid, to complex gas-washing devices, which increase the time the air and liquid are in contact with each other (see Figure 2.2.2).

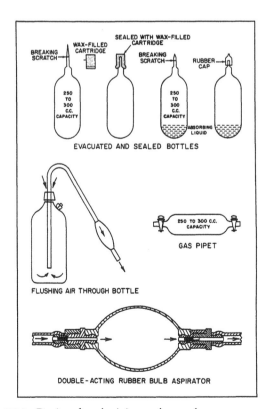

FIG. 2.2.1 Devices for obtaining grab samples.

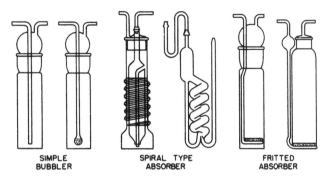

FIG. 2.2.2 Gas-absorbing vessels.

The impinger is probably the most widely used device. It is available in several sizes and configurations. An impinger consists of an entrance tube terminating in a small orifice that causes the velocity of the air passing through the orifice to increase. When this jet of air strikes a plate or the bottom of the sampling vessel at an optimal distance from the orifice, an intense impingement or bubbling action occurs. This impingement results in more efficient absorption of gases from the airstream than takes place if the air is simply bubbled through at low velocity. The two most frequently used impingement devices are the *standard impinger* and the *midget impinger* (see Figure 2.2.3). They are designed to operate at an airflow of 1 cubic foot per minute (cfm) and 0.1 cfm, or 28.3 and 2.8 lpm, respectively. Using such devices for sampling periods of 10

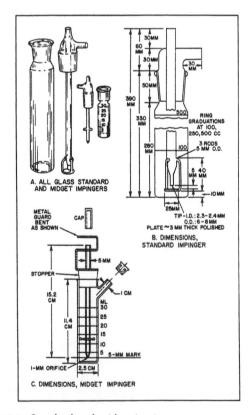

FIG. 2.2.3 Standard and midget impingers.

or 30 min results in a substantial amount of air passing through the devices, thus permitting low concentrations of trace substances to be determined with improved sensitivity and accuracy. Many relatively insoluble gases, such as nitrogen dioxide, are not quantitatively removed by passing through an impinger containing the usual sampling solutions.

The most useful sampling devices for absorbing trace gases from air are those in which a gas dispersion tube made of fritted or sintered glass, ceramic, or other material is immersed in a vessel containing the absorption liquid (see Figure 2.2.2). This device causes the gas stream to be broken into thousands of small bubbles, thus promoting contact between the gas and the liquid with resulting high-collection efficiencies. In general, fritted absorbers are more applicable to sampling gases and vapors than impingers and are not as dependent on flowrates as impingers. Fritted absorbers are available from scientific supply companies and come in various sizes suitable for many sampling tasks. Prefiltering the air prior to sampling with a fritted absorber is advisable to prevent the gradual accumulation of dirt within the pores of the frit.

The use of solid absorbents is not widely practiced in ambient air sampling because the quantity of absorbed gases is usually determined by gravimetric means. With the exception of carbon dioxide, few atmospheric gases lend themselves to this type of analysis.

ADSORPTION

Adsorption, by contrast with absorption, consists of the retention of gaseous substances by solid adsorbents which, in most cases, do not chemically combine with the gases. Instead, adsorptive forces hold the gases or vapors which can subsequently be removed unchanged. Any solid substance adsorbs a small amount of most gases; but to be useful as an adsorbent, a substance must have a large surface area and be able to concentrate a substantial amount of gas in a small volume of adsorbent.

Activated carbon or charcoal and activated silica gel are most widely used for this purpose. A small quantity of either adsorbent placed in a U-tube or other container through which air is passed quantitatively removes many vapors and gases from a large volume of air. These gases can then be taken to the laboratory where desorption removes the collected substances for analysis. Desorption commonly involves heating the adsorbent and collecting the effluent gases or eluting the collected substances with a suitable organic liquid.

For most organic vapors, subsequent analysis by GC, infrared (IR), or UV spectroscopy is most convenient. For some purposes, either silica gel or activated carbon is used; but the use of silica gel is not recommended because it also adsorbs water vapor, and a short sampling period in humid air can saturate the silica gel before sufficient contaminant is adsorbed. Because charcoal does not adsorb water, it can be used in humid environments for days or even weeks if the concentration of the contaminant is low.

The ease of sampling using adsorbents is offset somewhat by the difficulty of quantitatively desorbing samples for analysis. When published data are not available to predict the behavior of a new substance, the environmental engineer should perform tests in the laboratory to determine both the collection efficiency and the success of desorption procedures after sample collection.

FREEZE-OUT SAMPLING

Vapors or gases that condense at a low temperature can be removed from the sampled airstream by passage through a vessel immersed in a refrigerating liquid. Table 2.2.1 lists liquids commonly used for this purpose. Usually forming a sampling train in which two or three coolant liquids progressively lower the air temperature in its passage through the system is recommended.

All freeze-out systems are hampered somewhat by the accumulation of ice from water vapor and can eventually become plugged with ice. Flow rates through a freeze-out train are also limited because a sufficient residence time in the system is necessary for the air to be cooled to the required degrees. For these reasons and because of the inconvenience of assembling freeze-out sampling trains, they are not used for routine sampling purposes unless no other approach is feasible.

However, freeze-out sampling is an excellent means of collecting substances for research studies inasmuch as the low temperatures tend to arrest further chemical changes and ensure that the material being analyzed remains in the sampling container ready for analysis after warming. Analysis is usually conducted by GC, IR, or UV spectrophotometry or by mass spectrometry.

Particulate Matter Sampling

Particulate matter sampling methods include filtration, impingement and impaction, electrostatic precipitation, and thermal precipitation.

TABLE 2.2.1   COOLANT SOLUTIONS FOR FREEZE-OUT SAMPLING

Coolant	Temperature (°C)
Ice water	0
Ice and salt (NaCl)	−21
Dry ice and isopropyl alcohol	−78.5
Liquid air	−147
Liquid oxygen	−183
Liquid nitrogen	−196

FILTRATION

Passing air through a filter is the most convenient method to remove particulate matter (see Figure 2.2.4). Before filtration is used to obtain a sample, however, the purpose of the sample should be considered. Many filters collect particulates efficiently, but thereafter removing the collected matter may be impossible except by chemical treatment.

If samples are collected for the purpose of examining the particles and measuring their size or noting morphological characteristics, many filters are not suitable because the particles are imbedded in the fibrous web of the filter and cannot readily be viewed or removed. If the sample is collected for the purpose of performing a chemical analysis, the filter must not contain significant quantities of the substance. If the purpose of sampling is to collect an amount of particulate matter for weighing, then a filter which can be weighed with the required precision must be selected. Many filtration materials are hygroscopic and change weight appreciably in response to changes in the relative humidity.

Filters can be made of many substances, and almost any solid substance could probably be made into a filter. In practice, however, fibrous substances, such as cellulose or paper, fabrics, and a number of plastics or polymerized materials, are generally used. The most readily available filters are those made of cellulose or paper and used in chemical laboratories for filtering liquids. These filter papers come in a variety of sizes and range in efficiency from loose filters that remove only larger particles to papers that remove fine particles with high efficiency. All filters display similar behavior, and ordinarily a high collection efficiency is accompanied by increased resistance to airflow.

Certain filtration media are more suited to air sampling than most paper or fibrous filters. Of these, membrane filters have the greatest utility, and commercially available membrane filters combine high-collection efficiencies with low resistance to flow. Such filters are not made up of a fibrous mat but are usually composed of gels of cellulose esters or other polymeric substances that form a smooth surface of predictable characteristics.

Such filters contain many small holes, or pores, and can be made to exacting specifications so that their performance characteristics can be predicted. In addition, the filters usually have high chemical purity and are well suited to trace metal analyses. Some membrane filters can also be transparent and thus permit direct observation of collected particles with a microscope. Alternatively, the filters can be dissolved in an organic solvent, and the particles can be isolated and studied. Most membrane filters are not affected by relative humidity changes and can be weighed before and after use to obtain reliable gravimetric data.

Another kind of filter that is widely used in sampling ambient air is the fiberglass filter. These filters are originally made of glass fibers in an organic binder.

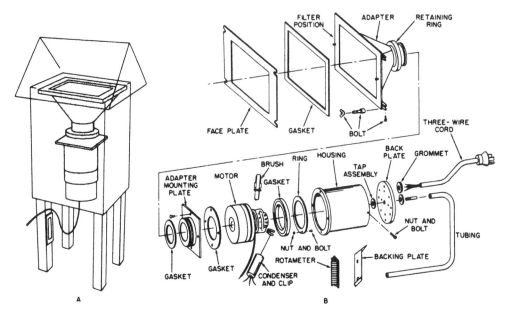

FIG. 2.2.4 High-volume sampler: (A) assembled sampler and shelter; (B) exploded view of typical high-volume air sampler.

Subsequently, the organic binder is removed by firing, leaving a web of glass that is efficient in collecting fine particles from the air. The principal advantages of using this type of filter are its low resistance to air flow and its unchanging weight regardless of relative humidity.

However, these filters are not well suited to particle size studies. In addition, they are not chemically pure, and the environmental engineer must ensure that the filter does not contribute the substance being analyzed in unknown quantities. In the United States, most data relating to suspended particulate matter in our cities have been obtained on filters made of fiberglass and used in conjunction with a sampling device referred to as a high-volume sampler (see Figure 2.2.4). Many other kinds of filters are available, but most sampling needs are well met by membrane or fiberglass filters.

IMPINGEMENT AND IMPACTION

The impingers previously described for sampling gases and vapors (see Figure 2.2.3) can also be used for the collection of particles and, in fact, were originally developed for that purpose. However, in ambient air sampling, they are not used because their collection efficiency is low and unpredictable for the fine particles present in ambient air. The low sampling rates also make them less attractive than filters for general air sampling, but instances do arise when impingers can be satisfactorily used. When impingers are used, the correct sampling rates must be maintained since the collection efficiency of impingers for particles varies when flow rates are not optimal.

Impactors are more widely used in ambient air sampling. In these devices, air is passed through small holes or orifices and made to impinge or impact against a solid surface. When these devices are constructed so that the air passing through one stage is subsequently directed onto another stage containing smaller holes, the resulting device is known as a *cascade impactor* and has the capability of separating particles according to sizes.

Various commercial devices are available. Figure 2.2.5 shows one impactor that is widely used and consists of

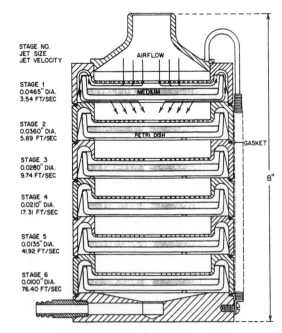

FIG. 2.2.5 Cascade impactor (Andersen sampler).

several layers of perforated plates through which the air must pass. Each plate contains a constant number of holes, but the hole size progressively decreases so that the same air volume passing through each stage impinges at an increased velocity. The result is that coarse particles are deposited on the first stage and successively finer particles are removed at each subsequent stage. Although these instruments do not achieve exact particle size fractions, they do perform predictably when the characteristics of the aerosol being sampled are known.

In use, an environmental engineer assembles a cascade impactor after scrupulously cleaning each stage and applying, if necessary, a sticky substance or a removable surface on which the particles are deposited. After a period of sampling during which time the volume of air is metered, the stages can be removed, and the total weight of each fraction is determined, as well as its chemical composition. Such information is sometimes more useful than a single weight or chemical analysis of the total suspended particulate matter without regard to its particle size.

ELECTROSTATIC PRECIPITATION

Particulate matter can be quantitatively removed from air by ESPs. Devices that operate on the same principal but are much larger are frequently used to remove particulate matter from stack gases prior to discharging into the atmosphere. Several commercially available ESPs can be used for air sampling, and all operate on the same general principle of passing the air between charged surfaces, imparting a charge to particles in the air, and collecting the particles on an oppositely charged surface or plate.

In one of widely used commercial devices (see Figure 2.2.6), a high-voltage discharge occurs along a central wire; the collecting electrode is a metallic cylinder placed around the central wire while the air is passing through the tube. An intense corona discharge takes place on the central wire; the particles entering the tube are charged and are promptly swept to the walls of the tube where they remain firmly attached. With this method, collecting a sample for subsequent weighing or chemical analysis and examining the particles and studying their size and shape is possible.

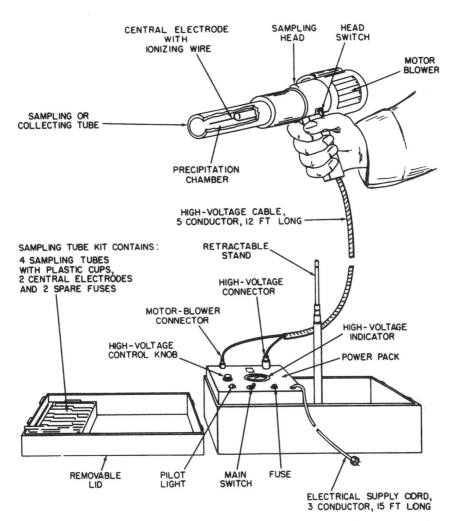

FIG. 2.2.6 Electrostatic precipitator. (Reprinted, with permission, from Mine Safety Appliances Company.)

However, the intense electrical forces can produce aggregates of particles that are different from those in the sampled air.

ESPs are not as widely used as filters for ambient air sampling because they are less convenient and tend to be heavy due to the power pack necessary to generate the high voltage. Nevertheless, they are excellent instruments for obtaining samples for subsequent analysis and sample, at high-flow rates and low resistance.

THERMAL PRECIPITATORS

Whenever a strong temperature gradient exists between two adjacent surfaces, particles tend to be deposited on the colder of these surfaces. Collecting aerosols by this means is termed *thermal precipitation,* and several commercial devices are available.

Because thermal forces are so weak, a large temperature difference must be maintained in a small area, and the air flow rate between the two surfaces must be low in order not to destroy the temperature gradient and to permit particles to be deposited before moving out of the collection area. As a result of these requirements, most devices use a heated wire as the source of the temperature differential and deposit a narrow ribbon of particles on the cold surface. Airflows are small, on the order of 10 to 25 ml per minute. At such low rates, the amount of material collected is normally insufficient for chemical analysis or weight determinations but is ample for examination by optical or electron microscopy.

Collection for particle size studies is the principal use for thermal precipitation units, and they are well suited to collecting samples for such investigations. Because the collecting forces are gentle, the particles are deposited unchanged. The microscopic examination gives information that can be translated into data concerning the number of particles and their morphological characteristics. The use of a small grid suitable for insertion into an electron microscope is also convenient as the collecting surface; this use eliminates additional manipulation of the sample prior to examination by electron microscopy.

Air Quality Monitoring Systems

Many options are available for the type of air quality information which can be collected, and the cost of air quality monitoring systems varies greatly. Only with a thorough understanding of the decisions which must be made based on the information received from the air quality monitoring system can an environmental engineer make an appropriate selection.

Regardless of the instruments used to measure air quality, the data are only as good as they are representative of the sampling site selected.

The simplest air quality monitors are static sensors that are exposed for a given length of time and are later ana-

lyzed in the laboratory. In some cases, these devices provide all the required information. More commonly, a system of automatic instruments measuring several different air quality parameters is used. When more than a few instruments are used, the signals from these instruments can be retained on magnetic tape rather than on recorder charts.

The most common errors in the design of air quality monitoring systems are poor site location and the acquisition of more data than necessary for the purpose of the installation.

PURPOSES OF MONITORING

The principal purpose of air quality monitoring is to acquire data for comparison to regulated standards. In the United States, standards have been promulgated by the federal government and by many states. Where possible, these standards are based on the physiological effect of the air pollutant on human health. The averaging time over which various concentration standards must be maintained differs for each pollutant.

Some air quality monitoring systems determine the impact of a single source or a concentrated group of sources of emissions on the surrounding area. In this case, determining the background level, the maximum ground-level concentration in the area, and the geographical extent of the air pollutant impact is important.

When the source is isolated, e.g., a single industrial plant in a rural area, the system design is straightforward. Utilizing the meteorological records available from nearby airports or government meteorological reporting stations, environmental engineers can prepare a wind rose to estimate the principal drift direction of the air pollutant from the source. Then they can perform dispersion calculations to estimate the location of the expected point of maximum ground-level concentration. As a general rule with stacks between 50 and 350 ft tall, this point of maximum concentration is approximately 10 stack heights downwind.

The air quality monitoring system should include at least one sensor at the point of expected maximum ground-level concentration. Additional sensors should be placed not less than 100 stack heights upwind (prevailing) to provide a background reading, and at least two sensors should be placed between 100 and 200 stack heights downwind to determine the extent of the travel of the pollutants from the source. If adequate resources are available, sampling at the intersection points of a rectilinear grid with its center at the source is recommended.

With such a system for an isolated source, environmental engineers can obtain adequate data in a one-year period to determine the impact of the source on the air quality of the area. Because of the variability in climatic conditions on an annual basis, few areas require less than one year of data collection to provide adequate information. If a study is performed to determine the effect of process changes on air quality, the study may have to continue for two to five years to develop information that is statistically reliable.

Some air quality monitoring is designed for the purpose of investigating complaints concerning an unidentified source. This situation usually happens in urban areas for odor complaints. In these cases, human observers use a triangulation technique to correlate the location of the observed odor with wind direction over several days. Plotting on a map can pinpoint the offending source in most cases. While this technique is not an air pollution monitoring system in an instrumental sense, it is a useful tool in certain situations.

Air pollution research calls for a completely different approach to air quality monitoring. Here, the purpose is to define some unknown variable or combination of variables. This variable can be either a new atmospheric phenomenon or the evaluation of a new air pollution sensor. In the former case, the most important consideration is the proper operation of the instrument used. In the latter case, the most important factor is the availability of a reference determination to compare the results of the new instrument against.

MONITORING IN URBAN AREAS

Urban areas are of major interest in air pollution monitoring in the United States since most of the population lives in these areas. The most sophisticated and expensive air quality monitoring systems in the United States are those for large cities (and one or two of the largest states) where data collection and analysis are centralized at a single location through the use of telemetry. Online computer facilities provide data reduction.

The three philosophies that can be used in the design of an urban air quality monitoring system are locating sensors on a uniform area basis (rectilinear grid), locating sensors at locations where pollutant concentrations are high, and locating sensors in proportion to the population distribution. The operation of these systems is nearly identical, but the interpretation of the results can be radically different.

The most easily designed systems are those where sensors are located uniformly on a geographical basis according to a rectilinear grid. Because adequate coverage of an urban area frequently requires at least 100 sensors, this concept is usually applied only with static or manual methods of air quality monitoring.

Locating air quality sensors at points of maximum concentration indicates the highest levels of air pollutants encountered throughout the area. Typically, these points include the central business district and the industrial areas on the periphery of the community. This type of data is useful when interpreted in the context of total system design. One or two sensors are usually placed in clean or background locations, so that the environmental engineer can estimate the average concentration of air pollutants over the entire area. The basis of this philosophy is that

if the concentrations in the dirtiest areas are below air quality standards, certainly the cleaner areas will have no problems.

The design of air quality monitoring systems based on population distribution means placing air quality sensors in the most populous areas. While this philosophy may not include all high-pollutant concentration areas in the urban region, it generally encompasses the central business district and is a measure of the air pollution levels to which most of the population are exposed. The average concentrations from this type of sampling network are an adequate description from a public health standpoint. This system design can, however, miss some localized high concentration areas.

Before the system is designed, the system designer and those responsible for interpreting the data must agree on the purpose of air quality monitoring.

SAMPLING SITE SELECTION

Once the initial layout has been developed for an air quality monitoring system, specific sampling sites must be located as close as practical to the ideal locations. The major considerations are the lack of obstruction from local interferences and the adequacy of the site to represent the air mass, accessibility, and security.

Local interferences cause major disruptions to air quality sensor sites. A sampler inlet placed at a sheltered interior corner of a building is not recommended because of poor air motion. Tall buildings or trees immediately adjacent to the sampling site can also invalidate most readings.

The selection of sampling sites in urban areas is complicated by the canyon effect of streets and the high density of pollutants, both gaseous and particulate, at street level. In order for the data to represent the air mass sampled, environmental engineers must again review the purpose of the study. If the data are collected to determine areawide pollutant averages, the sampler inlet might best be located in a city park, vacant lot, or other open area. If this location is not possible, the sampler inlet could be at the roof level of a one- or two-story building so that street-level effects are minimized. On the other hand, if the physiological impact of air pollutants is a prime consideration, the samplers should be at or near the breathing level of the people exposed. As a general rule, an elevation of 3 to 6 m above the ground is an optimum elevation.

The sampling site location can be different for the same pollutant depending on the purpose of the sampling. Carbon monoxide sampling is an example. The federally promulgated air quality standards for carbon monoxide include both an 8-hr and a 1-hr concentration limit. Maximum 1-hr concentrations are likely to be found in a high-traffic density, center-city location. People are not ordinarily exposed to these concentrations for 8-hr periods. When sampling for comparison with the 1-hr standard, the environmental engineer should locate the sensor within about 20 ft of a major traffic intersection. When sampling for comparison with the 8-hr standard, the engineer should locate the sensor near a major thoroughfare in either the center-city area or in the suburban area with the sampler less than 50 ft from the intersection. The reason for two different sampling site locations is to be consistent with the physiological effects of carbon monoxide exposure and the living pattern of most of the population. If only one site can be selected, the location described for the 8-hr averaging time is recommended.

When the sampling instruments are located inside a building and an air sample is drawn in from the outside, using a sampling pipe with a small blower is advantageous to bring outside air to the instrument inlets. This technique improves sampler line response time. An air velocity of approximately 700 ft per min in the pipe balances problems of gravitational and inertial deposition of particulate matter when particulates are sampled.

The sampling site should be accessible to operation and maintenance personnel. Since most air pollution monitoring sites are unattended much of the time, sample site security is a real consideration; the risk of vandalism is high in many areas.

STATIC METHODS OF AIR MONITORING

Static sensors used to monitor air quality require the minimum capital cost. While averaging times are in terms of weeks and sensitivity is low, in many cases, static monitors provide the most information for the amount of investment. Although inexpensive, they should not be rejected but should be considered as useful adjuncts to more sophisticated systems.

Dustfall Jars

The simplest of all air quality monitoring devices is the dustfall jar (see Figure 2.2.7). This device measures the fallout rate of coarse particulate matter, generally above about 10 μm in size. Dustfall and odor are two major reasons for citizen complaints concerning air pollution. Dustfall is offensive because it builds up on porches and automobiles and is highly visible and gritty to walk upon.

Dustfall seldom carries for distances greater than $\frac{1}{2}$ mi because these large particles are subject to strong gravitational effects. For this reason, dustfall stations must be spaced more closely than other air pollution sensors for a detailed study of an area.

Dustfall measurements in large cities in the United States in the 1920s and 1930s commonly indicated dustfall rates in hundreds of tn per sq mi per mon. These levels are considered excessive today, as evidenced by the dustfall standards of 25 to 30 tn per sq mi per mon promulgated by many of the states. While the measurement of low or moderate values of dustfall does not indicate freedom from air pollution problems, measured dustfall values in excess of

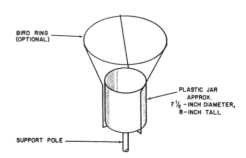

FIG. 2.2.7 Dustfall jar.

50 to 100 tn per sq mi per mon are an indication of excessive air pollution.

The large size of the particulate matter found in dustfall jars makes it amenable to chemical or physical analysis by such techniques as microscopy. These analyses are useful to identify specific sources.

Lead Peroxide Candles

For many years, sulfur dioxide levels have been determined through the use of lead peroxide candles. These devices, known as candles because they are a mixture of lead peroxide paste spread on a porcelain cylinder about the size and shape of a candle, are normally exposed for periods of 1 mon. Sulfur gases in the air react with lead peroxide to form lead sulfate. Environmental engineers analyze the sulfate according to standard laboratory procedures to indicate the atmospheric levels of sulfur gases during the period of exposure.

A modification of this technique that simplifies the laboratory procedure is to use a fiber filter cemented to the inside of a plastic petri dish (a flat-bottom dish with shallow walls used for biological cultures). The filter is saturated with an aqueous mixture of lead peroxide and a gel, commonly gum tragacanth, and is allowed to dry. These dishes or plates are exposed in an inverted position for periods of 1 to 4 wk.

The lead peroxide estimation of sulfur dioxide has inherent weaknesses. All sulfur gases, including reduced sulfur, react with lead peroxide to form lead sulfate. More importantly, the reactivity of lead peroxide depends on its particle size distribution. For this reason, the results from different investigators are not directly comparable. Nevertheless, a network of lead peroxide plates over an area provides a good indication of the exposure to sulfur gases during the exposure period. This technique is useful for determining the geographical extent of sulfur pollution.

Other Static Methods

Environmental engineers have modified the technique of using fiber filter cemented to a petri dish by using sodium carbonate rather than lead peroxide to measure sulfur gases. This method also indicates the concentration of other gases, including nitrogen oxides and chlorides. Engineers have measured relative levels of gaseous fluoride air pollution using larger filters, e.g., 3-in diameter, dipped in sodium carbonate, and placed in shelters to protect them from the rain. With all these static methods, the accuracy is low, and the data cannot be converted directly into ambient air concentrations. They do, however, provide a low-cost indicator of levels of pollution in an area.

Environmental engineers have evaluated the corrosive nature of the atmosphere using standardized steel exposure plates for extended periods to measure the corrosion rate. This method provides a gross indication of the corrosive nature of the atmosphere. As with other static samplers, the results are not directly related to the concentration of air pollutants.

Manual Analyses

Manual analyses for air quality measurements are those that require the sample first be collected and then analyzed in the laboratory. Manual instruments provide no automatic indication of pollution levels.

The manual air sampling instrument in widest use is the high-volume sampler. With this method, ambient air is drawn through a preweighed filter at a rate of approximately 50 atmospheric cubic feet per minute (acfm) for a period of 24 hr. The filter is then removed from the sampler, returned to the laboratory, and weighed. The weight gain, combined with the measured air volume through the sampler, allows the particulate mass concentration, expressed in micrograms per cubic meter, to be calculated.

Reference methods for nearly all gaseous air pollutants involve the use of a wet sampling train in which air is drawn through a collecting medium for a period of time. The exposed collecting medium is then returned to the laboratory for chemical analysis. Sampling trains have been developed that allow sampling of five or more gases simultaneously into separate bubblers. Sequential samplers, which automatically divert the airflow from one bubbler to another at preset time intervals, are also available.

These sampling methods can be accomplished with a modest initial investment; however, the manpower required to set out and pick up the samples, combined with the laboratory analysis, raises the total cost to a point where automated systems can be more economical for long-term studies.

Instrumental Analyses

As the need for accurate data that can be statistically reduced in a convenient manner increases, automated sampling systems become necessary. The elements of an automated system include the air-flow handling system, the sensors, the data transmission storage and display apparatus, and the data processing facility. The overall system is no more valuable than its weakest link.

SENSORS

The output reliability from an air quality sensor depends on its inherent accuracy, sensitivity, zero drift, and calibration. The inherent accuracy and sensitivity are a function of the the instrument's design and its operating principle. Zero drift can be either an electronic phenomenon or an indication of difficulties with the instrument. In instruments that use an optical path, lenses become dirty. In wet chemical analyzers, the flow rates of reagents can vary, changing both the zero and the span (range) of the instrument. Because of these potential problems, every instrument should have routine field calibration at an interval determined in field practice as reasonable for the sensor.

Environmental engineers calibrate an air quality sensor using either a standard gas mixture or a prepared, diluted gas mixture using permeation tubes. In some cases, they can currently sample the airstream entering the sensor by using a reference wet chemical technique.

The operator of air quality sensors should always have a supply of spare parts and tools to minimize downtime. At a minimum, operator training should include instruction to recognize the symptoms of equipment malfunction and vocabulary to describe the symptoms to the person responsible for instrument repair. Ideally, each operator should receive a short training session from the instrument manufacturer or someone trained in the use and maintenance of the instrument so that the operator can make repairs on site. Since this training is seldom possible in practice, recognition of the symptoms of malfunction becomes increasingly important.

DATA TRANSMISSION

The output signal from a continuous monitor in an air quality monitoring system is typically the input to a strip chart recorder, magnetic tape data storage, or an online data transmission system. The output of most air quality sensors is in analog form. This form is suitable for direct input to a strip chart recorder; but in automated systems, the analog signal is commonly converted to a digital signal. For those sensors that have linear output, the signal can go directly to the recorder or transmission system. For sensors with logarithmic output, it may be advantageous to convert this signal to a linear form.

Many early automated air quality monitoring systems in the United States had difficulty with the data transmission step in the system. In some cases, this difficulty was caused by attempts to overextend the lower range of the sensors so that the signal-to-noise ratio was unfavorable. In other cases, matching the sensor signal output to the data transmission system was poor. With developments and improvements in systems, these early difficulties were overcome.

TABLE 2.2.2 COMPOUNDS THAT CAN BE ANALYZED BY THE MICROPROCESSOR-CONTROLLED PORTABLE IR SPECTROMETER

Compound	Range of Calibration (ppm)	Compound	Range of Calibration (ppm)
Acetaldehyde	0 to 400	Ethylene oxide	0 to 10 and 0 to 100
Acetic acid	0 to 50	Ethyl ether	0 to 1000 and 0 to 2000
Acetone	0 to 2000	Fluorotrichloromethane (Freon 11)	0 to 2000
Acetonitrile	0 to 200	Formaldehyde	0 to 20
Acetophenone	0 to 100	Formic acid	0 to 20
Acetylene	0 to 200	Halothane	0 to 10 and 0 to 100
Acetylene tetrabromide	0 to 200	Heptane	0 to 1000
Acrylonitrile	0 to 20 and 0 to 100	Hexane	0 to 1000
Ammonia	0 to 100 and 0 to 500	Hydrazine	0 to 100
Aniline	0 to 20	Hydrogen cyanide	0 to 20
Benzaldehyde	0 to 500	Isoflurane	0 to 10 and 0 to 100
Benzene	0 to 50 and 0 to 200	Isopropyl alcohol	0 to 1000 and 0 to 2000
Benzyl chloride	0 to 100	Isopropyl ether	0 to 1000
Bromoform	0 to 10	Methane	0 to 100 and 0 to 1000
Butadiene	0 to 2000	Methoxyflurane	0 to 10 and 0 to 100
Butane	0 to 200 and 0 to 2000	Methyl acetate	0 to 500
2-Butanone (MEK)	0 to 250 and 0 to 1000	Methyl acetylene	0 to 1000 and 0 to 5000
Butyl acetate	0 to 300 and 0 to 600	Methyl acrylate	0 to 50
n-Butyl alcohol	0 to 200 and 0 to 1000	Methyl alcohol	0 to 500 and 0 to 1000
Carbon dioxide	0 to 2000	Methylamine	0 to 50
Carbon disulfide	0 to 50	Methyl bromide	0 to 50
Carbon monoxide	0 to 100 and 0 to 250	Methyl cellosolve	0 to 50
Carbon tetrachloride	0 to 20 and 0 to 200	Methyl chloride	0 to 200 and 0 to 1000
Chlorobenzene	0 to 150	Methyl chloroform	0 to 500
Chlorobromomethane	0 to 500	Methylene chloride	0 to 1000
Chlorodifluoromethane	0 to 1000	Methyl iodide	0 to 40
Chloroform	0 to 100 and 0 to 500	Methyl mercaptan	0 to 100
m-Cresol	0 to 20	Methyl methacrylate	0 to 250
Cumene	0 to 100	Morpholine	0 to 50
Cyclohexane	0 to 500	Nitrobenzene	0 to 20
Cyclopentane	0 to 500	Nitromethane	0 to 200
Diborane	0 to 10	Nitrous oxide	0 to 100 and 0 to 2000
m-Dichlorobenzene	0 to 150	Octane	0 to 100 and 0 to 1000
o-Dichlorobenzene	0 to 100	Pentane	0 to 1500
p-Dichlorobenzene	0 to 150	Perchloroethylene	0 to 200 and 0 to 500
Dichlorodifluoromethane (Freon 12)	0 to 5 and 0 to 800	Phosgene	0 to 5
		Propane	0 to 2000
1,1-Dichloroethane	0 to 200	n-Propyl alcohol	0 to 500
1,2-Dichloroethylene	0 to 500	Propylene oxide	0 to 200
Dichloroethyl ether	0 to 50	Pyridine	0 to 100
Dichloromonofluoromethane (Freon 21)	0 to 1000	Styrene	0 to 200 and 0 to 500
		Sulfur dioxide	0 to 100 and 0 to 250
Dichlorotetrafluoroethane (Freon 114)	0 to 1000	Sulfur hexafluoride	0 to 5 and 0 to 500
		1,1,2,2-Tetrachloro 1,2-difluoroethane (Freon 112)	0 to 2000
Diethylamine	0 to 50	1,1,2,2-Tetrachloroethane	0 to 50
Dimethylacetamide	0 to 50	Tetrahydrofuran	0 to 500
Dimethylamine	0 to 50	Toluene	0 to 1000
Dimethylformamide	0 to 50	Total hydrocarbons	0 to 1000
Dioxane	0 to 100 and 0 to 500	1,1,2-TCA	0 to 50
Enflurane	0 to 10 and 0 to 100	TCE	0 to 200 and 0 to 2000
Ethane	0 to 1000	1,1,2-Trichloro 1,2,2-TCA (Freon 113)	0 to 2000
Ethanolamine	0 to 100		
2-Ethoxyethyl acetate	0 to 200	Trifluoromonobromomethane (Freon 13B1)	0 to 1000
Ethyl acetate	0 to 400 and 0 to 1000		
Ethyl alcohol	0 to 1000 and 0 to 2000	Vinyl acetate	0 to 10
Ethylbenzene	0 to 200	Vinyl chloride	0 to 20
Ethyl chloride	0 to 1500	Vinylidene chloride	0 to 20
Ethylene	0 to 100	Xylene (Xylol)	0 to 200 and 0 to 2000
Ethylene dibromide	0 to 10 and 0 to 50		
Ethylene dichloride	0 to 100		

Source: Foxboro Co.

Online systems, i.e., those that provide an instantaneous readout of the dynamic situation monitored by the system, add an additional step to air quality monitoring systems. Data from continuous monitors can be stored on magnetic tape for later processing and statistical reduction. In an online system, this processing is done instantaneously. The added expense of this sophistication must be evaluated in terms of the purposes of the air quality monitoring system. When decisions with substantial community impact must be made within a short time, this real-time capability may be necessary.

DATA PROCESSING

The concentration of many air pollutants follows a log normal rather than a normal distribution. In a log normal distribution, a plot of the logarithm of the measured values more closely approximates the bell-shaped Gaussian distribution curve than does a plot of the numerical data. Suspended particulate concentrations are a prime example of this type of distribution. In this case, the geometric mean is the statistical parameter that best describes the population of data. The arithmetic mean is of limited value because it is dominated by a few occurrences of high values. The geometric mean, combined with the geometric standard deviation, completely describes a frequency distribution for a log normally distributed pollutant.

Environmental engineers should consider the averaging time over which sample results are reported in processing and interpreting air quality data. For sulfur dioxide, various agencies have promulgated air quality standards based on annual arithmetic average, monthly arithmetic average, weekly arithmetic average, 24-hr arithmetic average, 3-hr arithmetic average, and 1-hr arithmetic average concentrations. The output of a continuous analyzer can be averaged over nearly any discrete time interval. To reduce the computation time, an environmental engineer must consider the time interval over which continuous analyzer output is averaged to obtain a discrete input for the calculations. If a 1-hr average concentration is the shortest time interval of value in interpreting the study results, using a 1- or 2-min averaging time for input to the computation program is not economic.

Environmental engineers must exercise caution in using strip chart recorders to acquire air quality data. The experience of many organizations, both governmental and industrial, is that the reduction of data from strip charts is tedious. Many organizations decide that they do not really need all that data once they find a large backlog of unreduced strip charts. Two cautions are suggested by this experience. First, only data that is to be used should be collected. Second, magnetic tape data storage followed by computer processing has advantages.

The visual display of air quality data has considerable appeal to many nontechnical personnel. Long columns of numbers can be deceptive if only one or two important trends are shown. The use of bar charts or graphs is frequently advantageous even though they do not show the complete history of air quality over a time span.

Portable and Automatic Analyzers

Microprocessor-controlled spectrometers are available to measure concentrations of a variety of gases and vapors in ambient air. These units can be portable or permanently installed and can comply with environmental and occupational safety regulations. In the IR spectrometer design, an integral air pump draws ambient air into the test cell, operating at a flow rate of 0.88 acfm (25 l/min). The microprocessor selects the wavelengths for the components and the filter wheel in the analyzer allows the selected wavelengths to pass through the ambient air sample in the cell. The microprocessor automatically adjusts the path length through the cell to give the required sensitivity. Because of the folded-path-length design, the path length can be increased to 20 m (60 ft), and the resulting measurement sensitivity is better than 1 ppm in many cases.

As shown in Table 2.2.2, practically all organic and some inorganic vapors and gases can be monitored by these IR spectrometers. The advantage of the microprocessor-based operation is that the monitor is precalibrated for the analysis of over 100 Occupational Safety and Health Administration (OSHA)-cited compounds. The memory capacity of the microprocessor is sufficient to accommodate another ten user-selected and user-calibrated gases. Analysis time is minimized because the microprocessor automatically sets the measurement wavelengths and parameters for any compound in its memory. A general scan for a contaminant in the atmosphere takes about 5 min, while the analysis of a specific compound can be completed in just a few minutes. The portable units are battery-operated for 4 hr of continuous operation and are approved for use in hazardous areas.

—R.A. Herrick
R.G. Smith
Béla G. Lipták

2.3
STACK SAMPLING

Type of Sample
  Gas containing particulates

Standard Design Pressure
  Generally atmospheric or near-atmospheric

Standard Design Temperature
  −25 to 1500°F (−32 to 815°C)

Sample Velocity
  400 to 10,000 ft (120 to 3000 m) per minute

Materials of Construction
  316 or 304 stainless steel for pitot tubes; 304 or 316 stainless, quartz, or Incoloy for sample probes

Partial List of Suppliers
  Andersen Samplers, Inc.; Applied Automation/Hartmann & Braun Process Control Systems Div.; Bacharach, Inc.; Columbia Scientific Industries Corp.; Cosa Instruments; Gastech, Inc.; Mine Safety Appliance Co., Instrument Division; Sensidyne, Inc. Gas & Particulate Detection Systems; Scientific Glass & Instruments Inc.; Sierra Monitor Corp.; Teledyne Analytical Instrument, Teledyne Inc.

A complete EPA particulate sampling system (reference method 5) is comprised of the following four subsystems (U.S. EPA 1971; 1974; 1977; 1987; 1989; 1991; Morrow, Brief, and Bertrand 1972):

1. A pitot tube probe or pitobe assembly for temperature and velocity measurements and for sampling
2. A two-module sampling unit that consists of a separate heated compartment with provisions for a filter assembly and a separate ice-bath compartment for the impinger train and bubblers

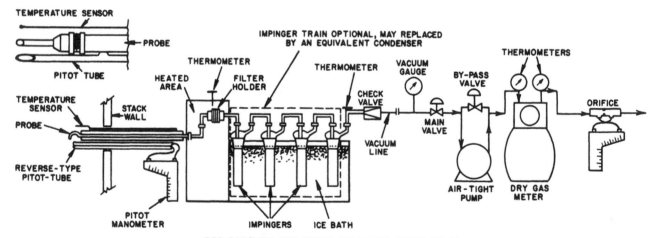

EPA PARTICULATE SAMPLING TRAIN (METHOD 5)
(FEDERAL REGISTER, VOL. 36, NOS. 234 AND 247)

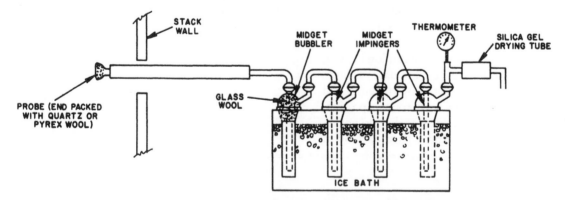

SAMPLING CASE FOR SO₂, SO₃ AND H₂SO₄ MIST (METHOD 8)

FIG. 2.3.1 (Top) EPA particulate sampling train (reference method 5); (Bottom) Sampling case for SO_2, SO_3, and H_2SO_4 mist (reference method 8). (Reprinted from *Federal Register* 36, nos. 234 and 247.)

3. An operating or control unit with a vacuum pump and a standard dry gas meter

4. An integrated, modular umbilical cord that connects the sample unit and pitobe to the control unit.

Figure 2.3.1 is a schematic of a U.S. EPA particulate sampling train (reference method 5). As shown in the figure, the system can be readily adapted for sampling sulfur dioxide (SO_2), sulfur trioxide (SO_3), and sulfuric acid (H_2SO_4) mist (reference method 8) (U.S. EPA 1971; 1977).

This section gives a detailed description of each of the four subsystems.

Pitot Tube Assembly

Procuring representative samples of particulates suspended in gas streams demands that the velocity at the entrance to the sampling probe be equal to the stream velocity at that point. The environmental engineer can equalize the velocities by regulating the rate of sample withdrawal so that the static pressure within the probe is equal to the static pressure in the fluid stream at the point of sampling. A specially designed pitot tube with means for measuring the pertinent pressures is used for such purposes. The engineer can maintain the pressure difference at zero by automatically controlling the sample drawoff rate.

Figure 2.3.2 shows the pitot tube manometer assembly for measuring stack gas velocity. The type S (Stauscheibe or reverse) pitot tube consists of two opposing openings, one facing upstream and the other downstream during the measurement. The difference between the impact pressure (measured against the gas flow) and the static pressure gives the velocity head.

Figure 2.3.3 illustrates the construction of the type S pitot tube. The external tubing diameter is normally between $\frac{3}{16}$ and $\frac{3}{8}$ in (4.8 and 9.5 mm). As shown in the figure, the distance is equal from the base of each leg of the tube to its respective face-opening planes. This distance (P_A and P_B) is between 1.05 and 1.50 times the external tube diameter. The face openings of the pitot tube should be aligned as shown.

Figure 2.3.4 shows the pitot tube combined with the sampling probe. The relative placement of these components eliminates the major aerodynamic interference effects. The probe nozzle has a bottom-hook or elbow design. It is made of seamless 316 stainless steel or glass with a sharp, tapered leading edge. The angle of taper should be less than 30°, and the taper should be on the outside to preserve a constant internal diameter. For the probe lining, either borosilicate or quartz glass probe liners are used for stack temperatures to approximately 900°F (482°C); quartz liners are used for temperatures between 900 and 1650°F (482 and 899°C). Although borosilicate or quartz glass probe linings are generally recommended, 316 stainless steel, Incoloy, or other corrosion-resistant metal can also be used.

INSTALLATION

The environmental engineer selects the specific points of the stack for sampling to ensure that the samples represent the material being discharged or controlled. The en-

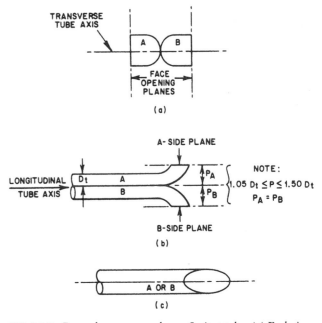

FIG. 2.3.3 Properly constructed type S pitot tube. (a) End view: face-opening planes perpendicular to transverse axis, (b) Top view: face-opening planes parallel to longitudinal axis, (c) Side view: both legs of equal length and center lines coincident; when viewed from both sides, baseline coefficient values of 0.84 can be assigned to pitot tubes constructed this way.

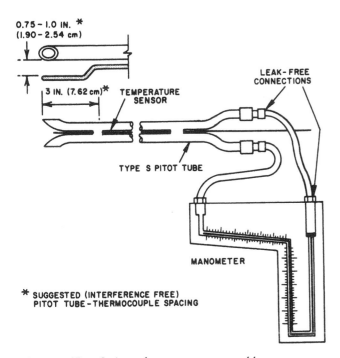

FIG. 2.3.2 Type S pitot tube manometer assembly.

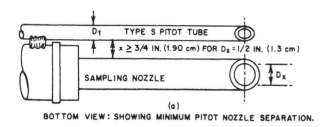

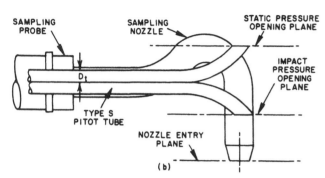

FIG. 2.3.4 Proper pitot tube with sampling probe nozzle configuration to prevent aerodynamic interference. (a) Bottom view: minimum pitot nozzle separation, (b) Side view: the impact pressure opening plane of the pitot tube located even with or above the nozzle entry plane so that the pitot tube does not interfere with gas flow streamlines approaching the nozzle.

gineer determines these points after examining the process or sources of emissions and their variation with time.

In general, the sampling point should be located at a distance equal to at least eight stack or duct diameters downstream and two diameters upstream from any source of flow disturbance, such as expansion, bend, contraction, valve, fitting, or visible flame. (Note: This eight-and-two-diameter criterion ensures the presence of stable, fully developed flow patterns at the test section.) For rectangular stacks, the equivalent diameter is calculated from the following equation:

$$\text{Equivalent diameter} = 2 \text{ (length} \times \text{width)/(length} + \text{width)}$$
$$2.3(1)$$

Next, provisions must be made to traverse the stack. The number of traverse points is 12. If the eight-and-two-diameter criterion is not met, the required number of traverse points depends on the sampling point distance from the nearest upstream and downstream disturbances. Figure 2.3.5 shows how to determine this number.

The cross-sectional layout and location of traverse points are as follows:

1. For circular stacks, the traverse points should be located on two perpendicular diameters as shown in Figure 2.3.6 and Table 2.3.1.
2. For rectangular stacks, the cross section is divided into as many equal rectangular areas as traverse points so that the length-to-width ratio of the elemental area is between one and two. The traverse points are located at the centroid of each equal area as shown in Figure 2.3.6.

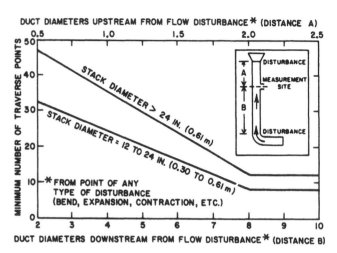

FIG. 2.3.5 Minimum number of traverse points for particulate traverses.

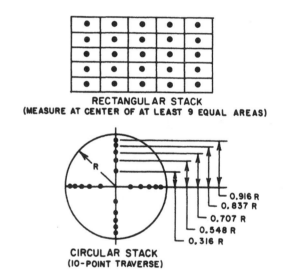

FIG. 2.3.6 Traverse point locations for velocity measurement or for multipoint sampling.

OPERATION

The environmental engineer measures the velocity head at various traverse points using the pitot tube assembly shown in Figure 2.3.2. The engineer collects the gas samples at a rate proportional to the stack gas velocity and analyzes them for carbon monoxide (CO), carbon dioxide (CO_2), and oxygen (O_2). Measuring the velocity head at a point in the flowing gas stream with both the type S pitot tube and a standard pitot tube with a known coefficient calibrates the pitot tube. Other data needed to calculate the volumetric flow are the stack temperature, stack and barometric pressures, and wet-bulb and dry-bulb temperatures of the gas sample at each traverse.

Table 2.3.2 gives the equations for converting pitot tube readings into velocity and mass flow and shows a typical data sheet for stack flow measurements (Morrow, Brief, and Bertrand 1972).

TABLE 2.3.1 LOCATION OF TRAVERSE POINTS IN CIRCULAR STACKS
(Percent of stack diameter from inside wall to traverse point)

Traverse Point Number on a Diameter	Number of Traverse Points on a Diameter											
	2	4	6	8	10	12	14	16	18	20	22	24
1	14.6	6.7	4.4	3.2	2.6	2.1	1.8	1.6	1.4	1.3	1.1	1.1
2	85.4	25.0	14.6	10.5	8.2	6.7	5.7	4.9	4.4	3.9	3.5	3.2
3		75.0	29.6	19.4	14.6	11.8	9.9	8.5	7.5	6.7	6.0	5.5
4		93.3	70.4	32.3	22.6	17.7	14.6	12.5	10.9	9.7	8.7	7.9
5			85.4	67.7	34.2	25.0	20.1	16.9	14.6	12.9	11.6	10.5
6			95.6	80.6	65.8	35.6	26.9	22.0	18.8	16.5	14.6	13.2
7				89.5	77.4	64.4	36.6	28.3	23.6	20.4	18.0	16.1
8				96.8	85.4	75.0	63.4	37.5	29.6	25.0	21.8	19.4
9					91.8	82.3	73.1	62.5	38.2	30.6	26.2	23.0
10					97.4	88.2	79.9	71.7	61.8	38.8	31.5	27.2
11						93.3	85.4	78.0	70.4	61.2	39.3	32.3
12						97.9	90.1	83.1	76.4	69.4	60.7	39.8
13							94.3	87.5	81.2	75.0	68.5	60.2
14							98.2	91.5	85.4	79.6	73.8	67.7
15								95.1	89.1	83.5	78.2	72.8
16								98.4	92.5	87.1	82.0	77.0
17									95.6	90.3	85.4	80.6
18									98.6	93.3	88.4	83.9
19										96.1	91.3	86.8
20										98.7	94.0	89.5
21											96.5	92.1
22											98.9	94.5
23												96.8
24												98.9

Based on the range of velocity heads, the environmental engineer selects a probe with a properly sized nozzle to maintain isokinetic sampling of particulate matter. As shown in Figure 2.3.7, a converging stream develops at the nozzle face if the sampling velocity is too high. Under this subisokinetic sampling condition, an excessive amount of lighter particles enters the probe. Because of the inertia effect, the heavier particles, especially those in the range of 3μ or greater, travel around the edge of the nozzle and are not collected. The result is a sample indicating an excessively high concentration of lighter particles, and the weight of the solid sample is in error on the low side.

Conversely, portions of the gas stream approaching at a higher velocity are deflected if the sampling velocity is below that of the flowing gas stream. Under this superisokinetic sampling condition, the lighter particles follow the deflected stream and are not collected, while the heavier particles, because of their inertia, continue into the probe. The result is a sample indicating a high concentration of heavier particles, and the weight of the solid sample is in error on the high side.

Isokinetic sampling requires precisely adjusting the sampling rate with the aid of the pitot tube manometer readings and nomographs such as APTD–0576 (Rom). If the pressure drop across the filter in the sampling unit becomes too high, making isokinetic sampling difficult to maintain, the filter can be replaced in the midst of a sample run.

Measuring the concentration of particulate matter requires a sampling time for each run of at least 60 min and

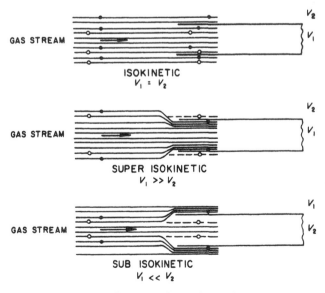

FIG. 2.3.7 Particle collection and sampling velocity.

TABLE 2.3.2 PITOT TUBE CALCULATION SHEET

Stack Volume Data

Stack No. .... Station .... Date .... Page ....

Name of Firm ....

Point	Position, in	Reading, H, in of H_2O	$\overline{H}$	Temperature t_b, °F	Velocity, V_b ft/sec
1					
2					
3					
4					
5					
6					
7					
8					
9					
10					
11					
12					
13					
14					
15					
16					
	Totals				
	Average				
	Absolute temperature, $T_s = t_s + 460 =$ °R.				

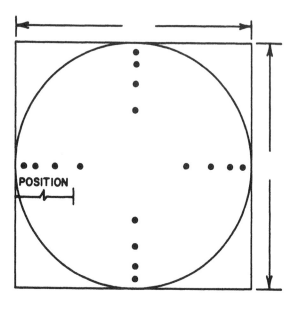

Dry-bulb temperature, $t_d =$ _____ °F   Barometer, $P_b =$ _____ in, Hg

Wet-bulb temperature, $t_u =$ _____ °F   Stack gage pressure = _____ in, H_2O

Absolute humidity, W = _____ lb H_2O/lb dry gas   Stack absolute pressure, $P_s =$ _____ $\dfrac{\text{in, }H_2O}{13.6} \pm P_b =$ _____ in, Hg

Stack area, $A_s =$ _____ sq ft   Pitot correction factor, $F_s =$ _____

Component	Volume Fraction, Dry Basis x	Molecular Weight	= Weight Fraction, Dry Basis
Carbon dioxide		44	=
Carbon monoxide		28	=
Oxygen		32	=
Nitrogen		28	=

Average dry gas molecular weight, M = _____

Specific gravity of stack gas, $G_s = \dfrac{0.62\,M\,(W+1)}{18+MW} = \dfrac{0.62 \times ____ \times ____}{18+_____} = $ _____

(Reference dry air at same conditions)

Velocity, $V_s = 2.9F_s\sqrt{\dfrac{29.92 \times T_s}{P_s \times G_s}}\sqrt{H} = 2.9 \times ____ \sqrt{\dfrac{29.92 \times __}{__ \times __}}\sqrt{H} = $ _____ ft/sec

Volume = _____ ft/sec × _____ sq. ft. × 60 _____ = _____ cfm

Standard volume = cfm × $\dfrac{530}{T_s} \times \dfrac{P_s}{29.92} = $ _____ × $\dfrac{530}{__} \times \dfrac{___}{29.92} = $ _____ scfm a minimum volume of 30 dry acfm (51 m³/hr) (U.S. EPA 1974).

Two-Module Sampling Unit

The two-module sampling unit has a separate heated compartment and ice-bath compartment.

HEATED COMPARTMENT

As shown in Figure 2.3.1, the probe is connected to the heated compartment that contains the filter holder and other particulate-collecting devices, such as the cyclone and flask. The filter holder is made of borosilicate glass, with a frit filter support and a silicone rubber gasket. The compartment is insulated and equipped with a heating system capable of maintaining the temperature around the filter holder during sampling at 248 ± 25°F (120 ± 14°C), or at another temperature as specified by the EPA. The thermometer should measure temperature to within 5.4°F (3°C). The compartment should have a circulating fan to minimize thermal gradients.

ICE-BATH COMPARTMENT

The ice-bath compartment contains the system's impingers and bubblers. The system for determining the stack gas moisture content consists of four impingers connected in series as shown in Figure 2.3.1. The first, third, and fourth impingers have the Greenburg–Smith design. The pressure drop is reduced when the tips are removed and replaced with a ½-in (12.5 mm) ID glass tube extending to ½ in (12.5 mm) from the bottom of the flask. The second impinger has the Greenburg–Smith design with a standard tip. During sampling for particulates, the first and second impingers are filled with 100 ml (3.4 oz) of distilled and deionized water. The third impinger is left dry to separate entrained water. The last impinger is filled with 200 to 300 g (7 to 10.5 oz) of precisely weighed silica gel (6 to 16 mesh) that has been dried at 350°F (177°C) for 2 hr to completely remove any remaining water. A thermometer, capable of measuring temperature to within 2°F (1.1°C), is placed at the outlet of the last impinger for monitoring purposes. Adding crushed ice during the run maintains the temperature of the gas leaving the last impinger at 60°F (16°C) or less.

Operating or Control Unit

As shown in Figure 2.3.1, the control unit consists of the system's vacuum pump, valves, switches, thermometers, and the totalizing dry gas meter and is connected by a vacuum line with the last Greenburg–Smith impinger. The pump intake vacuum is monitored with a vacuum gauge just after the quick disconnect. A bypass valve parallel with the vacuum pump provides fine control and permits recirculation of gases at a low-sampling rate so that the pump motor is not overloaded.

Downstream from the pump and bypass valve are thermometers, a dry gas meter, and calibrated orifice and inclined or vertical manometers. The calibrated orifice and inclined manometer indicate the instantaneous sampling rate. The totalizing dry gas meter gives an integrated gas volume. The average of the two temperatures on each side of the dry gas meter gives the temperature at which the sample is collected. The addition of atmospheric pressure to orifice pressure gives meter pressure.

Precise measurements require that the thermometers are capable of measuring the temperature to within 5.4°F (3°C); the dry gas meter is inaccurate to within 2% of the volume; the barometer is inaccurate within 0.25 mmHg (torr) (0.035 kPa); and the manometer is inaccurate to within 0.25 mmHg (torr) (0.035 kPa).

The umbilical cord is an integrated multiconductor assembly containing both pneumatic and electrical conductors. It connects the two-module sampling unit to the control unit, as well as connecting the pitot tube stack velocity signals to the manometers or differential pressure gauges.

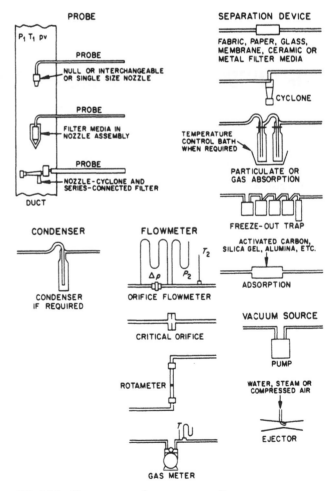

FIG. 2.3.8 Components of common sampling systems.

Sampling for Gases and Vapors

Some commonly used components in stack sampling systems are shown in Figure 2.3.8. If ball-and-socket joints and compression fittings are used, any arrangement of components is readily set up for field use. Environmental engineers select the stack sampling components on the basis of the source to be sampled, the substances involved, and the data needed.

Industrial hygienists developed a summary of sampling procedure outlines for specific substances (Vander Kolk 1980). After considering the complications that might arise from the presence of interfering substances in the gas samples, an environmental engineer should use the procedural outlines as a starting point in assembling a stack sampling system. The American Society for Testing and Materials (1971) provides other recommended sampling procedures for gases and vapors.

—David H.F. Liu

References

American Society for Testing and Materials. 1971. *Standards of methods for sampling and analysis of atmospheres.* Part 23.

Morrow, N.L., R.S. Brief, and R.R. Bertrand. 1972. Sampling and analyzing air pollution sources. *Chemical Engineering* 79, no. 2 (24 January): 84–98.

Rom, J.J. *Maintenance, calibration and operation of isokinetic source sampling equipment.* APTD-0576. U.S. EPA.

U.S. Environmental Protection Agency (EPA). 1971. Standards of performance for new stationary sources. *Federal Register* 36, no. 159 (17 August): 15,704–15,722.

———. 1974. Standards of performance for new stationary sources. *Federal Register* 30, no. 116 (14 June): 20,790–820,794.

———. 1977. Standards of performance for new stationary sources, Revision to reference method 1–8. *Federal Register* 42, no. 160 (18 August): 41,754–841,789.

———. 1987. Standards of performance for new stationary sources. *Federal Register* 52, no. 208 (28 October): 41,424–41,430.

———. 1989. Standards of performance for new stationary sources. *Federal Register* 54, no. 58 (28 March): 12,621–126,275.

———. 1990. Standards of performance for new stationary sources. *Federal Register* 55, no. 31 (14 February): 5,211–5,217.

———. 1991. Standards of performance for new stationary sources. *Federal Register* 56, no. 30 (13 February): 5,758–5,774.

Vander Kolk, A.L. 1980. *Michigan Department of Public Health, private communications.* (17 September).

2.4
CONTINUOUS EMISSION MONITORING

Requirements

More stringent clean air standards require more stringent monitoring of the release of pollutants into the atmosphere. The need is growing for reliable continuous emission monitoring (CEM) capabilities and for documenting the release amount from process plants. Therefore, a CEM system is an integral part of utility and industrial operations. For operators, collecting real-time emission data is the first step to attaining the nationally mandated reduction in SO_x and NO_x emissions. A company uses CEM to ensure compliance with the Acid Rain Program requirements of the CAAA.

A CEM system is defined by the U.S. EPA in Title 40, Part 60, Appendix B, Performance Specification 2 of the *Code of Federation Regulations,* as all equipment required to determine a gas concentration or emission rate. The regulation also defines a CEM system as consisting of subsystems that acquire, transport, and condition the sample, determine the concentration of the pollutant, and acquire and record the results. For the measurement of opacity, the specifics of the major subsystems are slightly different but basically the same.

The EPA has codified the standards of performance, equipment specifications, and installation and location specifications for the measurement of opacity, total reduced sulfur (TRS), sulfur dioxide, carbon dioxide, and hydrogen sulfide. These standards include requirements for the data recorder range, relative accuracy, calibration drift and frequency of calibration, test methods, and quarterly and yearly audits. The regulations also require opacity to be measured every 10 sec, the average to be recorded every 6 min, and pollutants to be measured a minimum of one cycle of sampling, analyzing, and data recording every 15 min. The readings from the gas analyzer must agree with a stack sampler to within 20% relative accuracy.

This section provides an overview of the CEM technology, the components of a proper analysis system, and some details in the reliable and accurate operation of CEM.

System Options

Figure 2.4.1 shows the various CEM options. All systems fall into one of two categories: in situ or extractive. Extractive units can be further classified into full-extractive (wet or dry basis) or dilution-extractive systems.

IN SITU SYSTEMS

An in situ system measures a gas as it passes by the analyzer in a stack. Figure 2.4.2 shows an in situ probe-type analyzer. The measurement cavity is placed directly into the sample flow system to measure the gas received on a wet basis.

The most commonly accepted in situ analyzer is the zirconium oxide (ZrO_2) oxygen analyzer. It is the most reliable method for measuring and controlling a combustion process. Since the introduction of ZrO_2 analyzers, many other gases have been measured in situ with light absorption instrumentation, such as UV and IR spectrometers. The gases that can be measured with such spectrometers include CO, CO_2, SO_2 and NO.

Most of the first analyzers designed to measure these gases have disappeared. Most noticeably, across-the-stack technology units are no longer used in this country due primarily to the promulgation of Title 40, Part 60,

Appendix F of the *Code of Federal Regulations* by the EPA in 1986. Appendix F requires that an analyzer be able to complete gas calibrations and that it be certified quarterly. An across-the-stack unit works like an opacity instrument in that it lacks the capability of using protocol gases for cylinder gas audits (CGA) as allowed by the EPA under Appendix F.

EXTRACTIVE SYSTEMS

Extractive CEM systems can be configured for either dry- or wet-basis measurements. Both configurations can achieve the CAAA-required 10% relative accuracy.

Dry-Extractive Systems

A standard extractive system (see Figure 2.4.3) extracts a gas sample from the stack and delivers it to an analyzer cabinet through heated sample lines. Filtration removes particulates at the sample probe. (However, uncontrolled condensation can block the sample line because it cannot trap and hold fines that pass through the stack filter.) The dry-extractive CEM system removes moisture through a combination of refrigeration, condensation, and permeation tube dryers that pass the sample through water-excluding membranes. This system helps keep the gas analyzer dry and removes any interferences caused by water.

To convert the volumetric concentration into a mass emission rate when using velocity-measuring flow moni-

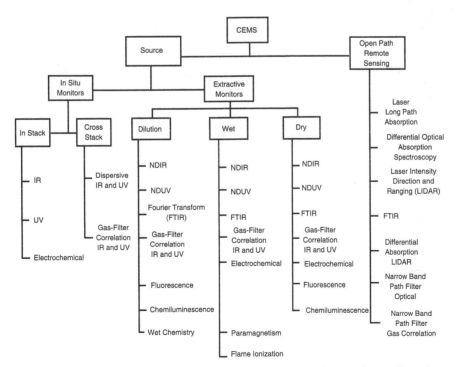

FIG. 2.4.1 CEM options. (Reprinted, with permission, from J. Schwartz, S. Sample, and R. McIlvaine, 1994, Continuous emission monitors—Issues and predictions, *Air & Waste*, Vol. 44 [January].)

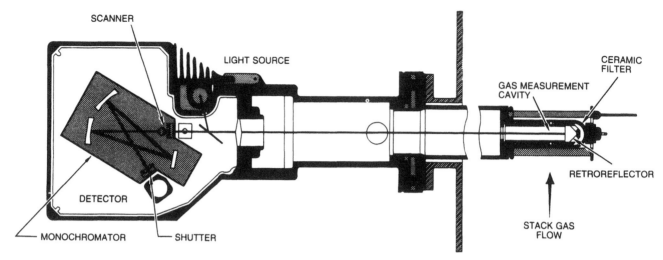

FIG. 2.4.2 Probe-type stack gas analyzer. (Reprinted, with permission, from Lear Siegler Inc.)

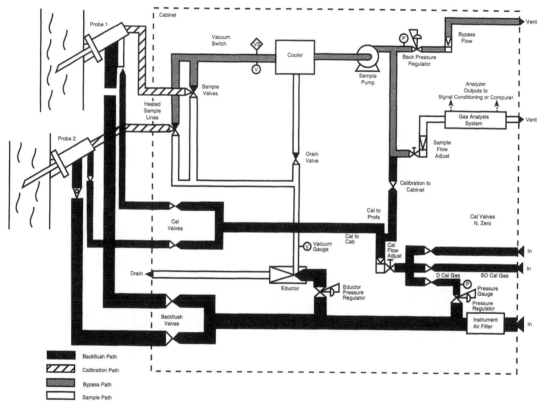

FIG. 2.4.3 Standard extractive system.

tors, the environmental engineer must measure or calculate the flue-gas moisture to account for the moisture removed before dry-extractive analysis. This additional measurement or calculation adds complexity to the system, which lowers the reliability and accuracy of dry-extractive CEM systems.

Operating an extractive system is complicated by the daily calibration, zero and span checks of the analyzers, and the need to backpurge the sample handling system to clear the probe and filters. These tasks are accomplished with additional lines and connections to the stack probe

and control valves. A controller system sequences the operation of all valves and controls gas flows.

The major advantages of extractive systems is their ability to share analyzers and spread the analyzer cost over several measurement points. An extractive system can be used to monitor two or eight points.

Wet-Extraction Systems

Wet-extraction CEM systems are similar to dry-extraction ones except that the sample is maintained hot, and moisture in the flue gas is retained throughout. Heat tracing is the critical component of a wet-extractive system. Sample temperatures must be maintained between 360° and 480°F to prevent acid gases from condensing within the sampling lines and analyzer.

Since nothing is done to the sample before analysis, wet-extractive CEM has the potential of being the most accurate measurement technique available. However, only a limited number of vendors supply wet-extractive CEM systems. In addition, increased flexibility and accuracy makes the wet-extractive CEM systems more costly than dry-extractive ones.

Dilution-Extractive Systems

The newest, accepted measurement system is dilution-extractive CEM. By precisely diluting the sample system at stack temperature with clean, dry (lower than −40°F dewpoint) instrument air, dilution-extractive systems eliminate the need for heat tracing and conditioning of extracted samples. Particulates are filtered out at the sample point. Thus, a dilution system measures all of the sample along with the water extracted with the sample on a wet basis. Figure 2.4.4 shows a schematic of a dilution-extractive system.

The dilution-extraction CEM system uses a stack dilution probe (see Figure 2.4.5) designed in Europe. A precisely metered quantity of flue gas is extracted through a critical orifice (or sonic orifice) mounted inside the probe. Dilution systems deliver the sample under pressure from the dilution air to the gas analyzers. Thus, the system pro-

tects the sample from any uncontrolled dilution from a leak in the sample line or system.

The environmental engineer selects the dilution ratios based on the expected water concentration in the flue gas and the lower limit of the ambient air temperature to avoid freezing. This kind of sample line is still considerably less expensive than the heated sample line for wet-extractive CEM.

When choosing the dilution ratio, the engineer must also consider the lowest pollutant concentration that can be detected by the monitoring device. These systems use a dilution ratio ranging from 12:1 to more than 700:1. Irrespective of the dilution ratio, the diluted sample must match the analyzer range.

SYSTEM SELECTION

The majority of the systems purchased for existing utility boilers are the dilution-extractive type. Beyond the utility industry, the dry-extractive, wet-extractive, and dilution-extractive systems are all viable choices. The accuracy, reliability, and cost differences must be considered in the selection of a system for a specific application. The CEM system selection process can also be limited by local regulatory agencies. These restrictions can include preapproval procedures that limit the type of system an agency accepts. For example, with some types of emission monitors, some states require that only dedicated, continuous measurement are used; that is no time-sharing. They can also define such terms as "continuous." This factor directly defines the requirements for a data acquisition system.

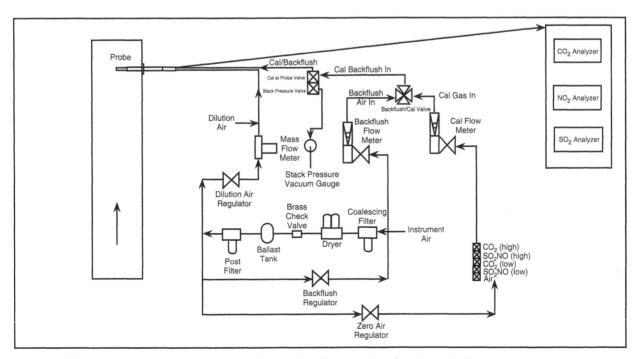

FIG. 2.4.4 Dilution-extractive system schematic. (Reprinted, with permission, from Lear Siegler.)

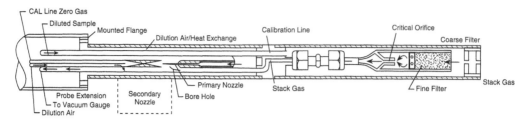

FIG. 2.4.5 Dilution probe.

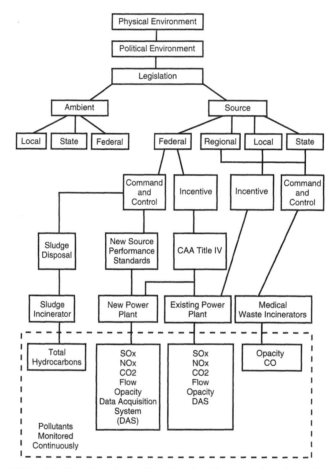

FIG. 2.4.6 Decision chain. (Reprinted, with permission, from J. Schwartz, S. Sample, and R. McIlvaine, 1994, Continuous emission monitors—Issues and predictions, *Air & Waste*, Vol. 44 [January].)

Continuous Emission Monitors

A complex combination of regulatory bodies determines which pollutants must be measured and in what manner the measurement must be made for a specific plant. Figure 2.4.6 shows the complexity of laws that affect the CEM design for a plant. For example, if the plant falls under the incentive program in Title IV of the 1990 CAA, this plant has an emission allowance in tons of SO_2 per year. It must continually monitor its mass flow and accurately record its total mass emissions in tons per year for the entire period. This requirement means that a data acquisition and reporting system must be incorporated into the CEM sys-

tem (see Figure 2.4.7). A plant in a nonattainment area or region can be required to reduce NO_x to a greater degree than that required in Title IV. This requirement could mean installing control equipment and measuring ammonia and other pollutants.

Table 2.4.1 shows the pollutants measured for various applications.

Table 2.4.2 summarizes the type of monitors and the concentration measurement range available for several CEM systems. Performance specifications for CO and O_2 are given in Table 2.4.3. Dilution-extractive systems prefer a pulse of chopped fluorescence for SO_2 analysis.

NO is generally measured using chemiluminescence (see

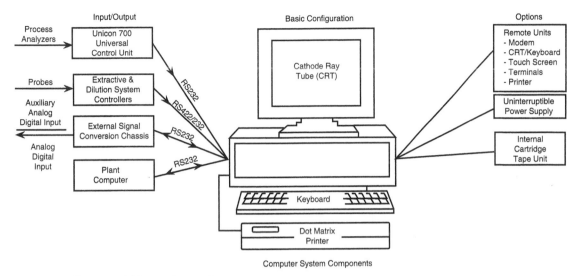

FIG. 2.4.7 Data acquisition and reporting computer system.

TABLE 2.4.1   MEASUREMENT REQUIREMENTS

	SO_2	*TRS*	NO_x	*THC*	*HCl*	*CO*	CO_2	O_2	NH_3	*Flow*	*Opacity*	*DAS*
Utility												
New gas			1			1		1	*			1
New coal	2		1			1	1	1	*	1	1	1
Part 75	1		1				1		*	1		1
Incinerators												
Municipal	2		1			2	1	2			1	1
Sewage				1								1
Hazardous	1		1	1	1	1		1			1	1
Pulp & Paper												
Recovery boiler		1						1			1	1/3
Lime kiln		1						1			1	1/3
Power boiler	1		1				†	1			1	1/3

Source: J. Schwartz, S. Sample, and R. McIlvaine, 1994, Continuous emission monitors—Issues and predictions, *Air & Waste*, Vol. 44 (January).
Notes: *Ammonia monitor required if selective catalytic reduction (SCR) or selective noncatalytic reduction (SNCR) is utilized.
†Can be substituted for O_2.

Figure 2.4.8). The term NO_x refers to all nitrogen oxides; however, in air pollution work, NO_x refers only to NO (about 95–98%) and NO_2. Because NO is essentially transparent in the visible and UV light region, it must be converted to NO_2 before it can be measured. NO can then react with ozone to form NO_2 with chemiluminescence. The light emitted can be measured photometrically as an indication of the reaction's extent. With excess ozone, the emission is in proportion to the amount of NO. Including NO_2 (which reacts slowly with ozone) involves bringing the sample gas in contact with a hot molybdenum catalyst. This technique converts all NO_2 to NO before it reacts with ozone.

LOCATION OF SYSTEMS

Title 40, Part 60 of the *Code of Federal Regulations* includes guidelines for instrument location within specific performance specifications. GEP dictates a location sepa-

rate from bends, exits, and entrances to prevent measurements in areas of high gas or material stratification. Environmental engineers can determine the degree of stratification of either the flowrate or pollution concentration by taking a series of preliminary measurements at a range of operating conditions. For gas measurements, certification testing confirms the proper location. Under the certification, readings from the gas analyzer must agree with the stack sampler measurements to within 20% relative accuracy (RA).

To convert the volumetric pollution concentration (ppmv) measured by the CEM system to the mass emission rate (lb/hr) required by the CAAA, an environmental engineer must measure the flowrate. Proper placement of the sensor in the flue-gas stream is a key requirement. The goal is to measure flow that represents the entire operation at various boiler loads. Measurement must compare favorably with the standard EPA reference methods to demonstrate the RA of the unit.

TABLE 2.4.2 SUMMARY OF CONTINUOUS EMISSION MONITORS

Pollutant	Monitor Type	Expected Concentration Range	Available Range[a]
O_2	Paramagnetic Electrocatalytic (e.g., zirconium oxide)	5–14%	0–25%
CO_2	NDIR[b]	2–12%	0–21%
CO	NDIR	0–100 ppmv	0–5000 ppmv
HCl	NDIR	0–50 ppmv	0–10000 ppmv
Opacity	Transmissometer	0–10%	0–100%
NO_x	Chemiluminescence	0–4000 ppmv	0–10000 ppmv
SO_2	Flame photometry Pulsed fluorescence NDUV[c]	0–4000 ppmv	0–5000 ppmv
SO_3	Colorimetric	0–100 ppmv	0–50 ppmv
Organic compounds	GC (FID)[d] GC (ECD)[e] GC (PID)[f] IR absorption UV absorption	0–50 ppmv	0–100 ppmv
HC	FID	0–50 ppmv	0–100 ppmv

Source: E.P. Podlenski et al., 1984, *Feasibility study for adapting present combustion source continuous monitoring systems to hazardous waste incinerators,* EPA-600/8-84-011a, NTIS PB 84-187814, U.S. EPA (March).

Notes: [a]For available instruments only. Higher ranges are possible through dilution.
[b]Nondispersion infrared.
[c]Nondispersion ultraviolet.
[d]Flame ionization detector.
[e]Electron capture detector.
[f]Photo-ionization detector.

TABLE 2.4.3   PERFORMANCE SPECIFICATIONS FOR CO AND O_2 MONITORS

Parameter	CO Monitors		O_2 Monitors
	Low Range	High Range	
Calibration drift (CD) 24 hr	≤6 ppm[a]	≤90 ppm	≤0.5% O_2
Calibration error (CE)	≤10 ppm[a]	≤150 ppm	≤0.5% O_2
Response time RA[b]	≤2 min ([c])	≤2 min ([c])	≤2 min (incorporated in CO and RA calculation)

Source: U.S. Environmental Protection Agency (EPA), 1990, *Method manual for compliance with BIF regulations,* EPA/530-SW-91-010, NTIS PB 90-120-006.

Notes: [a]For Tier II, CD and CE are ≤3% and ≤5% of twice the permit limit, respectively.
[b]Expressed as the sum of the mean absolute value plus the 95% confidence interval of a series of measurements.
[c]The greater of 10% of the performance test method (PTM) or 10 ppm.

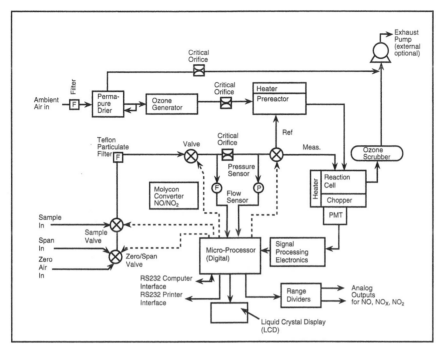

FIG. 2.4.8 System schematic of nitrogen oxides analyzer.

MAINTENANCE OF SYSTEMS

Because downtime can result in a fine, designing and engineering CEM systems to maximize reliability is essential. Certification follows start up and commissioning. Once a CEM system is running, users must devote appropriate resources to the ongoing maintenance and calibration requirements.

Proper maintenance procedures require periodically inspecting the probe and replacing the filter material. A good blowback system increases the time interval between these maintenance inspections.

Cochran, Ferguson, and Harris (1993) outline ways of enhancing the accuracy and reliability of CEM systems.

—David H.F. Liu

Reference

Cochran, J.R., A.W. Ferguson, and D.K. Harris. 1993. Pick the right continuous emissions monitor. *Chem. Eng.* (June).

2.5
REMOTE SENSING TECHNIQUES

PARTIAL LIST OF SUPPLIERS

Altech Systems Corp.; Ametek, Process and Analytical Instruments Div.; Extrel Corp.; The Foxboro., Environmental Monitoring Operations; KVB Analect, Inc.; MDA Scientific, Inc.; Nicolet Instrument Corp.; Radian Corp.; Rosemount Analytical, Inc.

The open-path, optical remote sensing techniques make new areas of monitoring possible. These techniques measure the interaction of light with pollutants in real time (see Figure 2.5.1). In an open-path configuration, path lengths—the distance between a light source and a reflecting device—can range to kilometers. Thus, these techniques are ideal for characterizing emission clouds wafting over process areas, storage tanks, and waste disposal sites. Open-path optical sensing systems can monitor gas concentrations at the fence lines of petrochemical plants and hazardous waste sites.

Wide-area systems are also being used indoors for emergency response and to demonstrate compliance with OSHA regulations. When used indoors, these systems can quickly detect routine or accidental releases of many chemical species. They are frequently fitted to control systems

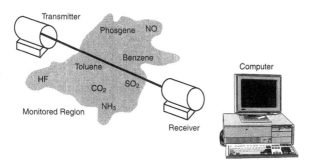

FIG. 2.5.1 Open-path monitoring system. A beam of light is transmitted from a transmitter through an area of contamination to a receiver. (Reprinted, with permission, from L. Nudo, 1992, Emerging technologies—Air, *Pollution Engineering* [15 April].)

and alarms to evacuate personnel when toxins are detected above a threshold limit.

Open-Path Optical Remote Sensing Systems

Table 2.5.1 describes several sensing techniques. Of the techniques listed, the FTIR and UV differential optical absorption spectroscopy (UV-DOAS) are the most widely publicized. They both measure multiple compounds in open areas in near real time.

These systems use a high-pressure xenon lamp in the UV-DOAS or a grey-body source in the FTIR to produce a broadband light that is collimated by a telescope into a narrow beam. This beam is then transmitted through the area of contamination to a receiver. The transmission path can range from several feet, in stacks, to 1 to 3 km, in ambient air over large areas (see Figure 2.5.1).

In route to the receiver, a portion of the light energy is absorbed by the pollutants. The light that reaches the receiver is sent to a spectrometer or interferometer, which generates a spectrum. The computer analyzes the spectrum by comparing it with a precalibrated reference spectrum for both the measured and the interfering components. Because each pollutant absorbs a unique pattern of wavelengths, missing light reveals the presence of a pollutant. The amount of missing light determines its concentration. Measurements are taken as often as every minute and averaged at any interval.

UV-DOAS systems are best suited to monitor criteria pollutants such as SO_2, NO, NO_2, O_3 and benzene, toluene, and xylene (BTX). FTIR systems are better suited to measure VOCs. FTIR systems also detect criteria pollutants but not with the same degree of sensitivity as UV-DOAS. These limitations stem from interferences of water vapor, which absorbs strongly in the IR region but does not absorb in the UV region. However, powerful software algorithms are becoming available to filter out the interference signals from water.

In general, more chemicals absorb light in the IR spectrum than in the UV spectrum. Thus, FTIR systems measure a broader range of compounds (more than fifty) than UV-DOAS systems, which detect about twenty-five (Nudo 1992). With the aid of a contractor, the EPA has developed reference IR spectra for approximately 100 of the 189 HAPs listed under Title III. The maximum number of HAPs to which FTIR might ultimately be applicable is about 130 (Schwartz, Sample, and McIlvaine 1994). The EPA plans to initiate its own investigation into this area. Additional investigation into the potential of FTIR in continuous applications is underway at Argonne National Labs.

TABLE 2.5.1   REMOTE SENSING TECHNIQUES FOR AIR ANALYSIS

Method	Operating Principle
FTIR	A broadband source allows collection and analysis of a full IR spectrum. This method compares standard spectra with field readings to determine constituents and concentrations
UV spectroscopy	UV spectra are collected over a limited absorption spectral region. This method uses the differential absorptions of the compounds in the air to determine the identity and concentrations of contaminants
Gas-filter correlation	A sample of the gas to be detected is used as a reference. This method compares the correlation between the spectrum of the sample gas to the gas in the measurement path to determine concentration
Filtered band-pass absorption	This method measures absorption of the gas in certain bands to detect composition and concentration
Laser absorption	This method uses one or more lasers to measure absorption at different wavelengths
Photoacoustic spectroscopy	The pressure change from the deactivation of excited molecules is measured in a closed acoustic chamber
LIDAR	This method measures molecular or aerosol backscatter by either differential absorption or Raman scattering to identify gases and determine their concentrations. Unlike the other methods listed, this method provides ranging information on measurements
Diode-laser spectroscopy	A developing technology for open-air use. This method scans a line feature of the gas spectrally to identify and quantify the compound

Source: G.E. Harris, M.R. Fuchs, and L.J. Holcombe, 1992, A guide to environmental testing, *Chem. Eng.* (November).

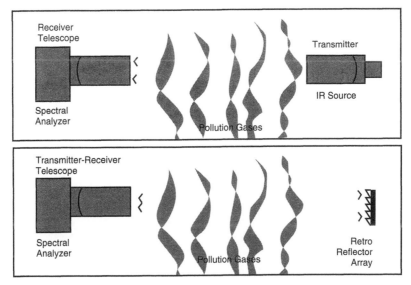

FIG. 2.5.2 Bistatic and unistatic systems. Bistatic systems (top), which place the IR source and receiver in different positions, require two power sources and precise alignment to record properly. Unistatic systems (bottom) are simpler to operate. Most installations use four separate bi- or unistatic source–receiver pairs—one for each side of a property. The next step may be to use one centrally located IR source on a tower, with retroreflectors around it.

UV-DOAS systems are more selective. They have a lower limit of detection and fewer problems with interferences from water vapor and carbon dioxide. For the best features of both systems, these complimentary systems can be integrated.

Instrumentation

The instrumentation used in remote sensing techniques includes bistatic systems and unistatic systems.

BISTATIC SYSTEMS

Early bistatic design (see the top of Figure 2.5.2) placed the light source and the receiver at opposite ends of the area to be monitored. Precise alignment, to assure that the receiver gets the full strength of the transmitted signal, is often a problem. Therefore, two separate electrical sources are usually required to operate these systems (Shelley 1991).

UNISTATIC SYSTEMS

A new unistatic design (see the bottom of Figure 2.5.2), developed by MDA Scientific, uses a single telescope that acts as both source and receiver. The IR beam traverses the area being monitored and returns to the receiver from an array of cubed mirrors, or retro-reflectors, on the opposite end of the field.

These clusters collect and refocus the beam and, unlike flat-faced mirrors, need not be precisely aligned to achieve near-perfect reflection along a light beam's original path.

This design makes the unistatic system easier to set up and reduces error during analysis.

LIMITATIONS

With open-path systems, sophisticated, meteorological modeling software is necessary to integrate information about regional topography and moment-to-moment changes in atmospheric conditions. Such modeling programs are still in the early stages. Therefore, discrete-point sampling is needed to confirm optical sensing results. Another shortcoming of using open-path systems is the assumption that the distribution of molecules in the beam is homogeneous. Path-averaging assumptions can underestimate a localized problem.

Despite some remaining problems, the use of perimeter monitoring is rising for documenting cumulative emissions over process facilities. The EPA is developing procedures to demonstrate equivalence with reference test methods for use with some toxic gases (Schwartz, Sample, and McIlvaine 1994). These procedures are required before regulatory agencies and industry can adopt optical techniques.

—David H.F. Liu

References

Nudo, L. 1992. Emerging technologies—Air. *Pollution Engineering* (15 April).

Schwartz, J., S. Sample, and R. McIlvaine. 1994. Continuous emission monitors—Issues and predictions. *Air & Waste,* Vol. 44 (January).

Shelley, S. 1991. On guard! *Chem. Eng.* (November): 31–39.

3

Pollutants: Minimization and Control

Samuel S. Cha | Karl T. Chuang | David H.F. Liu | Gurumurthy Ramachandran | Parker C. Reist | Alan R. Sanger

3.1
POLLUTION REDUCTION

The CAA of 1970 establishes primary and secondary standards for criteria pollutants. Primary standards protect human health, while secondary standards protect materials, crops, climate, visibility, and personal comfort. The standard for total suspended particulates (TSP) is an annual geometric mean of 75 $\mu g/m^3$ and a maximum 24 hr average of 260 $\mu g/m^3$. This standard is currently being reviewed by the U.S. EPA for possible change.

Prevention of air pollution from industrial operations starts within the factory or mill. Several alternatives are available to prevent the emission of a pollutant including:

- Selecting process inputs that do not contain the pollutant or its precursors
- Operating the process to minimize generation of the pollutant
- Replacing the process with one that does not generate the pollutant
- Using less of the product whose manufacture generates the pollutant
- Removing the pollutant from the process effluent.

Raw Material Substitution

Removing some pollutants involves simply substituting materials which perform equally well in the process but which discharge less harmful products to the environment. This method of air pollution reduction usually produces satisfactory control at a low cost.

Typical examples are the substitution of high-sulfur coal with low-sulfur coal in power plants. This substitution requires little technological change but results in a substantial pollution reduction. Changing to a fuel like natural gas or nuclear energy can eliminate all sulfur emissions as well as those of particulates and heavy metals. However, natural gas is more expensive and difficult to ship and store than coal, and many people prefer the known risks of coal pollution than the unknown risks of nuclear power. Coal gasification also greatly reduces sulfur emissions. Another example is substituting gasoline with ethanol or oxygenated fuels in internal combustion engines. This substitution reduces the O_3 pollution in urban areas. Alternative energy sources, such as wind or solar power, are being explored and may become economically feasible in the future.

Process Modification

Chemical and petroleum industries have changed dramatically by implementing automated operations, computerized process control, and completely enclosed systems that minimize the release of materials to the outside environment.

A process modification example is industry reducing the oxidation of SO_2 to SO_3 by reducing excess air from ~20% to <1% when burning coal, resulting in reduced sulfuric acid emissions. However, this process change has increased fly ash production.

Powders and granulated solids are widely used in industry. Handling these materials at locations such as transfer points or bagging and dumping operations generates dust that can affect worker health. Plinke et al. (1991) found that the amount of dust generated by an industrial process depends on the size distribution of the dust, the ratio of impaction forces that disperse the dust during material handling, and the cohesion properties of the dust, such as moisture content. The implication is that by modifying both the material handling processes and the properties of the powders, the dust generated can be reduced.

Similar approaches are being used in municipal trash incinerators which emit carcinogenic dioxins. By adjusting the temperature of incineration, dioxin emissions can be prevented.

Marketing Pollution Rights

Some economists view absolute injunctions and rigid limits as counterproductive. They prefer to rely on market mechanisms to balance costs and effects and to reduce pollution. Corporations can be allowed to offset emissions by buying, selling, and banking pollution rights from other factories at an expected savings of $2 to 3 billion per year. This savings may make economic sense for industry and even protect the environment in some instances, but it may be disastrous in some localities.

Demand Modification

Demand modification and lifestyle changes are another way to reduce pollution. To combat smog-causing emissions, some districts in California have proposed substantial lifestyle changes. Aerosol hair sprays, deodorants, charcoal lighter fluid, gasoline-powered lawn mowers, and drive-through burger stands could be banned. Clean-burning, oxygenated fuels or electric motors would be required for all vehicles. Car-pooling would be encouraged, and parking lots would be restricted.

Gas Cleaning Equipment

Industry controls air pollution with equipment that removes contaminants at the end of the manufacturing process. Many such devices exist and are described in the next sections. These devices display two characteristics: the size and cost of the equipment increase with the volume of gas cleaned per unit of time, and the cost of removing a contaminant rises exponentially with the degree of removal. The degree of pollutant removal is described as collection efficiency, while the operating cost depends on the pressure drop across the unit and its flow capacity.

—*Gurumurthy Ramachandran*

Reference

Plinke, M.A.E., D. Leith, D.B. Holstein, and M.G. Boundy. 1991. Experimental examination of factors that affect dust generation. *Am. Ind. Hyg. Assoc. J.* 52, no. 12:521–528.

3.2
PARTICULATE CONTROLS

Control Equipment

If waste reduction is not possible, either by process or material change, waste particulate material must be removed from the process air stream. Selecting the type of control equipment depends largely on the characteristics of the particulate material to be removed. Factors such as the physical form of the particulates (whether solid or liquid), particle size and size distribution, density and porosity of the particulates, and particle shape (spheres, fibers, or plates) affect particle behavior.

Control equipment can be divided into three broad classes: gravitational and inertial collectors, electrostatic precipitators (ESPs), and filters. Included in the category of gravitational and inertial collectors are wet and dry scrubbers, gravity settling chambers, and cyclone collectors. ESPs include both single and two-stage units. Finally, filters encompass a variety of media ranging from woven and nonwoven fabric, fiber, and paper media. In addition to media, filters are often classified according to the method of cleaning.

Particulate Size

Although particulate material varies in both size and composition, certain particle types are usually associated with a range of particle sizes. For example, particulate material produced by crushing or grinding or resuspended from settled dusts is generally made up of particles with diameters larger than 1 μm. On the other hand, particulate material produced by condensation or gas-phase chemical reactions is comprised of many small particles, all much smaller than 1 μm in diameter. Figure 3.2.1 illustrates typical particle diameters for a variety of substances, including particulates such as beach sand and pollens. For comparison, this figure also includes the size range for types of electromagnetic radiation, estimates of gas molecule diameters, size ranges for fogs, mists and raindrops, and ranges for inspirable particles.

Other Factors

Generally, particle size is the most important factor in the selection of a collector since the range of particle sizes to be collected strongly affects the expected cost and efficiency. In some instances, other factors such as serviceability or the pressure drop across the collector are the most important factors. Table 3.2.1 lists the characteristics of various particulate removal equipment and compares some of these features. For the most part, the collection devices that are the least expensive to install and operate are also the least efficient.

Table 3.2.2 also compares the collection characteristics and total annual costs of various air cleaning equip-

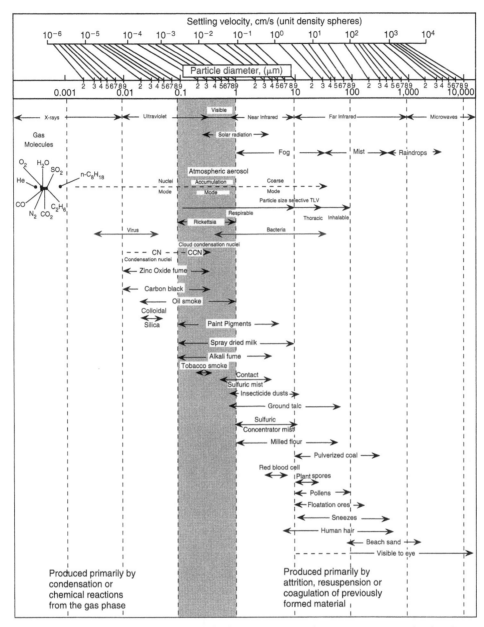

FIG. 3.2.1 Molecular and aerosol particle diameters, copyright © P.C. Reist. Molecular diameters calculated from viscosity data. See Bird, Stewart, and Lightfoot, 1960, *Transport phenomona,* Wiley. (Adapted from Lapple, 1961, *Stanford Research Institute Journal,* 3d quarter; and J.S. Eckert and R.F. Strigle, Jr., 1974, *JAPCA* 24:961–965.)

TABLE 3.2.1 CHARACTERISTICS OF AIR AND GAS CLEANING METHODS AND EQUIPMENT FOR COLLECTING AEROSOL PARTICLES

Characteristics	Gravitational and Inertial Collectors — Dry	Gravitational and Inertial Collectors — Wet, Low Energy	Gravitational and Inertial Collectors — Wet, High Energy	ESPs — 2-Stage Low Voltage	ESPs — 1-Stage High Voltage	Filters — Fabric, Inside Collector (Reverse Gas Cleaning)	Filters — Fabric, Outside Collector	Filters — Ventilation	Filters — Absolute
Type	Settling Chamber / Cyclone Chamber	Wetted Cyclone / Impingement & Entrainment	Venturi / Disintegrator	Electronic Air Cleaner / Furnace Air Cleaner	Wire and Plate / Wire and Tube	Shaker Cleaning	Pulse-Jet Cleaning	Pleated Paper Media	Paper Deep Bed
Contaminants	Crushing, grinding, machining	Crushing, grinding, machining	Metallurgical fumes	Room air, oil mist	Fly ash acid mist	All dry dusts	All dry dusts	Room air	Precleaned room air
Loadings (g/m³)	0.1–100	0.1–100	0.1–100	<0.1	0.1–10	0.1–20	0.1–20	<0.01	<0.001
Overall efficiency, %	High for >10 μm	High for >10 μm	High for >1 μm	High for >0.5 μm	High for >0.5 μm	High for all sizes	High for all sizes	High for >5 μm	High for all sizes
Pressure drop (mm Hg)	4–20	4–20	15–200	4–10	4–20	5–20	5–20	2–10	2–4
Initial cost	Low	Moderate	Moderate	Moderate	High	Moderate to high	Moderate	Low	Moderate
Operating cost	Moderate	Moderate	High	Low	Moderate	Moderate	Moderate	Low	High
Serviceability	Good (erosion)	Good (corrosion)	Good (corrosion)	Poor (shorts)	Fair (shorts)	Fair to good	Fair to good (blinding)	Fair	Fair to poor
Limitations	Efficiency	Disposal	Operating cost	Loading	Resistivity	Temperature	Temperature	Loading	Loading

Source: M.W. First and D. Leith, 1994, Personal communication.

TABLE 3.2.2 COMPARISON OF PARTICULATE REMOVAL SYSTEMS

Unit	Collection Characteristics			Space Required	Cost[a] $/yr/m^3
	0.1–1 μm	1–10 μm	10–50 μm		
Standard cyclone	Poor	Poor	Good	Large	7
High-efficiency cyclone	Poor	Fair	Good	Moderate	11
Baghouse (cotton)	Fair	Good	Excellent	Large	14
Baghouse (dacron, nylon, orlon)	Fair	Good	Excellent	Large	17
Baghouse (glass fiber)	Fair	Good	Good	Large	21
ESP	Excellent	Excellent	Good	Large	21
Dry scrubber	Fair	Good	Good	Large	21
Baghouse (teflon)	Fair	Good	Excellent	Large	23
Impingement scrubber	Fair	Good	Good	Moderate	23
Spray tower	Fair	Good	Good	Large	25
Venturi scrubber	Good	Good	Excellent	Small	56

Source: A.C. Stern, R.W. Boubel, D.B. Turner, and D.L. Fox, 1984, *Fundamentals of air pollution*, 2d ed., 426.
[a]Includes water and power cost, maintenance cost, operating cost, capital, and insurance costs (1984 dollars).

ment. These cost comparisons, although representative of air cleaners as a whole, may not reflect actual cost differences for various types of air cleaning equipment when applied to specific applications. Thus, these tables should be used as guides only and should not be relied upon for accurate estimates in a specific situation.

—Parker C. Reist

3.3
PARTICULATE CONTROLS: DRY COLLECTORS

Gravity Settling Chambers

FEATURE SUMMARY

Pressure Drop
Less than 0.5 in of water

Operating Temperatures
Up to 1000°C

Applications
Precleaners for removing dry dust produced by grinding in cement and lime kilns, grain elevators, rock crushers, milling operations, and thermal coal dryers

Dust Particle Sizes
Greater than 50 μm

Settling chambers serve as preliminary screening devices for more efficient control devices. These chambers remove large particles from gas streams by gravity. The chambers slow the gas sufficiently to provide enough time for particles to collect (Leith, Dirgo, and Davis 1986). Figure 3.3.1 shows a chamber of length L, width W, and height H. A mass balance for an infinitesimal slice dL yields:

$$V_g HWC = V_g HW(C - dC) + WdL(-V_{ts})C \quad \text{3.3(1)}$$

where:

V_g = the uniform gas velocity
V_{ts} = the particle terminal settling velocity
C = particle concentration in the slice dL

This equation assumes that the particles are well-mixed in every plane perpendicular to the direction of air flow due to lateral turbulence. This assumption is reasonable because even at low velocities, gas flow in industrial-scale settling chambers is turbulent. Rearranging Equation 3.3(1) and integrating between the two ends of the chamber yields the collection efficiency of the chamber as:

$$\eta = 1 - \exp\left[\frac{-V_{ts}L}{V_g H}\right] \quad \text{3.3(2)}$$

The terminal settling velocity is as follows:

$$V_{ts} = \frac{\rho_p d^2 C_c g}{18\mu} \quad \text{3.3(3)}$$

where:

ρ_p = the particle density
d = the particle diameter
μ = the viscosity of the medium
C_c = the Cunningham slip correction factor

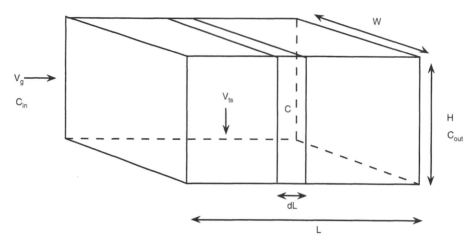

FIG. 3.3.1 Schematic diagram of settling chamber with complete lateral mixing due to turbulence.

Equation 3.3(2) shows that settling chambers have high collection efficiency for low gas velocities, high terminal settling velocities, and large ratios of chamber length to height. Thus, gas velocities are usually below 3 m/sec, which has the added benefit of preventing reentrainment. A large L/H ratio ensures a long residence time and a short vertical distance for the particle to travel to be collected. The equation yields high terminal settling velocities for large particle sizes, therefore, these devices efficiently remove particles greater than 50 μm in diameter.

Settling chambers are characterized by low capital costs and low pressure drop (~0.5 in water gauge [w.g.]). They can be used under temperature and pressure extremes (~1000°C and ~100 atm). Settling chambers are used as precleaners to remove dry dusts produced by grinding, e.g. in coal dryers, grain elevators, and rock crushers.

Cyclones

FEATURE SUMMARY

Types of Designs
Reverse flow, or straight through, with tangential, scroll, and swirl vane entries

Gas Flow Rates
50 to 50,000 m³/hr

Pressure Drop
Between 0.5 and 8.0 in of water

Operating Temperatures
Up to 1000°C

Applications
Cement and lime kilns, grain elevators, milling operations, thermal coal dryers, and detergent manufacturing

Dust Particle Sizes
Greater than 5 μm

Partial List of Suppliers
Advanced Combustion Systems Inc.; Alfa-Laval Separation Inc.; Bayliss-Trema Inc.; Beckert and Heister Inc.; Clean Gas Systems Inc.; Dresser Industries Inc.; Ducon Environmental Systems; Emtrol Corp.; Fisher-Klosterman Inc.; Hough International Inc.; HVAC Filters Inc.; Interel Environmental Technologies Inc.; Joy Technologies Inc.; Quality Solids Separation Co.; Thiel Air Technologies Corp.; United Air Specialists Inc.; Wheelabrator Inc.

Cyclones operate by accelerating particle-laden gas in a vortex from which particles are removed by centrifugal force. One of the most widely used dust collecting devices, cyclones are inexpensive to construct and easy to maintain because they do not have any moving parts. Although cyclones are inefficient for collecting particles smaller than 5 μm in diameter, they operate with low to moderate pressure drops (0.5 to 8.0 in w.g.).

Cyclones can be constructed to withstand dust concentrations as high as 2000 g/m³, gas temperatures as high as 1000°C, pressures up to 1000 atm, and corrosivity. Such conditions are encountered in the pressurized, fluidized-bed combustion of coal, where cyclones are an economic control option. In addition, cyclones are used to control emissions from cement and lime kilns, grain elevators, grain drying and milling operations, thermal coal dryers, and detergent manufacturing.

Various cyclone designs have been proposed; however the reverse-flow cyclone is the type most commonly used for industrial gas cleaning. Figure 3.3.2 shows the dimensions of a reverse-flow cyclone. Dust-laden gas enters the cyclone at the top; the tangential inlet causes the gas to spin. After entry, the gas forms a vortex with a high tangential velocity that gives particles in the gas a high centrifugal force, moving them to the walls for collection. Below the bottom of the gas exit duct, the spinning gas gradually migrates to the cyclone axis and moves up and out the gas exit. Thus, the cyclone has an outer vortex moving downward and an inner vortex flowing upward (Dirgo and Leith 1986). Collected dust descends to the duct outlet at the bottom of the cone.

Cyclone dimensions are usually expressed as multiples of the diameter D. The dimension ratios a/D, b/D, D_e/D,

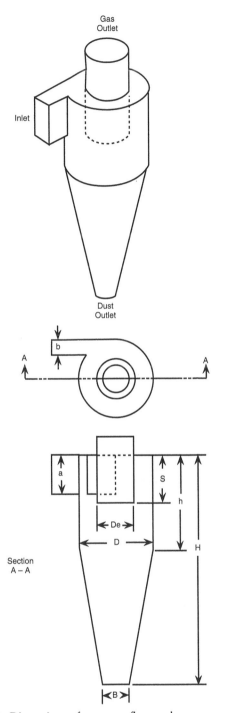

FIG. 3.3.2 Dimensions of a reverse-flow cyclone.

S/D, h/D, H/D, and B/D allow cyclones which differ in size to be compared. Table 3.3.1 gives some standard designs.

GAS FLOW PATTERNS

Understanding and characterizing the performance of a cyclone requires knowledge of the gas flow patterns within. The overall gas motion consists of two vortices: an outer vortex moving down and an inner vortex moving up.

According to Barth (1956), the boundary between the outer and inner vortices, called the *cyclone core,* is the cylindrical extension of the gas exit duct.

Gas flow is three dimensional and can be described in terms of three velocity components—tangential, radial, and axial. Ter Linden (1949) made the first systematic measurements of the velocity field inside a cyclone. In recent years, laser doppler velocimetry (Kirch and Loffler 1987) and digital imaging (Kessler and Leith 1991) have yielded more accurate measurements of the three velocity components.

TANGENTIAL GAS VELOCITY

Early theoretical treatments of cyclone flow patterns (Lissman 1930; Rosin, Rammler, and Intelman 1932) described the relationship between the tangential gas velocity V_t and the distance r from the cyclone axis as:

$$V_t r^n = \text{constant} \qquad 3.3(4)$$

The vortex exponent n is +1 for an ideal liquid and −1 for rotation as a solid body, while the usual range for n is from 0.5 to 0.9 (Ter Linden 1949; Shepherd and Lapple 1939; First 1950).

Equation 3.3(4) implies that tangential velocity increases from the walls to the cyclone axis where it reaches infinity. Actually, in the outer vortex, tangential velocity increases with decreasing radius to reach a maximum at the radius of the central core. In the inner vortex, the tangential velocity decreases with decreasing radius. Figure 3.3.3 shows Iozia and Leith's (1989) anemometer measurements of tangential velocity. Measurements by Ter Linden (1949) show similar velocity profiles.

RADIAL AND AXIAL GAS VELOCITY

Radial gas velocity is the most difficult velocity component to measure experimentally. Figure 3.3.4 shows Kessler and Leith's (1991) measurements of radial velocity profiles in a cyclone. Despite some uncertainty in the measurements, some trends can be seen. Radial flow velocity increases toward the center of the cyclone due to the conservation of mass principle. A tendency also exists for the maximum radial velocity in each cross section to decrease with decreasing cross-sectional height.

Figure 3.3.5 shows the axial gas velocity in a reverse-flow cyclone. The gas flows downward near the cyclone wall and upward near the cyclone axis. The downward velocity near the wall is largely responsible for transporting dust from the cyclone wall to the dust outlet.

MODELING GAS FLOWS

The fluid dynamics of a cyclone are complex, and modeling the detailed flow pattern involves solving the strongly coupled, nonlinear, partial differential equations of the

TABLE 3.3.1 STANDARD DESIGNS FOR REVERSE FLOW CYCLONES

Source	Duty	D	a/D	b/D	D_e/D	S/D	h/D	H/D	B/D	ΔH	Q/D^2 (m/hr)
Stairmand[a]	High efficiency	1	0.50	0.20	0.50	0.50	1.5	4.0	0.375	5.4	5500
Swift[b]	High efficiency	1	0.44	0.21	0.40	0.50	1.4	3.9	0.40	9.2	4940
Lapple[c]	General purpose	1	0.50	0.25	0.50	0.625	2.0	4.0	0.25	8.0	6860
Swift[b]	General purpose	1	0.50	0.25	0.50	0.60	1.75	3.75	0.40	7.6	6680
Stern et al.[d]	Consensus	1	0.45	0.20	0.50	0.625	0.75	2.0	—	—	—
Stairmand[a]	High through-put	1	0.75	0.375	0.75	0.875	1.5	4.0	0.375	7.2	16500
Swift[b]	High through-put	1	0.80	0.35	0.75	0.85	1.7	3.7	0.40	7.0	12500

Source: D. Leith, 1984, Cyclones, in *Handbook of powder science and technology,* edited by M.A. Fayed and L. Otten (Van Nostrand Reinhold Co.).
Notes: [a]Stairmand, C.J. 1951. The design and performance of cyclone separators. *Trans. Instn. Chem. Engrs.* 29:356.
[b]Swift, P. 1969. Dust controls in industry. *Steam and Heating Engineer* 38:453.
[c]Lapple, C. 1951. Processes use many collector types. *Chemical Engineering* 58:144.
[d]Stern, A.C., K.J. Kaplan, and P.D. Bush. 1956. *Cyclone dust collectors.* New York: American Petroleum Institute.

conservation of mass and momentum. Boysan, Ayers, and Swithenbank (1982) have developed a mathematical model of gas flow based on the continuity and momentum conservation principles, accounting for the anisotropic nature of turbulence and its dissipation rate. The velocity and pressure profiles of their model agree remarkably with the experimental measurements of Ter Linden (1949). However, as they suggest, the information obtained from such modeling is more detailed than that required by a process design engineer.

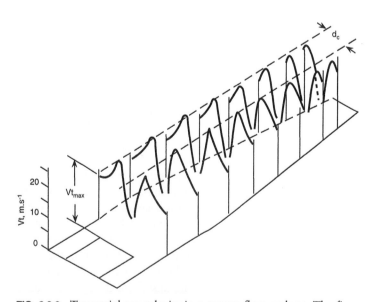

FIG. 3.3.3 Tangential gas velocity in a reverse-flow cyclone. The figure is a vertical cross section of half the cyclone. Each of the nodes represents tangential gas velocity. (Reprinted, with permission, from D.L. Iozia and D. Leith, 1989, Effect of cyclone dimensions on gas flow pattern and collection efficiency, *Aerosol Sci. and Technol.* 10:491.)

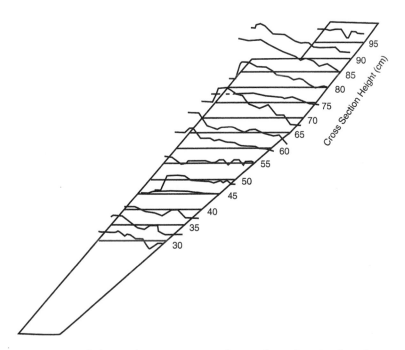

FIG. 3.3.4 Radial gas velocity in a reverse-flow cyclone. (Reprinted, with permission, from M. Kessler and D. Leith, 1991, Flow measurement and efficiency modeling of cyclones for particle collection, *Aerosol Sci. and Technol.* 15.)

COLLECTION EFFICIENCY

Collection efficiency η is defined as the fraction of particles of any size that is collected by the cyclone. The plot of collection efficiency against particle diameter is called *a fractional* or *grade efficiency curve*.

Determining the path of the particles in the gas flow field is essential to estimate the collection efficiency. When all external forces except drag force are neglected, the equation of motion for a small particle is expressed in a Lagrangian frame of reference as:

$$\frac{du_p}{dt} = -\left(\frac{1}{8m_p} C_D \pi d^2 \rho_p\right)(u_p - u - u')|u_p - u - u'|$$

$$\frac{dv_p}{dt} = \frac{w_p^2}{r} - \left(\frac{1}{8m_p} C_D \pi d^2 \rho_p\right)(v_p - v - v')|v_p - v - v'|$$

$$3.3(5)$$

$$\frac{dw_p}{dt} = 2\frac{w_p}{r}v_p - \left(\frac{1}{8m_p} C_D \pi d^2 \rho_p\right)$$

$$\times (w_p - w - w')|w_p - w - w'|$$

where:

u_p, v_p, w_p	= the components of the particle velocities in the z, r, and θ directions
u, v, w	= the gas velocities in the same directions
$u', v',$ and w'	= the fluctuating components of the gas velocities
d_p	= the particle diameter
ρ_p	= the particle density
C_D	= the drag coefficient
m_p	= the particle mass

In addition to Equations 3.3(5), the following equation applies:

$$\frac{dz}{dt} = u_p; \quad \frac{dr}{dt} = v_p; \quad \text{and} \quad \frac{d\theta}{dt} = \frac{w_p}{r} \qquad 3.3(6)$$

Boysan, Ayers, and Swithenbank (1982) used a stochastic approach whereby the values of u', v', and w', which pre-

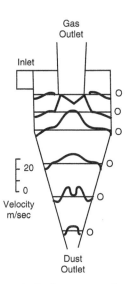

FIG. 3.3.5 Axial gas velocity in a reverse-flow cyclone.

vail during the lifetime of a fluid eddy in which the particle is traversing, are sampled assuming that these values possess a Gaussian probability distribution. They used the lifetime of the fluid eddy as a time interval over which the gas velocity remains constant. This assumption allowed the direct solution of the equations of motion to obtain local, closed-form solutions. They obtained the values of z, r, and θ by a simple stepwise integration of the equation of the trajectory. Many random trajectories were evaluated, and a grade efficiency curve was constructed based on these results. The results agreed with the experimental results of Stairmand (1951).

Kessler and Leith (1991) approximated the drag force term using Stokes' law drag for spheres. They solved the equations of motion numerically using backward difference formulas to solve the stiff system of differential equations. They modeled the collection efficiency for a single, arbitrarily sized particle by running several simulations for that particle. For each particle size, the ratio of collected particles to total particles described the mean of a binomial distribution for the probability distribution. Figure 3.3.6 shows this curve, three other grade efficiency models (Barth [1956], Dietz [1981], and Iozia and Leith [1990]), and the direct measurements of Iozia and Leith.

Early collection efficiency treatments balanced the centrifugal and drag forces in the vortex to calculate a *critical particle* size that was collected with 50 or 100% efficiency. The tangential velocity of the gas and the particle were assumed to be equal, and the tangential velocity was given by the vortex exponent law of Equation 3.3(4). As previously shown, the maximum tangential velocity, V_{max} occurs at the edge of the central core. The average inward

radial velocity at the core edge is as follows:

$$V_r = -\frac{Q}{2\pi r_{core}(H - S)}$$

3.3(7)

For particles of the critical diameter, the centrifugal and drag forces balance, and these static particles remain suspended at the edge of the core. Larger particles move to the wall and are collected, while smaller particles flow into the core and out of the cyclone. Barth (1956) and Stairmand (1951) used different assumptions about V_{max} to develop equations for the critical diameter. Collection efficiencies for other particle sizes were obtained from a separation curve—a plot that relates efficiency to the ratio of particle diameter to the critical diameter. However, theories of the critical particle type fail to account for turbulence in the cyclone (Leith, Dirgo, and Davis 1986); moreover, a single separation curve is not universally valid.

Later theories attempted to account for turbulence and predict the entire fractional efficiency curve (Dietz 1981; Leith and Licht 1972; Beeckmans 1973). Of these, the theory by Leith and Licht (1972) has been frequently used; however, some of the assumptions on which the theory is based have been invalidated. As with most efficiency theories, the interactions between particles is not taken into account. The assumption of complete turbulent mixing of aerosol particles in any lateral plane has been proven untrue.

Iozia and Leith (1989) used experimental data to develop an equation to predict the *cut diameter* d_{50}, which is the particle size collected with 50% efficiency. Their empirical equations are as follows:

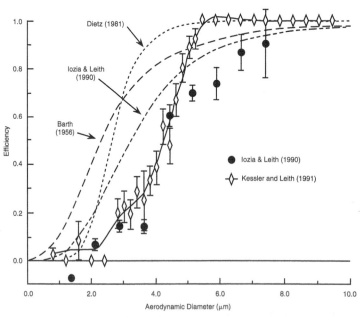

FIG. 3.3.6 Measured collection efficiency from studies by Barth, Dietz, and Iozia and Leith and modeled collection efficiency by Kessler and Leith.

$$d_{50} = \left(\frac{9\mu Q}{\pi z_c \rho_p V_{max}^2} \right)^{1/2} \qquad 3.3(8)$$

where:

$$V_{max} = 6.1 V_i \left(\frac{ab}{D^2} \right)^{0.61} \left(\frac{D_e}{D} \right)^{-0.74} \left(\frac{H}{D} \right)^{-0.33} \qquad 3.3(9)$$

where V_i is the inlet gas velocity, and z_c is the core length which depends on core diameter d_c as follows:

$$d_c = 0.52D \left(\frac{ab}{D^2} \right)^{-0.25} \left(\frac{D_e}{D} \right)^{1.53} \qquad 3.3(10)$$

When $d_c < B$, the core intercepts the cyclone wall, and the following equation applies:

$$z_c = (H - S) - \left(\frac{H - h}{\frac{D}{B} - 1} \right) \left(\frac{d_c}{B} - 1 \right) \qquad 3.3(11)$$

When $d_c < B$, the core extends to the bottom of the cyclone, and the following equation applies:

$$z_c = (H - S) \qquad 3.3(12)$$

Iozia and Leith (1989, 1990) also developed the following logistic equation for the fractional efficiency curve:

$$\eta = \frac{1}{1 + \left(\dfrac{d_{50}}{d} \right)^{\beta}} \qquad 3.3(13)$$

where the parameter β is estimated from the cut size d_{50}, and the geometry of the cyclone is as follows:

$$\ln(\beta) = 0.62 - 0.87 \ln(d_{50}(\text{cm}))$$
$$+ 5.21 \ln\left(\frac{ab}{D^2} \right) + 1.05 \left(\ln\left(\frac{ab}{d^2} \right) \right)^2 \qquad 3.3(14)$$

The preceding model describes collection efficiencies significantly better than other theories.

PRESSURE DROP

A cyclone pressure drop results from the following factors (Shepherd and Lapple 1939):

1. Loss of pressure due to gas expansion as it enters the cyclone
2. Loss of pressure due to vortex formation
3. Loss of pressure due to wall friction
4. Regain of rotational kinetic energy as pressure energy

Factors 1, 2, and 3 are probably the most important. Iinoya (1953) showed that increasing wall roughness decreases the pressure drop across the cyclone probably by inhibiting vortex formation. If this assumption is true, then energy consumption due to vortex formation is more important than wall friction. First (1950) found that wall friction is not an important contributor to pressure drop.

Devices such as an inlet vane, an inner wall extension of the tangential gas entry within the cyclone body to a position close to the gas exit duct, and a cross baffle in the gas outlet lower the pressure drop (Leith 1984). However, such devices work by suppressing vortex formation and decrease collection efficiency as well. Designing cyclones for low pressure drop without attaching devices that also lower collection efficiency is possible.

Cyclone pressure drop has traditionally been expressed as the number of inlet velocity (v_i) heads ΔH. The following equation converts velocity heads to pressure drop ΔP:

$$\Delta P = \Delta H \left(\frac{1}{2} \rho_g v_i^2 \right) \qquad 3.3(15)$$

The value of ΔH is constant for a cyclone design (i.e., cyclone dimension ratios), while ΔP varies with operating conditions.

Many analytical expressions for determining ΔH from cyclone geometry are available (Barth 1956; First 1950; Shepherd and Lapple 1940; Stairmand 1949; Alexander 1949). None of these expressions predicts pressure drop accurately for a range of cyclone designs; predictions differ from measured values by more than a factor of two (Dirgo 1988). Further, evaluations of these models by different investigators produced conflicting conclusions as to which models work best.

Dirgo (1988) and Ramachandran et al. (1991) developed an empirical model for predicting pressure drop, which was developed through statistical analysis of pressure drop data for ninety-eight cyclone designs. They used stepwise and backward regression to develop the following expression for ΔH based on cyclone dimension ratios:

$$\Delta H = 20 \left(\frac{ab}{D_e^2} \right) \left\{ \frac{\left(\dfrac{S}{D} \right)}{\left(\dfrac{H}{D} \right) \left(\dfrac{h}{D} \right) \left(\dfrac{B}{D} \right)} \right\}^{1/3} \qquad 3.3(16)$$

This model made better predictions of pressure drop than the models of Barth (1956), First (1950), Shepherd and Lapple (1940), Stairmand (1949), and Alexander (1949).

DUST LOADING

Increasing inlet dust loading, C_i (g/cm^3), increases collection efficiency and decreases pressure drop. Briggs (1946) quantified the influence of dust loading on pressure drop as follows:

$$\Delta P_{dusty} = \frac{\Delta P_{clean}}{1 + 0.0086 C_i^{1/2}} \qquad 3.3(17)$$

The following equation (Whiton 1932) gives the effect on efficiency of changing the inlet loading from C_{i1} to C_{i2}:

$$\frac{100 - \eta_1}{100 - \eta_2} = \left(\frac{C_{i2}}{C_{i1}}\right)^{0.182}$$ 3.3(18)

CYCLONE DESIGN OPTIMIZATION

Cyclone design usually consists of choosing an accepted standard design or a manufacturer's proprietary model that meets cleanup requirements at a reasonable pressure drop. However, investigators have developed analytical procedures to optimize cyclone design, trading collection efficiency against pressure drop.

Leith and Mehta (1973) describe a procedure to find the dimensions of a cyclone with maximum efficiency for a given diameter, gas flow, and pressure drop. Dirgo and Leith (1985) developed an iterative procedure to improve cyclone design. While holding the cyclone diameter constant, their method alters one cyclone dimension and then searches for a second dimension to change to yield the greatest collection efficiency at the same pressure drop.

They selected the gas outlet diameter D_e as the primary dimension to vary. This variation changed the pressure drop. Then, they varied the inlet height a, inlet width b, and gas outlet duct length S, one at a time, to bring pressure drop back to the original value. They predicted the d_{50} for each new design from theory. The new design with the lowest d_{50} became the baseline for the next iteration. In the next iteration, they varied D_e again with the three other dimensions to find the second dimension change that most reduced d_{50}. They continued iterations until the predicted reduction in d_{50} from one iteration to the next was less than one nanometer.

Ramachandran et al. (1991) used this approach with the efficiency theory of Iozia and Leith, Equations 3.3(8)–3.3(14), and the pressure drop theory of Dirgo, Equation 3.3(16), to develop optimization curves. These curves predict the minimum d_{50} and the dimension ratios of the optimized cyclone for a given pressure drop (see Figures 3.3.7 and 3.3.8).

To design a cyclone, the design engineer must obtain the inlet dust concentration and size distribution and other design criteria such as gas flow rate, temperature, and particle density, preferably by stack sampling. When a cyclone is designed for a plant to be constructed, stack testing is impossible, and the design must be based on data obtained from similar plants. Once the size distribution of the dust is known, the design engineer chooses a value of d_{50}. For each size range, the engineer calculates the collection efficiency using Equation 3.3(13). The overall efficiency is calculated from the following equation:

$$\eta_{overall} = \sum_i f_i \eta_i$$ 3.3(19)

where f_i is the fraction of particles in the ith size range. By trial and error, the engineer chooses a value of d_{50} to obtain the required overall efficiency. This d_{50} is located on the optimization curve (e.g., for H = 5D) of Figure 3.3.7, and the pressure drop corresponding to this d_{50} is found. Figure 3.3.8 gives the cyclone dimension ratios. The cyclone diameter is determined from the following equation:

$$D_2 = D_1 \left(\frac{\rho_{p2} Q_2}{\rho_{p1} Q_1}\right)^{1/3}$$ 3.3(20)

where D_1, ρ_{p1}, and Q_1 are the cyclone diameter (0.254 m), particle density (1000 kg/m³), and flow (0.094 m³/sec) of the cyclone optimized in Figure 3.3.8; and ρ_{p2} and Q_2 are the corresponding values for the system being designed. The following equation gives the design pressure drop:

$$\Delta P_2 = \Delta P_1 \left(\frac{Q_2 D_1^2}{Q_1 D_2^2}\right)^2$$ 3.3(21)

where ΔP_1 is the pressure drop from Figure 3.3.7 corresponding to the chosen d_{50}.

If the pressure drop ΔP_2 is too high, then the design engineer should explore other options, such as choosing a taller cyclone as the starting point or reducing the flow through the cyclone by installing additional cyclones in parallel. These options, however, increase capital costs. The design engineer has to balance these various factors.

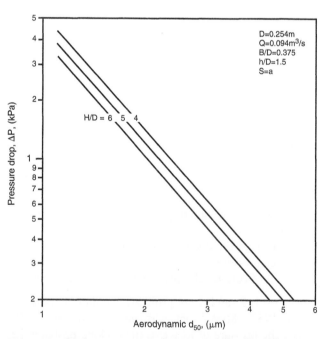

FIG. 3.3.7 Optimum pressure drops versus d_{50} for different cyclone heights. (Reprinted, with permission, from G. Ramachandran, D. Leith, J. Dirgo, and H. Feldman, 1991, Cyclone optimization based on a new empirical model for pressure drop, *Aerosol Sci. and Technol.* 15.)

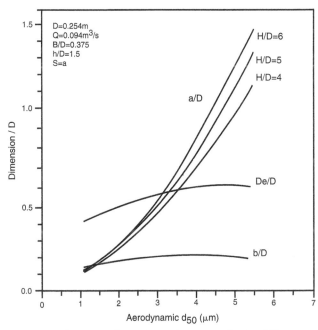

FIG. 3.3.8 Optimum dimension ratios versus d_{50} for different cyclone heights. (Reprinted, with permission, from Ramachandran et al. 1991.)

Filters

FEATURE SUMMARY

Types of Filter Media
    Paper, woven fabric with a low air/cloth ratio, felted fabric with high air/cloth ratio

Fibers
    Glass, Teflon, Nomex, dacron, orlon, nylon, wool, cotton, and dynel

Cleaning Methods
    Reverse flow, pulse jet, and various shaking methods including manual, mechanical, and sonic

Partial List of Suppliers
    Aeropulse, Inc.; Airguard Industries Inc.; Air Sentry, Inc.; Alco Industries Inc.; Amco Engineering Co.; BGI Inc.; Dresser Industries Inc.; Dustex Corp.; Emtrol Corp.; Environmental Elements Inc.; Esstee Mfg. Co., Inc.; Flex-Cleen Corp.; Fuller Co.; General Filters Inc.; Great Lakes Filter, Filpaco Div.; Hough International, Inc.; HVAC Filters Inc.; Interel Environmental Technologies Inc.; Kimbell-Bishard; Mikropul Environmental Systems; North American Filter Co.; Perry Equipment Corp.; Purolator Products Air Filtration Co.; Science Applications International Co.; Stamm International Group; Standard Filter Corp.; TFC Corp.; Tri-Dim Filter Corp.; Trion Inc.; United McGill Corp.; Wheelabrator Air Pollution Control; Zelcron Industries, Inc.; Zurn Industries Inc.

FIBER FILTERS

Fiber bed filters collect particles within the depth of the filter so that a dust cake does not build up on the filter surface. Fiber sizes range from less than 1 μm to several hundred μm. The materials used for fibrous filters include cellulose, glass, quartz, and plastic fibers. Fiber beds are used as furnace filters, air conditioning filters, and car air filters. They are also used for sampling in air pollution and industrial hygiene measurements. The collected particles cannot generally be cleaned from the bed, so these filters must be replaced when the pressure drop becomes too high.

Filtration Theory

The basis of this theory is the capture of particles by a single fiber. The single-fiber efficiency η_{fiber} is defined as the ratio of the number of particles striking the fiber to the number of particles that would strike the fiber if streamlines were not diverted around the fiber. If a fiber of diameter d_f collects all particles contained in a layer of thickness y, then the single-fiber efficiency is y/d_f (see Figure 3.3.9).

The general approach involves finding the velocity field around an isolated fiber, calculating the total collection efficiency of the isolated fiber due to the mechanisms of particle deposition, expressing the influence of neighboring fibers (interference effects) by means of empirical corrections, and finally obtaining the overall efficiency of a filter composed of many fibers. The following equation relates the overall efficiency of a filter composed of many fibers in a bed E to the single-fiber efficiency:

$$E = 1 - \exp\left[\frac{-4\eta_{\text{fiber}}\alpha L}{\pi d_f(1 - \alpha)}\right] \qquad 3.3(22)$$

where:

α = the solidity or packing density of the filter
L = the filter thickness
d_f = the fiber diameter

Note than even if $\eta_{\text{fiber}} = 1$, the total filter efficiency can be low if α is low or the filter thickness is small. The derivation of this equation assumes that particles are well-mixed in every plane perpendicular to the gas flow direction, no particles are reentrained into the gas stream, all fibers are perpendicular to the gas flow direction, and the gas has a uniform velocity.

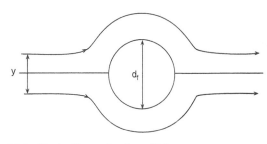

FIG. 3.3.9 Single-fiber collection efficiency.

Filtration Mechanisms

As air passes through a filter, the trajectories of particles deviate from the streamlines due to various mechanisms, most important of which are interception, inertial impaction, diffusion, gravity, and electrostatic forces. The radiometric phenomena such as thermophoresis and diffusiophoresis are usually negligible. The single-fiber efficiency η_{fiber} is approximated as the sum of the efficiencies due to each of the preceding mechanisms acting individually as follows:

$$\eta_{fiber} = \eta_{inter} + \eta_{impaction} + \eta_{diffusion} + \eta_{gravity} + \eta_{elec} \quad 3.3(23)$$

The maximum value of the single-fiber efficiency is 1.0; so if several of the mechanisms have large efficiencies, their sum cannot exceed unity.

Interception

Particles are collected when the streamline brings the particle center to within one particle radius from the fiber surface. This effect is described by a dimensionless interception parameter R, which is the ratio of particle diameter d_p to fiber diameter as follows:

$$R = \frac{d_p}{d_f} \quad 3.3(24)$$

Assuming Kuwabara flow, η_{inter} is given by the following equation:

$$\eta_{inter} = \frac{1 + R}{2Ku} \left[2\ln(1 + R) - 1 \right.$$
$$\left. + \alpha + \left(\frac{1}{1 + R}\right)^2 \left(1 - \frac{\alpha}{2}\right) - \frac{\alpha}{2}(1 + R)^2 \right] \quad 3.3(25)$$

where Ku is the Kuwabara hydrodynamic factor given by the following equation:

$$Ku = -\frac{1}{2} \ln \alpha - \frac{3}{4} + \alpha - \frac{\alpha^2}{4} \quad 3.3(26)$$

Interception is independent of the flow velocity around a fiber.

Inertial Impaction

A particle, because of its inertia, may be unable to adjust to the rapidly changing curvature of streamlines in the vicinity of a fiber and may cross the streamlines to hit the fiber. The dimensionless number used to describe this effect is the Stokes number, which is the ratio of particle stopping distance to fiber diameter as follows:

$$Stk = \frac{\tau U}{d_f}, \quad \text{where } \tau = \frac{1}{18} d_p^2 \frac{(\rho_p - \rho_m)}{\mu} C_c \quad 3.3(27)$$

where:

C_c = the Cunningham slip correction factor
U = the free-stream velocity (cm/sec)
μ = the viscosity of the medium (poise)
ρ_p and ρ_m = the densities of the particle and medium, respectively (gm/cc)

The slip correction factor is given by the equation $C_c = 1 + [2\lambda(1.257)/d_p]$ where λ is the mean free path of gas molecules and is equal to 0.071 μm for air molecules at 25°C and $d_p > 2\lambda$.

Stechkina, Kirsch, and Fuch (1969) calculated the inertial impaction efficiency for particles using the Kuwabara flow field as follows:

$$\eta_{impaction} = \frac{1}{(2Ku)^2}$$
$$\times [(29.6 - 28\alpha^{0.62})R^2 - 27.5R^{2.8}]Stk \quad 3.3(28)$$

Inertial impaction is a significant collection mechanism for large particles, high gas velocities, and small-diameter fibers.

Brownian Diffusion

Very small particles generally do not follow streamlines but wobble randomly due to collisions with molecules of gas. Random Brownian motion causes incidental contact between a particle and the fiber. At low velocities, particles spend more time near fiber surfaces, thus enhancing diffusional collection. A dimensionless parameter called the Peclet number Pe is defined as a ratio of convective transport to diffusive transport as follows:

$$Pe = \frac{d_c U}{D} \quad 3.3(29)$$

where:

d_c = the characteristic length of the collecting medium
U = the average gas velocity
D = the diffusion coefficient of the particles as follows:

$$D = \frac{kT C_c}{3\pi\mu d_p} \quad 3.3(30)$$

where:

k = the Boltzman constant (1.38×10^{-16} erg/°Kelvin)
T = the absolute temperature
μ = the viscosity of the medium (poise)
d_p = the particle diameter

Lee and Liu (1982a,b) use a boundary layer model commonly used in heat and mass transfer analysis with a flow field around multiple cylinders that account for flow interference due to neighboring fibers as follows:

$$\eta_{diffusion} = 2.58 \frac{1 - \alpha}{Ku} Pe^{-2/3} \quad 3.3(31)$$

Gravitational Settling

Particles deviate from streamlines when the settling velocity due to gravity is sufficiently large. When flow is downward, gravity increases collection; conversely, when flow is upward, gravity causes particles to move away from the collector resulting in a negative contribution. Gravity is important for large particles at low flow velocities. The dimensionless parameter for this mechanism is as follows:

$$Gr = \frac{V_{ts}}{U} \qquad 3.3(32)$$

where U is the free stream velocity and V_{ts} is the settling velocity. The single-filtration efficiency due to gravity is as follows:

$$\eta_{gravity} = \frac{Gr}{1 + Gr} \qquad 3.3(33)$$

Electrostatic Forces

Aerosol particles can acquire an electrostatic charge either during their generation or during flow through a gas stream. Likewise, the filter fibers can acquire a charge due to the friction caused by a gas stream passing over them. The following three cases can occur:

1. Charged Particle–Charged Fiber

$$\eta_{elec} = \frac{4Qq}{3\mu dd_f V_g} \qquad 3.3(34)$$

where Q is the charge per unit length of the fiber and q is the charge on the particle.
2. Charged Fiber–Neutral Particle

$$\eta_{elec} = \frac{4}{3}\left(\frac{\varepsilon-1}{\varepsilon-1}\right)\left(\frac{d^2Q^2}{d_f^3 \mu V_g}\right) \qquad 3.3(35)$$

where   is the dielectric constant of the particle.
3. Charged Particle–Neutral Fiber

$$\eta_{elec} = \sqrt{\frac{\varepsilon-1}{\varepsilon-1}\left(\frac{q^2}{3\pi\mu dd_f^2 V_g(2-\ln Re)}\right)^{1/2}} \qquad 3.3(36)$$

where μ is the viscosity of the medium and Re is the Reynold's number given by $(d_f V_g \rho_g)/\mu$.

Pressure Drop of Fibrous Filters

The following equation gives the pressure drop for fibrous filters (Lee and Ramamurthi 1993):

$$\Delta P = \frac{16\eta\alpha UL}{Kud_f^2} \qquad 3.3(37)$$

The measured pressure drop across a filter is used with Equation 3.3(37) to determine the effective fiber diameter d_f for use in the filtration efficiency theory, previously described (Lee and Ramamurthi 1993). Predicting the pressure drop for real filters is not straightforward. Comparing the measured pressure drop and calculated pressure drop based on an ideal flow field indicates how uniformly the media structure elements, such as fibers or pores, are arranged.

FABRIC FILTERS (BAGHOUSES)

Baghouses separate fly ash from flue gas in separate compartments containing tube-shaped or pocket-shaped bags or fabric filters. Baghouses are effective in controlling both total and fine particulate matter. They can filter fly ash at collection efficiencies of 99.9% on pulverized, coal-fired utility boilers. Other baghouse applications include building material dust removal, grain processing, oil mist recovery in workplace environments, soap powders, dry chemical recovery, talc dust recovery, dry food processing, pneumatic conveying, and metal dust recovery.

The main parameters in baghouse design are the pressure drop and air-to-cloth ratio. Pressure drop is important because higher pressure drops imply that more energy is required to pull gas through the system. The *air-to-cloth ratio* determines the unit size and thus, capital cost. This ratio is the result of dividing the volume flow of gas received by a baghouse by the total area of the filtering cloth and is usually expressed as acfm/ft^2. This ratio is also referred to as the *face velocity*. Higher air-to-cloth ratios mean less fabric, therefore less capital cost. However, higher ratios can lead to high pressure drops forcing energy costs up. Also, more frequent bag cleanings may be required, increasing downtime. Fabric filters are classified by their cleaning method or the direction of gas flow and hence the location of the dust deposit.

Inside Collectors

During filtration, as shown in Figure 3.3.10, dusty gas passes upward into tubular or pocket-shaped bags that are closed at the top. Tubular bags are typically 10 m tall and 300 mm in diameter. A dust cake builds on the inside bag surface during filtration. Clean gas passes out through the filter housing. Filtration velocities are about 10 mm/sec (2 cfm/ft^2). Many bags in a compartment act in parallel, and a fabric filter usually comprises several compartments.

Two cleaning techniques are used with inside collectors: reverse flow and shaking. In reverse flow, as shown in Figure 3.3.10, filtered gas from the outlets of other compartments is forced backward through the bags. Between the support rings, this reverse flow causes the bags to partially collapse causing the dust cake on the inside to deform, crack, and partially dislodge. A framework of rings keeps the bags open during cleaning. After a few minutes, the reverse gas stops, and filtration resumes. Cleaning is usually performed offline. Typical air-to-cloth ratios are 2 cfm/ft^2, and dust cake weights range from 0.5 to 1.5 lb/ft^2. The pressure drop across the fabric and dust cake with reverse flow is 0.5 to 1.0 in w.g. Some inside collectors used for oil mist recovery are not cleaned but are replaced when they become saturated with oil.

When equilibrium is established, the forces acting to remove the residual dust cake must equal those tending to retain it (Carr and Smith 1984). The removing forces are mechanical flexing and deformation, aerodynamic pressure, gravity, erosion, and acceleration by snapping the bags. Forces acting to retain the dust cake are adhesion and cohesion. *Adhesion* refers to the binding forces between particles and fibers when they contact each other; whereas *cohesion* refers to the bonding forces that exist between the collected particles. Forces acting to remove the residual dust cake increase with thickness or weight of the dust cake, while those tending to retain it decrease or remain the same. Thus, when equilibrium is achieved, a residual dust cake is established.

In shaking, the bag tops are connected to an oscillatory arm, causing the dust on the inside bag surfaces to separate from the fabric and fall into the hopper. Shaking is usually used in conjunction with reverse-gas cleaning. Although bags in shake and deflate units contain no anti-collapse rings, they retain a nominally circular cross section because the reverse-gas flow is low. The shaking force is applied to the tops of the bags causing them to sway and generating traveling waves in them. Deformation of the bags is significant in dislodging the dust cake.

Bags made from woven fabrics are generally used for these filters. Most large filters cleaned by reverse flow collect fly ash at coal-fired plants (Noll and Patel 1979). Bags are usually made of glass fiber to withstand the hot flue gases. Bag lifetime has exceeded 20,000 hr for woven bags. The long life is due to the low filtration velocity and infrequent cleaning (Humphries and Madden 1981). Filter problems in the utility industry are caused by improper bag specifications, installation, or tensioning. Other problems are related to *bag blinding,* a gradual, irreversible increase in pressure drop.

Outside Collectors

As shown in Figure 3.3.11, in outside collectors, dusty gas flows radially inward through cylindrical bags held open by a metal frame inside them. The bags are typically 3 m tall and 200 mm in diameter. Dust collects on the outside bag surfaces. Clean gas passes out of the top of each bag to a plenum.

Cleaning outside collectors usually involves injecting a pulse of compressed air at the outlet of each bag. This pulse snaps the bag open and drives the collected dust away from the bag surface into the hopper. Because pulse-jet cleaning takes a fraction of a second, it can be done online without interrupting the gas flow to a compartment.

Pulse-jet filters generally use thick felt fabrics to reduce dust penetration even when the dust cake is not thick. Filtration velocities through a pulse-jet filter are several times higher than through a reverse gas filter; therefore, pulse-jet filters can be smaller and less expensive. Operating at high filtration velocities can lead to an excessive pressure drop, dust penetration, and fabric wear. These filters are most commonly used to control industrial dust.

Pulse-jet cleaning can be ineffective when the dust is fine. The particles are driven beneath the fabric surface and are not easily removed. As fine dust accumulates below the surface, the pressure drop across the filter gradually increases. If the pressure drop becomes too high, the blinded bags must be replaced. With online cleaning of pulse-jet filters, as little as 1% of the dust on a bag can fall to the hopper after each cleaning pulse (Leith, First, and Feldman 1977).

Pressure Drop

The pressure drop is the sum of the pressure drop across the filter housing and across the dust-laden fabric (Leith and Allen 1986). The pressure drop across the housing is

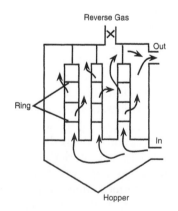
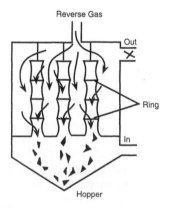

FIG. 3.3.10 Reverse-gas flow cleaning in inside collectors.

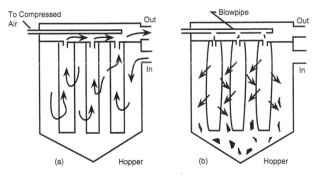

FIG. 3.3.11 Pulse-jet cleaning in outside collectors.

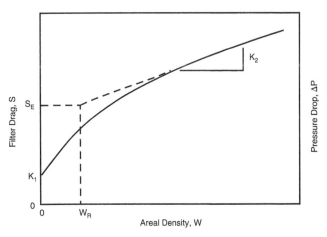

FIG. 3.3.12 Pressure drop and drag for a fabric filter versus areal density of the dust deposit.

proportional to the square of the gas-flow rate due to turbulence. The pressure drop across the dust-laden fabric is the sum of the pressure drop across the clean fabric and the pressure drop across the dust cake as follows:

$$\Delta P = \Delta P_f + \Delta P_d = K_1 v + K_2 vw \qquad 3.3(38)$$

where:

v = the filtration velocity
K_1 = the flow resistance of the clean fabric
K_2 = the specific resistance of the dust deposit
w = the fabric dust areal density

K_1 is related to Frazier permeability, which is the flow through a fabric in cfm/ft^2 of fabric when the pressure drop across the fabric is 0.5 in w.g. as follows:

$$K_1(Pa\ s\ m^{-1}) = \frac{24590}{\text{Frazier Permeability(cfm/ft}^2 \text{ at 0.5 in w.g.)}}$$

$$3.3(39)$$

The following equation determines the dust loading adding to the fabric between cleanings w_0:

$$w_0 = C_{in}vt \qquad 3.3(40)$$

where:

C_{in} = the dust inlet concentration
t = the time between cleanings

Dividing Equation 3.3(38) by v gives the filter drag as follows:

$$S = \frac{\Delta P}{v} = K_1 + K_2 w \qquad 3.3(41)$$

Figure 3.3.12 shows a plot of drag S versus w. Equation 3.3(41) assumes that the increase in drag with increasing areal density is linear; however, the increase is linear only after a dust cake has formed. Therefore, Equation 3.3(41) can be rewritten as follows in terms of the extrapolated residual areal dust density after cleaning w_R and the effective drag S_E.

$$S = S_E + K_2(w - w_R) \qquad 3.3(42)$$

Evaluation of Specific Resistance K_2

The Kozeny–Carman relationship (Billings and Wilder 1970; Carman 1956) is often used to describe pressure drop across a dust deposit. Rudnick and First (1978) showed that the Happel (1958) cell model corrected for slip flow gives better K_2 estimates than the Kozeny–Carman relationship for a dust cake with no interaction between the fabric and cake. All theoretical models for K_2 are sensitive to dust deposit porosity and dust particle size distributions. Porosity cannot usually be estimated correctly and varies with filtration velocity, humidity, and other factors. Whenever possible, K_2 should be measured rather than calculated from theory. Leith and Allen (1986) suggest that the dust collected on a membrane filter and K_2 should be calculated from the increase in pressure drop $(\Delta P_2 - \Delta P_1)$ with filter weight gain $(M_2 - M_1)$ as follows:

$$K_2 = \left(\frac{A}{v}\right)\left[\frac{(\Delta P_2 - \Delta P_1)}{(M_2 - M_1)}\right] \qquad 3.3(43)$$

where A is the surface area of the membrane filter.

Cleaning Fabric Filters

Dennis and Wilder (1975) present a comprehensive study on cleaning mechanisms. For a dust cake to be removed from a fabric, the force applied to the cake must be greater than the forces that bond the cake to the fabric. The major factors that combine to create both adhesion and cohesion are van der Waals, coulombic and induced dipole electrostatic forces, and chemical reactions between the gas and the dust. Capillary forces caused by surface tension are important under high relative humidity conditions.

The following equation calculates the drag across a single bag:

$$S = \frac{1}{\sum_{i=1}^{n} \left(\frac{a_i}{S_i}\right)} \qquad 3.3(44)$$

where a_i is the ith fraction of the total bag area and S_i is the drag through that area (Stephan, Walsh, and Herrick 1960). Substituting the values a_i and S_i for the cleaned and uncleaned portions in Equation 3.3(44) calculates the drag for the entire filter. When the local filtration velocity through each bag area (which depends on local drag) is used, the additional dust collected on each area over time can be determined. This collection in turn changes the local drag. This procedure allows an iterative prediction of pressure drop versus time.

Cleaning Inside Collectors

The factors responsible for cake removal in inside collectors have received little attention. Cake rupture due to fabric flexing is the primary cleaning mechanism. The other main mechanism is the normal stress to the cake from reverse pressurization.

Shaker filters use a motor and an eccentric cam to produce the cleaning action. Cleaning depends on factors such as bag shape and support, rigidity, bag tension, shaker frequency, and shaker amplitude. The following equation gives the acceleration a developed by the shaker arm:

$$a = 4\pi^2 f^2 l_d \qquad 3.3(45)$$

where:

   f = the frequency of shaking
   l_d = the amplitude of the sinusoidal motion

Modeling the transfer of motion to the bags is analogous to a vibrating string, whereby a wave travels down the bag and is reflected from its end. If the reflected wave reaches the arm at the same time that the next wave is produced, resonance occurs; this time is when the transfer of cleaning energy to the bag is optimum.

Pulse-Jet Cleaning

Leith and Ellenbecker (1980b) use Equation 3.3(38) as the basis of a model to predict pressure drop across a pulse-jet cleaned filter. Their model assumes that the fraction of the dust deposit removed by a cleaning pulse is proportional to the separation force applied and that impulse and conservation of momentum determine bag motion as follows:

$$\Delta P = K_v v^2 + \frac{P_s + K_1 v - \sqrt{\left(P_s - K_1 v\right) - 4w_o v\left(\frac{K_2}{K_3}\right)}}{2}$$

$$3.3(46)$$

where K_v depends on the pressure drop characteristics of the venturi at the top of each bag. Typically, $K_v = 57500$ Pa s^2/m^2. The value P_s is the pressure within the bag generated by the cleaning pulse, and it depends on venturi design as well as pulse pressure P as follows:

$$P_s(Pa) = 164[P(kPa)]^{0.6} \qquad 3.3(47)$$

The values K_1 and w_o are from Equations 3.3(39) and 3.3(40), respectively. The term (K_2/K_3) is a constant that depends on the interaction of the dust and the fabric.

An environmental engineer can use Equation 3.3(46) to estimate the pressure drop of an existing filter when the operating conditions change. When the pressure drop under the initial operating conditions is known, the engineer can use Equation 3.3(46) to determine (K_2/K_3). Then, using this value of (K_2/K_3), the engineer can determine the pressure drop for other operating conditions.

Collection Efficiency

The collection efficiency of a fabric filter depends on the interactions between the fabric, the dust, and the cleaning method. However, a clear understanding of these factors is not available, and existing models cannot be generalized beyond the data sets used in their development.

Impaction, diffusion, and interception in a dust cake are effective. Essentially all incoming dust is collected through a pulse-jet filter, with or without a dust cake. Nevertheless, some dust penetrates a fabric filter. It does so because gas bypasses the filter by flowing through pinholes in the dust cake or because the filter fails to retain the dust previously collected (seepage).

Pinholes form in woven fabrics at yarn intersections. With dust loading, some pinholes are bridged, while the gas velocity increases through open pinholes. Incoming particles bounce through pinholes rather than collect. A disproportionately high fraction of gas flows through unbridged pinholes due to their lower resistance. As the fabric flexes during cleaning, some dust particles dislodge and recollect deeper in the fabric. After several cycles of dislodgment and recollection, the particles completely pass through the filter. Particle penetration by this mechanism is called *seepage*.

Dennis et al. (1977) and Dennis and Klemm (1979) developed a model for predicting effluent fly ash concentration with fabric loading for woven glass fabrics. However, their models contain empirical constants that may be inappropriate for other fabrics. Leith and Ellenbecker (1982) showed that seepage is the primary mechanism for penetration through a pulse-jet-cleaned felt fabric. They developed a model (Leith and Ellenbecker 1980a) for outlet flux N assuming that all incoming dust is collected by the filter and that seepage of previously collected particles through the fabric accounts for all the dust emitted once the filter is conditioned. Seepage occurs as the bag strikes its supporting cage at the end of a cleaning pulse. The impact dislodges particles from the filter; the particles are then carried into the outlet gas stream. The following equation calculates the outlet flux:

$$N = \frac{kw^2v}{t} \qquad 3.3(48)$$

where:

- N = the outlet flux
- w = the areal density of the dust deposit
- v = the filtration velocity
- t = the time between cleaning pulses to each bag
- k = a constant that depends on factors such as dust characteristics, fabric types, and length of filter service

Figure 3.3.13 plots the outlet flux measured in laboratory experiments against the flux predicted using Equation 3.3(48) with k = 0.002 m − s/kg. These data are for different felt surfaces with different filtration velocities and different dusts. However, all data cluster about the same line.

Figure 3.3.14 plots the mass outlet flux against particle diameter for a pulse-jet filter. Comparatively little flux results from the seepage of small particles because of their small mass. The flux due to large particles is also little because they do not pass through the filter. Intermediate size particles contribute the most to the flux since they are large enough to have appreciable mass and small enough to seep

through. However, no theory exists to predict outlet flux as a function of particle size.

However, Christopher, Leith, and Symons (1990) evaluate mass penetration as a function of particle size. They developed the following empirical equation:

$$N = k\left(\frac{w^2v}{t}\right)^n \qquad 3.3(49)$$

where the exponent n is given by:

$$n = 0.916(d_p)^{0.257} \qquad 3.3(50)$$

where d_p is the particle diameter in μm.

The following equation gives model constant k:

$$k = a(d_p)^{4.82} \qquad 3.3(51)$$

where a = 4.45 × 10^{-6} for Ptfe-laminated fabric and a = 2.28 × 10^{-5} for untreated polyester felt fabric.

Bag Design

Bags are chosen for temperature and chemical resistance, mechanical stability, and the ability to collect the dust cake and then allow it to be easily removed. In utility applications, fiberglass bags are used exclusively. For low temperature applications (<280°F), acrylic is an economical alternative.

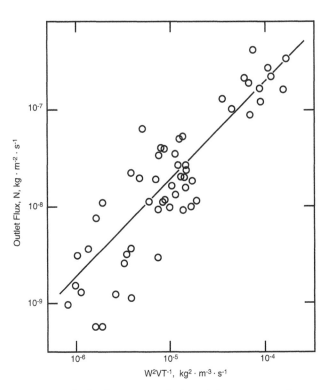

FIG. 3.3.13 Outlet flux versus w^2 v/t for laboratory experiments with a pulse-jet filter.

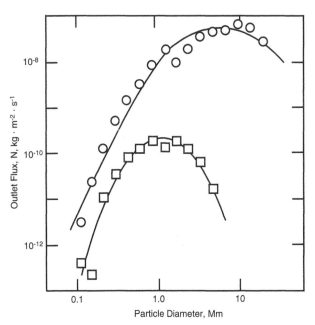

FIG. 3.3.14 Outlet flux versus particle diameter for a pulse-jet filter at 50 mm/sec velocity collecting granite dust on Ptfe-laminated bags (squares) and at 100 mm/sec velocity collecting fly ash on polyester felt with untreated surface (circles).

TABLE 3.3.2 RELATIVE PROPERTIES OF POPULAR FIBERS

Resistance to Dry Heat		Resistance to Moist Heat		Resistance to Abrasion		Relative Tensile Strength	Specific Gravity
Glass	550°	Glass	E	Nylon	E	Nylon	1.14
Teflon	400°	Teflon	E	Dacron	E	Polypro	0.90
Nomex	400°	Nomex*	E	Polypro	E	Dacron	1.38
Dacron	275°	Orlon	G	Nomex	E	Glass	2.54
Orlon	260°	Nylon*	G	Orlon	G	Polyeth	0.92
Rayon	200°	Rayon	G	Dynel	G	Rayon	1.52
Polypro	200°	Cotton	G	Rayon	G	Nomex	1.38
Nylon	200°	Wool	F	Cotton	G	Cotton	1.50
Wool	200°	Dacron*	F	Acetate	G	Dynel	1.30
Cotton	180°	Polypro	F	Polyeth	G	Orlon	1.14
Dynel	160°	Polyeth	F	Wool	F	Wool	1.32
		Dynel	F	Teflon	F	Teflon	2.10
		Acetate	F	Glass	P	Acetate	1.33

Resistance to Mineral Acids		Resistance to Alkalies		Relative Staple Fiber Cost		Relative Filament Fiber Cost	
Glass†	E	Teflon	E	Teflon	7.0	Teflon	54.
Teflon	E	Polypro	E	Nomex	3.2	Nomex	16.
Polypro	E	Nylon	G	Glass	6.	Acrylic	5.
Polyeth	G	Nomex	G	Orlon (Acrylic)	5.	Dacron	3.
Orlon	G	Cotton	G	Wool	5.	Nylon	3.
Dacron	G	Dynel	G	Nylon	5.	Polypro	2.8
Dynel	G	Polyeth	G	Dynel	4.	Polyeth	2.3
Nomex	F	Dacron	F	Dacron	3.5	Rayon	1.6
Nylon	P	Orlon	F	Acetate	2.0	Acetate	1.4
Rayon	P	Rayon	F	Rayon	1.5	Glass	1.0
Cotton	P	Acetate	P	Cotton	1.0		
Acetate	P	Glass	P				
Wool	G	Wool	P				

Notes: *These fibers degrade in hot, moist atmospheres; Dacron is affected most, Nomex next, nylon the least.
†Glass is destroyed by gaseous HF at dew point temperatures.
These ratings are only a general guide: E = Excellent, G = Good, F = Fair, P = Poor.

Bag fabric research focuses on two broad objectives: improving performance and economics. Different weaves and finishes for fiberglass are being studied, as well as alternative fabrics such as Teflon, Nomex, Ryton, and felted glass (see Table 3.3.2).

Bag weave is significant because different weaves produce different types of filter cakes. *Warp* is the system of yarns running lengthwise in a fabric, and *fill* is the system running crosswise. The style, weight, thickness, porosity, and strength are factors to consider in weave selection. Finishes are a lubricant against abrasion and protect against acid and alkali attack. Teflon, silicon, and graphite are the most popular finishes.

—*Gurumurthy Ramachandran*

References

Alexander, R. Mc K. 1949. Fundamentals of cyclone design and operation. *Proc. Australas Inst. Mining Metall.*, n.s., 203:152–153.

Barth, W. 1956. Design and layout of the cyclone separator on the basis of new investigations. *Brennstoff-Warme-Kraft* 8:1–9.

Beeckmans, J.M. 1973. A two-dimensional turbulent diffusion model of the reverse flow cyclone. *J. Aerosol Sci.* 4:329.

Billings, C.E., and J. Wilder. 1970. *Fabric filter cleaning studies.* ATD-0690, PB-200-648. Springfield, Va.: NTIS.

Boysan, F., W.H. Ayers, and J. Swithenbank. 1982. A fundamental mathematical modeling approach to cyclone design. *Trans. I. Chem. Eng.* 60:222–230.

Briggs, L.W. 1946. Effect of dust concentration on cyclone performance. *Trans. Amer. Inst. Chem. Eng.* 42:511.

Carman, P.C. 1956. *Flow of gases through porous media.* New York: Academic Press.

Carr, R.C., and W.B. Smith. 1984. Fabric filter technology for utility coal-fired power plants. *JAPCA* 43, no. 1:79–89.

Christopher, P.C., D. Leith, and M.J. Symons. 1990. Outlet mass flux from a pulse-jet cleaned fabric filter: Testing a theoretical model. *Aerosol Sci. and Technol.* 13:426–433.

Dennis, R., R.W. Can, D.W. Cooper, R.R. Hall, V. Hampl, H.A. Klemm, J.E. Longley, and R.W. Stern. 1977. *Filtration model for coal fly ash with glass fabrics.* EPA report, EPA-600/7-77-084. Springfield, Va.: NTIS.

Dennis, R., and H.A. Klemm. 1979. A model for coal fly ash filtration. *J. Air Pollut. Control Assoc.* 29:230.

Dennis, R., and J. Wilder. 1975. *Fabric filter cleaning studies.* EPA-650/2-75-009. Springfield, Va.: NTIS.

Dietz, P.W. 1981. Collection efficiency of cyclone separators. *A.I.Ch.E. Journal* 27:888.

Dirgo, J. 1988. Relationships between cyclone dimensions and performance. Sc.D. thesis, Harvard University, Cambridge, Mass.

Dirgo, J.A., and D. Leith. 1985. Performance of theoretically optimized cyclones. *Filtr. and Sep.* 22:119–125.

Dirgo, J.A., and D. Leith. 1986. Cyclones. Vol. 4 in *Encyclopedia of fluid mechanics,* edited by N.P. Cheremisinoff. Houston: Gulf Publishing Co.

First, M.W. 1950. Fundamental factors in the design of cyclone dust collectors. Sc.D. thesis, Harvard Univ., Cambridge, Mass.

Happel, J. 1958. Viscous flow in multiparticle systems: Slow motion of fluids relative to beds of spherical particles. *A.I.Ch.E. Journal* 4:197.

Humphries, W., and J.J. Madden. 1981. Fabric filtration for coal fired boilers: Nature of fabric failures in pulse-jet filters. *Filtr. and Sep.* 18:503.

Iinoya, K. 1953. Study on the cyclone. Vol. 5 in *Memoirs of the faculty of engineering,* Nagoya Univ. (September).

Iozia, D.L., and D. Leith. 1989. Effect of cyclone dimensions on gas flow pattern and collection efficiency. *Aerosol Sci. and Technol.* 10:491.

Iozia, D.L., and D. Leith. 1990. The logistic function and cyclone fractional efficiency. *Aerosol Sci. and Technol.* 12:598.

Kessler, M., and D. Leith. 1991. Flow measurement and efficiency modeling of cyclones for particle collection. *Aerosol Sci. and Technol.* 15:8–18.

Kirch, R., and F. Loffler. 1987. Measurements of the flow field in a gas cyclone with the aid of a two-component laser doppler velocimeter. *ICALEO '87 Proc., Optical Methods in Flow and Particle Diagnostics, Laser Inst. of America* 63:28–36.

Lee, K.W., and B.Y.H. Liu. 1982a. Experimental study of aerosol filtration by fibrous filters. *Aerosol Sci. and Technol.* 1:35–46.

Lee, K.W., and B.Y.H. Liu. 1982b. Theoretical study of aerosol filtration by fibrous filters. *Aerosol Sci. and Technol.* 1:147–161.

Lee, K.W., and M. Ramamurthi. 1993. Filter collection. In *Aerosol measurement, principles, techniques, and applications,* edited by K. Willeke and P.A. Baron. Van Nostrand Reinhold Co.

Leith, D. 1984. Cyclones. In *Handbook of powder science and technology,* edited by M.A. Fayed and L. Otten. Van Nostrand Reinhold Co.

Leith, D., and R.W.K. Allen. 1986. Dust filtration by fabric filters. In *Progress in filtration and separation,* edited by R.J. Wakeman. Elsevier.

Leith, D., J.A. Dirgo, and W.T. Davis. 1986. Control devices: Application, centrifugal force and gravity, filtration and dry flue gas scrubbing. Vol. 7 in *Air pollution,* edited by A. Stern. New York: Academic Press, Inc.

Leith, D., and M.J. Ellenbecker. 1980a. Theory for penetration in a pulse jet cleaned fabric filter. *J. Air Pollut. Control Assoc.* 30:877.

———. 1980b. Theory for pressure drop in a pulse-jet cleaned fabric filter. *Atmos. Environ.* 14:845.

———. 1982. Dust emission characteristics of pulse jet cleaned fabric filters. *Aerosol Sci. and Technol.* 1:401.

Leith, D., M.W. First, and H. Feldman. 1977. Performance of a pulse-jet filter at high filtration velocity II. Filter cake redeposition. *J. Air Pollut. Control Assoc.* 27:636.

Leith, D., and W. Licht. 1972. The collection efficiency of cyclone type particle collectors—A new theoretical approach. *AIChE Symposium Ser.* 68:196–206.

Leith, D., and D. Mehta. 1973. Cyclone performance and design. *Atmos. Environ.* 7:527.

Lissman, M.A. 1930. An analysis of mechanical methods of dust collection. *Chem. Met. Eng.* 37:630.

Noll, K.E., and M. Patel. 1979. Evaluation of performance data from fabric filter collectors on coal-fired boilers. *Filtr. Sep.* 16:230.

Ramachandran, G., D. Leith, J. Dirgo, and H. Feldman. 1991. Cyclone optimization based on a new empirical model for pressure drop. *Aerosol Sci. and Technol.* 15:135–148.

Rosin, P., E. Rammler, and W. Intelman. 1932. Grundhagen und greuzen der zyklonentstaubung. *Zeit. Ver. Deutsch. Ing. Z.* 76:433.

Rudnick, S., and M.W. First. 1978. *Third Symposium on Fabric Filters for Particulate Collection.* Edited by N. Surprenant. EPA-600/7-78-087, 251. Research Triangle Park, N.C.: U.S. EPA.

Shepherd, C.B., and C.E. Lapple. 1939. Flow pattern and pressure drop in cyclone dust collectors. *Ind. Eng. Chem.* 31:972.

Shepherd, C.B., and C.E. Lapple. 1940. Flow pattern and pressure drop in cyclone dust collectors. *Ind. Eng. Chem.* 32:1246.

Stairmand, C.J. 1949. Pressure in cyclone separators. *Engineering* (London) 168:409.

Stairmand, C.J. 1951. The design and performance of cyclone separators. *Trans. Instn. Chem. Engrs.* 29:356.

Stechkina, I.B., A.A. Kirsch, and N.A. Fuch. 1969. Studies in fibrous aerosol filters—IV. Calculation of aerosol deposition in model filter in the range of maximum penetration. *Ann. Occup. Hyg.* 12:1–8.

Ter Linden, A.J. 1949. Investigations into cyclone dust collectors. *Proc. Inst. Mech. Engrs.* (London) 160:233–240.

Whiton, L.C. 1932. Performance characteristics of cyclone dust collectors. *Chem. Met. Eng.* 39:150.

3.4
PARTICULATE CONTROLS:ELECTROSTATIC PRECIPITATORS

FEATURE SUMMARY

Types of Designs
Single- or two-stage

Applications
Single-stage units are used in coal-fired power plants to remove fly ash, cement kiln dust, lead smelter fumes, tar, and pulp and paper alkali salts. Two-stage units are used for air conditioning applications.

Flue Gas Limitations
Flow up to 4×10^6 acfm, temperature up to 800°C, gas velocity up to 10 ft/sec, pressure drop is under 1 in w.g.

Treatment Time
2–10 sec

Power Requirement
Up to 17.5 W per m³/min

Dust Particle Size Range
0.01 μm to greater than 1000 μm

Collection Efficiency
99.5 to 99.99%

Partial List of Suppliers
Air Cleaning Specialists, Inc.; Air Pol Inc.; Air Quality Engineering, Inc.; ASEA Brown Boveri Inc.; Babcock and Wilcox, Power Generation Group; Belco Technologies Corp.; Beltran Associates, Inc.; Dresser Industries Inc.; Ducon Environmental Systems; GE Company; Joy Environmental Technologies, Inc.; North American Pollution Control Systems; Research-Cottrell Companies; Scientific Technologies, Inc.; United Air Specialists, Inc.; Universal Air Precipitator Corp.; Wheelabrator Air Pollution Control.

Electrostatic precipitation uses the forces of an electric field on electrically charged particles to separate solid or liquid aerosols from a gas stream. The aerosol is deliberately charged and passed through an electric field causing the particles to migrate toward an oppositely charged electrode which acts as a collection surface. Gravity or rapping the collector electrode removes the particles from the precipitator. Various physical configurations are used in the charging, collection, and removal processes.

ESPs are characterized by high efficiencies, even for small particles. They can handle large gas volumes with low pressure drops and can be designed for a range of temperatures. On the other hand, they involve high capital costs, take up a lot of space, and are not flexible after installation to changes in operating conditions. They may not work on particles with high electrical resistivity.

Commercial ESPs accomplish charging using a high-voltage, direct-current corona surrounding a highly charged electrode, such as a wire. The large potential gra-

dient near the electrode causes a corona discharge comprising electrons. The gas molecules become ionized with charges of the same sign as the wire electrode. These ions then collide with and attach to the aerosol particles, thereby charging the particles. Two electrodes charge the particles, and two electrodes collect the particles, with an electric field between each pair.

When the same set of electrodes is used for both charging and collecting, the precipitator is called a *single-stage precipitator*. Rapping cleans the collecting electrodes; thus, these precipitators have the advantage of continuous operation. The discharge electrode consists of a wire suspended from an insulator and held in position by a weight at the bottom. A power source supplies a large direct current (DC) voltage (~50 kV) which can be either steady or pulsed.

Parallel plate ESPs are more widely used in industry. Here, gas flows between two vertical parallel plates with several vertical wires suspended between them. The wires are held in place by weights attached at the bottom. These wires constitute the charging electrodes, and the plates are the collecting electrodes. Rapping the plate removes the collected dust; the dust gathers in a dust hopper at the bottom. Figure 3.4.1 shows a single-stage, parallel plate precipitator with accessories such as hoppers, rappers, wire weights, and distribution baffles for the gas.

If different sets of electrodes are used for charging and collecting, the precipitator is called a *two-stage precipitator* (see Figure 3.4.2). Cleaning involves removing the collecting plates and washing them or washing the collecting plates in place. In a two-stage ESP, the charging section is short, providing a short residence time, and the collection section is five or more times longer to provide sufficient time for collection (Crawford 1976). Two-stage ESPs are used in air conditioning applications, while single-stage ESPs are used in industrial applications where dust loadings are higher and space is available.

Corona Generation

When the potential difference between the wire and plate electrodes increases, a voltage is reached where an electrical breakdown of the gas occurs near the wire. When gas molecules get excited, one or more of the electrons can shift to a higher energy level. This state is transient; once the excitation has ceased, the molecule reverts to its ground state, the reby releasing energy. Part of this energy con-

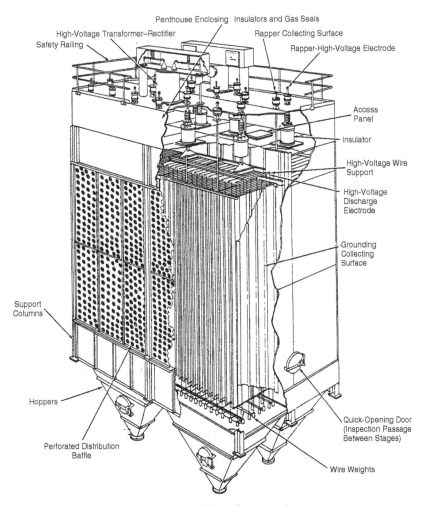

FIG. 3.4.1   Single-stage, parallel plate ESP with accessories.

verts to light. The bluish glow around the wire is the corona discharge. A different situation occurs when an electron or ion imparts additional energy on an excited molecule. This process causes a cascade or avalanche effect described next.

The space between the wire and the plate can be divided into an active and a passive zone (see Figure 3.4.3). In the active zone, defined by the corona glow discharge, electrons leave the wire electrode and impact gas molecules, thereby ionizing the molecules. The additional free electrons also accelerate and ionize more gas molecules. This avalanche process continues until the electric field decreases to the point that the released electrons do not acquire sufficient energy for ionization. The behavior of these charged particles depends on the polarity of the electrodes; a negative corona is formed if the discharge electrode is negative, and a positive corona is formed if the discharge electrode is positive.

In a negative corona, positive ions are attracted toward the negative wire electrode, and electrons are attracted toward the positive plate or cylinder electrode. Beyond the corona glow region, the electric field diminishes rapidly, and if electronegative gases are present, the gas molecules

become ionized by electron impact. The negative ions move toward the plate electrode. In the passive zone, these ions attach themselves to aerosol particles and serve as the principal means for charging the aerosol. The ion concentration is typically 10^7 to 10^9 ions/cm^3.

The corona current, and therefore the charge density in the space between the electrodes, depends on factors such as the ionic mobility, whether the gas is electropositive or electronegative, and whether the corona is positive or negative. If the gas is electropositive with low electron affinity like N_2, H_2, or an inert gas, its molecules absorb few electrons. Thus, the current is predominantly electronic. Due to the higher mobility of electrons, the corona current is high. Conversely, an electronegative gas like O_2 has high electron affinity and absorbs electrons easily. Here, the current is due to negative ions, and thus the corona current is low due to the lower mobility of the gas ions.

When a corona is negative, the free electrons leaving the active zone are transformed into negative ions with a substantially lower mobility on their way to the plates. The negative charge carriers thus cover the first part of their path as fast, free electrons and the second part as slower ions; their average mobility is lower than that of free elec-

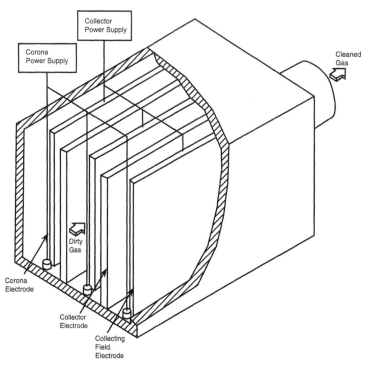

FIG. 3.4.2 Parallel plate, two-stage ESP.

trons but higher than that of the large ions. On the other hand, when a corona is positive, the positive charge carriers are large, slow ions by origin and retain this form throughout their motion. Consequently, a negative corona always has a higher corona current than a positive corona for an applied voltage.

Negative coronas are more commonly used in industrial applications, while positive coronas are used for cleaning air in inhabited spaces. A negative corona is accompanied by ozone generation and, therefore, is usually not used for cleaning air in inhabited spaces. However, most industrial gas-cleaning precipitators use a negative corona because of its superior electrical characteristics which increase efficiency at the temperatures at which they are used.

Current–Voltage Relationships

For a wire-in-cylinder configuration, the current–voltage relationship can be derived from Poisson's equation as follows:

$$\nabla^2 V = -\frac{4\pi}{\epsilon_0} \rho_s \qquad 3.4(1)$$

where:

ρ_s = the space charge per unit volume
ϵ_0 = the permittivity of the medium.

The equations in this treatment follow the electrostatic system of units (ESU). The value V is the electric potential which is related to the electric field $\vec{E}$ as follows:

$$-\nabla V = \vec{E} \qquad 3.4(2)$$

Thus, the relationship can be expressed as follows:

$$\nabla^2 V = \frac{\partial^2 V}{\partial r^2} + \frac{1}{r}\frac{\partial V}{\partial r} \qquad 3.4(3)$$

In the presence of a charge sufficient to produce corona discharge, the preceding equations can be expressed as:

$$\frac{dE}{dr} + \frac{1}{r}E = -\frac{4\pi}{\epsilon_0}\rho_s \qquad 3.4(4)$$

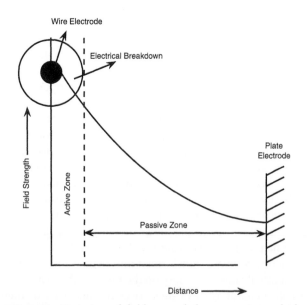

FIG. 3.4.3 Variation of field strength between wire and plate electrodes.

Assuming that the wire acts as an ion source, the current i applied to the wire is used to maintain the space charge as follows:

$$i = 2\pi r \rho_s Z E \qquad 3.4(5)$$

where Z is the ion mobility. Substituting Equation 3.4(5) in 3.4(4) and solving the resultant differential equation yields the following equation:

$$E = \left(\frac{2i}{Z} + \frac{C^2}{r^2}\right)^{1/2} \qquad 3.4(6)$$

where C is a constant of integration which depends on corona voltage, current, and the inner and outer cylinder diameters and r is the radial distance from the wire. For large values of i and r, Equation 3.4(6) reduces to E = $(2i/Z)^{1/2}$, implying a constant field strength over most of the cross section away from the inner electrode.

The following equation gives an approximation of the corona current i (White 1963):

$$i = V(V - V_0)\frac{2Z}{r_0^2 \ln\left[\dfrac{r_0}{r_i}\right]} \qquad 3.4(7)$$

where:

V   = the operating voltage
V_0 = the corona starting voltage
r_0 and r_i = the radii of the cylinder and wire, respectively.

The value V_0 can be estimated by use of the following equation:

$$V_0 = 100\delta f r_i \left(1 + \frac{0.3}{r_i}\right)\ln\left[\frac{r_0}{r_i}\right] \qquad 3.4(8)$$

where δ is a correction factor for temperature and pressure as follows:

$$\delta = \frac{293}{T}\frac{P}{760} \qquad 3.4(9)$$

The factor f is a wire roughness factor, usually between 0.5 and 0.7 (White 1963).

Equation 3.4(8) shows that reducing the size of the corona wire decreases the applied voltage necessary to initiate corona V_0. Decreasing the corona starting voltage increases the corona current for an applied voltage.

An analysis similar to that shown for wire-in-cylinder configuration can be applied to the parallel plate and wire electrode configuration. The equations are not as simple because the symmetry of the cylindrical precipitator simplifies the mathematics. The plate-type precipitator also has an additional degree of freedom—corona wire spacing. Qualitatively, if the corona wires are spaced close together, the system approaches the field configuration of a parallel plate capacitor, which yields a constant field strength in the interelectrode space. For an applied voltage, this configuration reduces the electric field near the corona wire. Therefore, using smaller wires or spacing the wires farther apart increases the current density in a region.

In the presence of aerosol particles, the particles get charged with the same sign as the wire. This particle charging decreases the voltage gradient, and therefore the corona starting voltage has a higher value. Another effect is that the electric field is distorted and the electric field near the collector electrode increases due to image forces.

Particle Charging

Two mechanisms are responsible for charging aerosol particles in an ESP: field charging and diffusion charging. Both mechanisms are active; however, each becomes significant for particles in different size ranges. Field charging is the dominant mechanism for particles with a diameter greater than 1 μm, while diffusion charging predominates for particles with a diameter less than 0.2 μm. In the intermediate region, both mechanisms contribute a significant charge.

FIELD CHARGING

The presence of particles with a dielectric constant greater than unity causes a localized deformation in the electric field (see Figure 3.4.4). Gas ions travel along electric field lines, and because the lines intercept the particle matter, the ions collide with these particles and charge them. When the particle reaches a *saturation charge*, additional ions are repelled and charging stops. The amount of charge q on a particle is the product of the number of charges n and the electronic charge e (q = ne). The following equation gives the rate of particle charging:

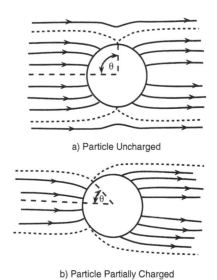

a) Particle Uncharged

b) Particle Partially Charged

FIG. 3.4.4 Distortion of an electric field around an aerosol particle.

$$\frac{d\left(\frac{n}{n_s}\right)}{dt} = \pi N_i e Z \left(1 - \frac{n}{n_s}\right)^2 \qquad 3.4(10)$$

On integration, this equation yields the number of charges on a particle as a function of time as follows:

$$n = n_s \left(\frac{t}{t + \tau}\right) \qquad 3.4(11)$$

where n_s is the saturation charge given by $(3/ + 2)$ $(d^2E/4e)$. The saturation charge is the maximum charge that can be placed on a particle of diameter d by a field strength E. The value is the dielectric constant of the particle. The value τ_e is a time constant for the rapidity of charging and is equal to $(1/\pi e Z_i N_i)$, where Z_i is the ion mobility and N_i is the ion concentration.

DIFFUSION CHARGING

Particles are charged by unipolar ions in the absence of an electric field. The collision of ions and particles occurs due to the random thermal motion of ions. If every ion that collides with a particle is retained, the rate of charging with respect to time is as follows:

$$\frac{dn}{dt} = \frac{\pi}{4} d^2 \bar{c} N_i \exp\left[-\frac{2ne^2}{dkT}\right] \qquad 3.4(12)$$

For an initially uncharged particle, integrating this equation gives the number of charges gained due to diffusion charging as follows:

$$n = \left(\frac{dkT}{2e^2}\right) \ln\left[1 + \frac{\pi d \bar{c} e^2 N_i}{2kT} t\right] \qquad 3.4(13)$$

where:

k = the Boltzman constant
T = the absolute temperature
e = the charge of one electron

Here, $\bar{c}$ is the mean thermal speed of the ions and for a Maxwellian distribution is given by $\sqrt{8kT/\pi m}$, where m is the ionic mass.

When both field and diffusion charging are significant, adding the values of the n calculated with Equations 3.4(11) and 3.4(13) is not satisfactory; adding the charging rates due to field and diffusion charging is better and gives an overall rate. This process yields a nonlinear differential equation with no analytic solution and therefore has to be solved numerically. An overall theory of combined charging agrees with experimental values but is computationally cumbersome.

Migration Velocity

The following equation determines the velocity of a charged particle suspended in a gas and under the influence of an electric field by equating the electric and Stokes drag forces on the particle:

$$V_p = \frac{neEC_c}{3\pi\mu d} \qquad 3.4(14)$$

Here, V_p is the particle migration velocity toward the collector electrode, and E is the collector electric field. The value for n depends on whether field or diffusion charging is predominant. Particles charged by field charging reach a steady-state migration velocity, whereas the migration velocity for small particles charged by diffusion charging continues to increase throughout the time that the dust is retained in the precipitator.

ESP Efficiency

The most common efficiency theory is the Deutsch–Andersen equation. The theory assumes that particles are at their terminal electrical drift velocity and well-mixed in every plane perpendicular to the gas flow direction due to lateral turbulence. The theory also assumes plug flow through the ESP, no reentrainment of particles from the collector plates, and uniform gas velocity throughout the cross section. Collection efficiency is expressed as follows:

$$\eta = 1 - \exp\left[-\frac{V_p A}{Q_g}\right] \qquad 3.4(15)$$

where A is the total collecting area of the precipitator and Q_g is the volumetric gas flow. The derivation of this equation is identical to that for gravity settling chambers except that the terminal settling velocity under a gravitational field is replaced with terminal drift velocity in an electric field.

The precipitator performance can differ from theoretical predictions due to deviations from assumptions made in the theories. Several factors can change the collection efficiency of an idealized precipitator. These factors include particle agglomeration, back corona, uneven gas flow, and rapping reentrainment.

Cooperman (1984) states that the Deutsch–Andersen equation neglects the role of mixing (diffusional and large-scale eddy) forces in the precipitator. These forces can account for the differences between observed and theoretical migration velocities and the apparent increase in migration velocity with increasing gas velocity. He postulates that the difference in particle concentration along the precipitator length produces a mixing force that results in a particle velocity through the precipitator that is greater than the gas velocity. At low velocities, the effect is pronounced, but it is masked at higher velocities. He presents a more general theory for predicting collection efficiency by solving the mass balance equation as follows:

$$D_1 \frac{\partial^2 C}{\partial x^2} + D_2 \frac{\partial^2 C}{\partial y^2} - V_g \frac{\partial C}{\partial x} + V_p \frac{\partial C}{\partial x} = 0 \qquad 3.4(16)$$

where:

C = the particle concentration
D_1 and D_2 = the longitudinal and transverse mixing coefficients, respectively

V_g = the gas velocity in the x direction
V_p = the migration velocity in the y direction
(see Figure 3.4.5)

The initial conditions chosen are $C(0,y) = C_0$, and $C(x,y) \to 0$ as $y \to 0$, where C_0 is the particle concentration at the inlet. The first condition implies a constant concentration across the inlet, while the second condition states that an infinitely long precipitator has zero penetration.

The boundary conditions are $D_2\, \partial C/\partial y + V_p C = 0$ at the center plane, and $D_2\, \partial C/\partial y - f V_p C = 0$ at the boundary layer. The first condition states that the diffusional flux balances the flux due to migration velocity, and no net particle flux occurs across the center plane. The second condition contains a reentrainment factor f. This factor is an empirical parameter indicating what fraction of the particles that enter the boundary layer are later reentrained back into the gas stream. The complete analytic solution is fairly complicated, and therefore only the first term of the solution is used; but it yields sufficient accuracy as follows:

$$\eta = 1 - \exp\left[\left(\frac{V_g}{2D_1}\right) - \sqrt{\left(\frac{V_g}{2D_1}\right)^2 + \left(\frac{D_2}{D_1}\right)\left(\frac{\lambda_1}{b}\right)^2 + \frac{V_p^2}{4D_1 D_2}}\,\right] L$$

3.4(17)

$$\tan \lambda_1 = \frac{\beta}{\chi}\, \frac{2\lambda_1(1-f)}{\lambda_1^2 - (1-f)\frac{\beta^2}{\chi^2}}$$

$$\beta = \frac{bV_p}{2D_1}; \quad \chi = \frac{D_2}{D_1}$$

where b is the distance from the center line to the collector plate and λ_1 is the smallest positive solution to the preceding transcedental equation.

The Deutsch–Andersen equation and other efficiency theories are limiting cases of this general theory. To use this exact theory, an environmental engineer has to first decide on appropriate mixing coefficients. For a large D_2 (i.e., large transverse mixing), the equation for efficiency reduces to the following equation:

$$\eta = 1 - \exp\left[\left(\frac{V_g b}{2D_1}\right) - \sqrt{\left(\frac{V_g b}{2D_1}\right)^2 + 2(1-f) + \frac{V_p b}{D_1}}\,\right]\frac{L}{b}$$

3.4(18)

Expanding the square root term as a Taylor series up to the first three terms and rearranging gives the following equation:

$$\eta = 1 - \exp\left[-(1-f)\left(1 - \frac{V_p}{V_g^2}\right)\frac{D_1(1-f)}{b}\,\frac{V_p A_c}{Q_g}\right]$$

3.4(19)

This equation is analogous to the Deutsch–Andersen equation except that two modifying factors are added. The $(1-f)$ factor corrects for reentrainment, and the second factor is related to the diffusional transport in the gas flow direction that modifies the gas velocity. Cooperman (1984) suggests a value of 18,000 cm^2/sec for D_1.

Thus, efficiency depends on longitudinal mixing as well as reentrainment.

DUST RESISTIVITY

A major factor affecting ESP performance is the electrical resistivity of dust. If the resistivity is low (high conductivity), the charge on a particle leaks away quickly as particles collect on the plate. Thus, reentrainment becomes possible. As van der Waals' forces may be insufficient to bind the particles, the particles hop or creep through the precipitator. Conversely, if the particle resistivity is high, a charge builds on the collected dust. This charge reduces field strength, reducing ionization and the migration velocity of particles through the gas.

A plot of resistivity versus temperature shows a maximum resistivity between 250 and 350°F (see Figure 3.4.6). This relationship is unfortunate because operators cannot reduce the ESP temperature below 250°F without risking condensation of H_2SO_4 on the plate surfaces. Increasing

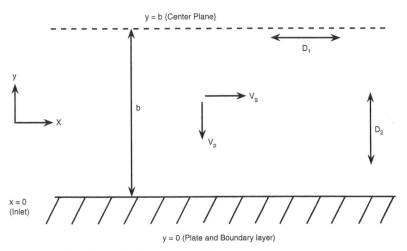

FIG. 3.4.5 Top view of wire-plate precipitator.

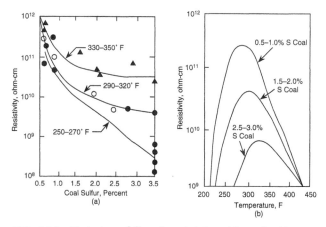

FIG. 3.4.6 Variation of fly ash resistivity with coal sulfur content and flue-gas temperature.

the temperature above 350°F results in excessive heat loss from the stack.

Resistivity decreases with increased sulfur content in coal because of the increased adsorption of conductive gases on the fly ash. Resistivity changes were responsible for the increased fly ash emissions when power plants switched from high-sulfur coal to low-sulfur coal to reduce SO_2 emissions in the United States.

In some cases of high resistivity in dust, adding a conditioning agent to the effluent gases substantially reduces resistivity and enhances particle collection. Examples are adding SO_3 to gas from power generator boilers and NH_3 to gas from catalytic cracking units used in petroleum refining. The overall efficiency can increase from 80% before injection to 99% after injection.

PRECIPITATOR DESIGN

Precipitator design involves determining the sizing and electrical parameters for an installation. The most important parameters are the precipitation rate (migration velocity), specific collecting area, and specific corona power (White 1984). In addition, the design includes ancillary factors such as rappers to shake the dust loose from the plates, automatic control systems, measures for insuring high-quality gas flow, dust removal systems, provisions for structural and heat insulation, and performance monitoring systems.

The design engineer should determine the size distribution of the dust to be collected. Based on this information, the engineer can calculate the migration velocity (also known as the precipitation rate) V_p using Equation 3.4(14) for each size fraction. The engineer calculates the number of charges on a particle n using Equation 3.4(11) or 3.4(13), depending on whether field or diffusion charging is predominant. Diffusion charging is the dominant charging mechanism for particles less than 0.2 μm, while field charging is predominant for particles greater than 1 μm. For particles of intermediate sizes, both mechanisms are

significant. The engineer can also calculate V_p empirically from pilot-scale or full-scale precipitator tests. The value V_p also varies with each installation depending on resistivity, gas flow quality, reentrainment losses, and sectionalization. Therefore, each precipitator manufacturer has a file of experience to aid design engineers in selecting a value of V_p. A high migration velocity value indicates high performance.

After selecting a precipitation rate, the design engineer uses the Deutsch–Andersen relationship, Equation 3.7(15) or 3.7(18), to determine the collecting surface area required to achieve a given efficiency when handling a given gas flow rate. If Equation 3.7(18) is used, the engineer can choose the value of the reentrainment factor f empirically from pilot-scale studies or previous experience or set it to zero as an initial guess. The quantity A/Q_g is called the specific collection area.

The corona power ratio is P_c/Q_g, where P_c is the useful corona power. The design engineer determines the power required for an application on an empirical basis. The power requirements are related to the collection efficiency and the gas volume handled. Figure 3.4.7 plots the collection efficiency versus the corona power ratio. At high efficiencies, large increments of corona power are required for small increments in efficiency. The precipitation rate (migration velocity) is related to corona power as follows:

$$V_p = \frac{kP_c}{A} \qquad 3.4(20)$$

where k is an empirical constant that depends on the application. The Deutsch–Andersen equation can therefore be expressed as follows:

$$\eta = 1 - \exp\left[-\frac{kP_c}{Q}\right] \qquad 3.4(21)$$

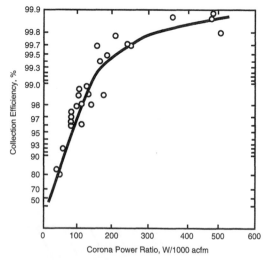

FIG. 3.4.7 Collection efficiency as a function of the corona power ratio.

TABLE 3.4.1   RANGES FOR ESP DESIGN PARAMETERS

Parameter	Range of Values
Precipitation rate V_p	1.0–10 m/min
Channel width D	15–40 cm
Specific collection area $\dfrac{\text{Plate Area}}{\text{Gas Flow}}$	0.25–2.1 m²/(m³/min)
Gas velocity u	1.2–2.5 m/sec
Aspect ratio $R = \dfrac{\text{Duct Length}}{\text{Height}}$	0.5–1.5 (Not less than 1 for η > 99%)
Corona power ratio $\dfrac{\text{Corona Power}}{\text{Gas Flow}} = \dfrac{P_c}{Q}$	1.75–17.5 W/(m³/min)
Corona current ratio $\dfrac{\text{Corona Current}}{\text{Plate Area}}$	50–750 μA/m²
Plate area per electrical set A_s	460–7400 m²
Number of electrical sections N_s	
a. In the direction of gas flow	2–8
b. Total number of sections	1–10 bus sections/(1000 m³/min)
Spacing between sections	0.5–2 m
L_{en}, L_{ex}	2–3 m
Plate Height; Length	8–15 m; 1–3 m

Power Density versus Resistivity Ash Resistivity, ohm-cm	Power Density, W/m²
10^4–10^7	43
10^7–10^8	32
10^9–10^{11}	27
10^{11}	22
10^{12}	16
10^{13}	10.8

Source: C.D. Cooper and F.C. Alley, 1986, *Air pollution control: A design approach* (Boston: PWS Publishers).

This relationship gives the corona power necessary for a given precipitation efficiency, independent of the precipitator design. The maximum corona power capability of the electrical sets should be considerably higher than the useful corona power.

The corona electrodes in large precipitators are subdivided into multiple groups or sections, referred to as bus sections. These sections are individually powered by separate rectifier sets to reduce the effects of sparking and better match the corona voltages and currents to the electrical characteristics of the gas and dust. The degree of sectionalization is expressed as the number of sections per 1000 m³/min. This parameter is a measure of the precipitator's ability to absorb corona power from the electrical sets. ESP performance improves with sectionalization. This improvement may be due to better electrode alignment and accurate spacing (White 1977). Sectionalization also implies that the unit is operational even if a few bus sections are taken offline.

Table 3.4.1 gives the range of some design parameters. A design engineer can specify the basic geometry of an ESP using the information in the table.

The following equation determines the total number of channels in parallel or the number of electrical sections in a direction perpendicular to that of gas flow N_d:

$$N_d = \frac{Q}{uDH} \qquad 3.4(22)$$

where:

Q = the volumetric gas flow rate
u = the linear gas velocity
D = the channel length (plate separation)
H = the plate height

The overall width of the precipitator is N_dD. The overall length of the precipitator is as follows:

$$L_0 = N_sL_p + (N_s - 1)L_s + L_{en} + L_{ex} \qquad 3.4(23)$$

where:

N_s = the number of electrical sections in the direction of flow
L_p = the plate length

L_s = the spacing between the electrical sections

L_{en} and L_{ex} = the entrance and exit section lengths, respectively.

The following equation determines the number of electrical sections N_s:

$$N_s = \frac{RH}{L_p} \qquad 3.4(24)$$

where R is the aspect ratio.

When the number of ducts and sections have been specified, the design engineer can use the following equation to calculate the actual area:

$$A_a = 2HL_pN_sN_d \qquad 3.4(25)$$

The preceding procedure provides a rational basis for determining the plate area, total power, and degree of sectionalization required.

—*Gurumurthy Ramachandran*

References

Cooperman, G. 1984. A unified efficiency theory for electrostatic precipitators. *Atmos. Environ.* 18, no. 2:277–285.

Crawford, M. 1976. Air pollution control theory. McGraw Hill, Inc.

White, H.J. 1963. Industrial electrostatic precipitation. New York: Addison Wesley.

———. 1977. Electrostatic precipitation of fly ash—Parts I, II, III, and IV. *JAPCA* 27 (January–April).

———. 1984. The art and science of electrostatic precipitation. *JAPCA* 34, no. 11:1163–1167.

3.5
PARTICULATE CONTROLS: WET COLLECTORS

FEATURE SUMMARY

Types of Designs
Low or high energy

Limitations
Low efficiency and liquid waste disposal for low-energy designs; operating costs for high-energy designs

Loadings
0.1 to 100 g/m³

Pressure Drop
Low energy—0.5 to 2 cm H_2O; high energy—2 to 100 cm H_2O

Overall Efficiency
Low energy—high for >10 μm; high energy—high for >1 μm

Partial List of Suppliers
Aerodyne Development Corp; Air-Cure Environmental Inc./Ceilcote Air Pollution Control; Air Pol Inc.; Beco Engineering Co.; Ceilcote Co.; Croll Reynolds Co. Inc.; Fairchild International; Hild Floor Machine Co. Inc.; Joy Technologies Inc./Joy Environmental Equipment Co.; Lurgi Corp USA; Merck & Co. Inc./Calgon Corp; Safety Railway Service Corp/Entoleter Inc.; Sonic Environmental Systems; Spendrup & Associates Inc./Spendrup Fan Co.; Svedala Industries Inc./Allis Mineral Systems; Kennedy Van Saun; Wheelabrator Air Pollution Control; Zurn Industries Inc./Air Systems Div.

A particulate scrubber or wet collector is a device in which water or some other solvent is used in conjunction with inertial, diffusion, or other forces to remove particulate matter from the air or gases. The scrubbing process partially mimics natural processes where dust-laden air is cleaned by rain, snow, or fog. The first industrial scrubbers attempted to duplicate this natural cleaning, with dusty air ascending through a rain of liquid droplets in a large, vertical tube. Subsequent developments reduced the space requirements for scrubbers.

The first patent for a particle scrubber design was issued in Germany in 1892, and the first gas scrubber with rotating elements was patented about 1900. Venturi scrubbers were developed just after World War II, mainly in the United States, and represented a breakthrough in scrubber design (Batel 1976).

General Description

In the scrubbing process, gas containing dust particles contacts a scrubbing liquid (often water). Here, some of the dust particles are captured by the scrubbing liquid. The scrubbing liquid can also condense on the dust particles, forming liquid droplets which are more easily removed than the dry dust particles. Finally, the mist droplets, which may or may not contain dust particles, are removed in an entrainment separator. Figure 3.5.1 schematically shows the process. The primary collection mechanisms for particle removal are inertial impaction, interception, diffusion or diffusiophoresis, and electrical attraction. At times, other mechanisms, such as those listed in Table 3.5.1, are involved in collection. However, in all cases, the dominant mechanism is inertial impaction followed by diffusion.

Scrubbers can be divided into three categories. In the first category, the scrubbing liquid is a spray, and collection occurs when particles are embedded by impaction in the scrubbing liquid surface (see part a in Figure 3.5.2). In the second category, collection occurs by impingement on a wetted surface (see part b of Figure 3.5.2). Finally, in the third category, particle-laden air is bubbled through the

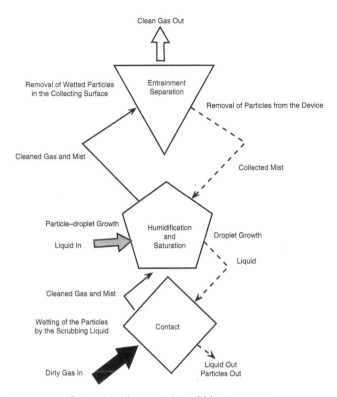

FIG. 3.5.1 Schematic diagram of scrubbing process.

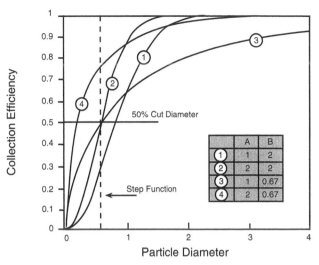

FIG. 3.5.3 Typical grade efficiency curves as computed from Efficiency $= \exp(Ad_{pa}^{B})$.

scrubbing liquid, and particles are removed by impingement (see part c of Figure 3.5.2).

The main factor influencing a scrubber's efficiency is the size of the dust particles being removed. Scrubber collection efficiency also varies with scrubber design and operating conditions. Figure 3.5.3 shows examples of typical grade efficiency curves. Because of the efficiency variation due to particle size, collection efficiency is often expressed as a function of one particle size, the *50% cut diameter,* that is, the particle diameter for which collection efficiency is 50% (Calvert et al. 1972; Calvert 1974, 1977). The rationale for this definition is that the grade efficiency curve for all scrubbing is steepest at the 50% point, and for estimating purposes this curve is considered to be a step function—all particles greater than this size are collected, all smaller are not (see dotted line on Figure 3.5.3). Because collection efficiency is also a function of

particle density, the concept of the aerodynamic particle diameter must be considered. The aerodynamic diameter of an aerosol particle is the diameter of a unit density sphere (density $= 1$ g/cm^3) which has the same settling velocity as the particle. With this definition, the relationship of a spherical particle of diameter d_s and density ρ to its aerodynamic diameter d_a is expressed as:

$$d_a = d_s(\rho C_c)^{1/2} \qquad 3.5(1)$$

For particles with diameters of several micrometers or less, a slip correction term C_c must be included in the aerodynamic diameter conversion as follows:

$$C_c = 1 + \frac{2\lambda}{d_s}\left[1.257 + 0.400 \exp\left(-\frac{1.1d_s}{2\lambda}\right)\right] \qquad 3.5(2)$$

The term d_s is the dust particle diameter, and λ is the mean free path of gas molecules. For air at 20°C, λ has a value of 0.0687 μm.

Scrubber Types

Over the years, many scrubber designs have been developed based on the three preceding scrubbing processes. Some representative examples are described next.

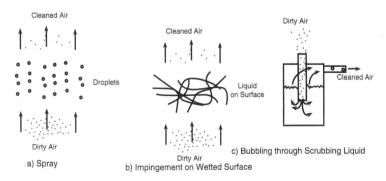

FIG. 3.5.2 Schematic illustration of various scrubbing mechanisms.

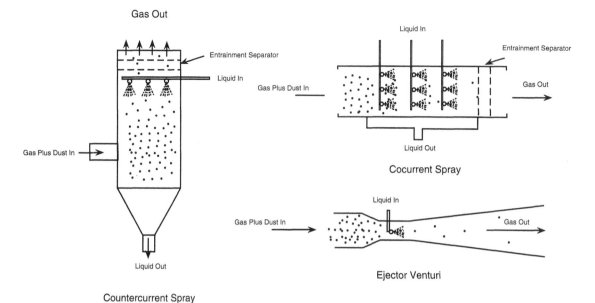

FIG. 3.5.4 Example of preformed spray unit.

SPRAY COLLECTORS—TYPE I

With spray-type units, particles are collected on liquid droplets within the scrubbing unit. The liquid droplets are formed either by atomization, where the flow rate of the scrubbing liquid and pressure in the atomizing nozzle control the droplet size and number, or by the moving gas stream, which atomizes and accelerates the resulting droplets.

Preformed Spray

With a preformed spray, droplets are formed by a nozzle and are then sprayed through the dust-laden air (see Figure 3.5.4). Removal of the dust is primarily by impaction, although diffusion and diffusiophoresis can also contribute to particle collection. Spray scrubbers, taking advantage of gravitational settling, achieve cut diameters around 2 μm, with high-velocity sprays capable of reducing this lower limit to about 0.7 μm (aerodynamic diameter) (Calvert et al. 1972).

Gas-Atomized Spray

With a gas-atomized spray, high-velocity, particle-laden gas atomizes liquid into drops, and the resulting turbulence and velocity difference between droplets and particles enhances particle–droplet collisions and hence particle removal (see Figure 3.5.5). Typical gas velocities range from 60 to 120 m/sec (200–400 f/sec), and the pressure drop in these units is relatively high. A typical example of this type of unit is the venturi scrubber. Venturi scrubbers have achieved cut diameters down to 0.2 μm (Calvert 1977).

Centrifugal Scrubbers

With centrifugal scrubbers, the scrubbing liquid is sprayed into the gas stream at the same time the unit is imparting a spinning motion to the mixture of particles and droplets (see Figure 3.5.6). The result is a centrifugal deposition of the particle–droplet mixture on the outer walls of the scrubber. Depending on the amount of scrubbing liquid used, the collection efficiency is good for particles down to about 1 to 2 μm diameter (aerodynamic) (Calvert et al. 1972). The tangential velocities in these units should not

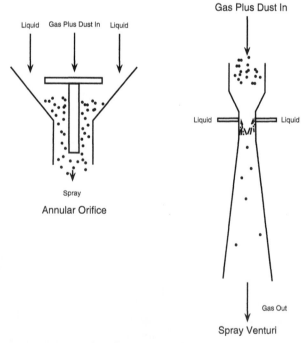

FIG. 3.5.5 Examples of gas-atomized spray unit.

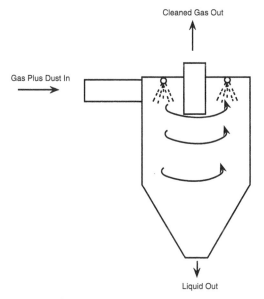

FIG. 3.5.6 Centrifugal scrubbers.

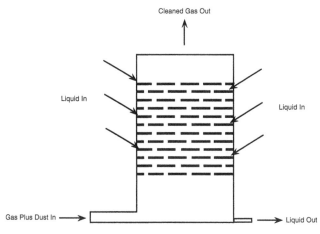

FIG. 3.5.7 Plate scrubbers.

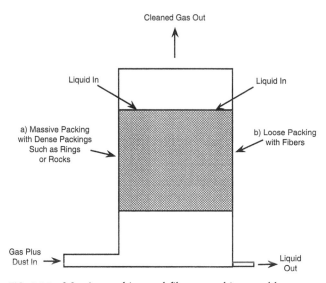

FIG. 3.5.8 Massive packing and fibrous packing scrubbers.

exceed 30 m/sec to prevent reentrainment of the droplet–particle mixture.

IMPINGEMENT ON A WETTED SURFACE— TYPE II

With impingement on a wetted surface, collection occurs by impingement. These scrubbers include plate scrubbers and scrubbers with massive packing or fibrous packing.

Plate Scrubbers

Plate scrubbers have plates or trays mounted within a vertical tower at right angles to the axis of the tower (see Figure 3.5.7). Gas flows from the bottom of the tower up through slots, holes, or other perforations in the plates where mixing with the scrubbing liquid occurs. Collection efficiency increases as the cut diameter decreases. A cut diameter of about 1 μm (aerodynamic diameter) is typical for $\frac{1}{8}$-in holes in a sieve plate (Calvert et al. 1972). Calvert (1977) points out that increasing the number of plates does not necessarily increase collection efficiency. Once particles around the size of the cut diameter are removed, adding more plates does little to increase collection efficiency.

Massive Packing

Packed-bed towers packed with crushed rock or various ring- or saddle-shaped packings are often used as gas scrubbers (see a in Figure 3.5.8). The gas–liquid contact can be crossflow, concurrent, or countercurrent. Collection efficiency rises as packing size falls. According to Calvert (1977), a cut diameter of about 1.5 μm (aerodynamic) can be achieved with columns packed with 1-in Berl saddles or Raschig rings. Smaller packings give higher efficiencies; however, the packing shape appears to have little importance.

Fibrous Packing

Beds of fibers (see part b in Figure 3.5.8) are also efficient at removing particles in gas scrubbers. Fibers can be made from materials such as steel, plastic, or even spun glass. With small diameter fibers, efficient operation is achieved. Efficiency increases as the fiber diameter decreases and also as gas velocity increases. The collection is primarily by impaction and interception. Diffusion is important for small particles, although to increase diffusional collection, lower gas velocities are necessary. Cut diameters can be as low as 1.0 to 2.0 μm (aerodynamic) or in some cases as low as 0.5 μm (aerodynamic) (Calvert et al. 1972). A major difficulty with fibrous beds is that they are prone to plugging; even so, they are widely used.

BUBBLING THROUGH SCRUBBING LIQUID—TYPE III

In scrubbers that bubble gas through a scrubbing liquid, the particulate-containing gas is subjected to turbulent mixing with the scrubbing liquid. Droplets form by shear forces, or the dust particles are impacted directly onto the scrubbing liquid. Droplets are also formed by motor-driven impellers (that are either submerged or in the free air of the scrubber). These impellers serve not only to form small droplets but also to enhance impaction of the dust particles.

Baffle and Secondary-Flow Scrubbers

These units remove particulates from an air stream by continually changing the flow direction and velocity as the gas flows through the unit (see Figure 3.5.9). This motion results in intimate particle mixing in the gas and the spray droplets of scrubbing liquid. Zig-zag baffles or louvers are examples of how flow direction and velocity are altered internally. With these units, cut diameters as low as 5 to 10 μm can be achieved with low pressure drops. Plugging can be a problem with some heavy particle loadings.

Impingement and Entrainment Scrubbers

With this type of scrubber design, the particle–gas mixture is bubbled through or skimmed over the scrubbing liquid surface which atomizes some droplets and mixes the particles with the scrubbing liquid (see Figure 3.5.10). Both particle impaction on the liquid surface and on the atomized drops and some diffusion contribute to particle collection, which is effective with high-velocity entrainment for cut diameters down to 0.5 μm (aerody-

namic diameter). The pressure drop for this class of device is high.

Mechanically Aided Scrubbers

With mechanically aided scrubbers, a motor-driven device in the scrubber produces spray droplets and mixes the incoming gas and scrubbing liquid more intimately (see Figure 3.5.11). The motor-driven device can be a fan with the scrubbing liquid introduced into the fan rotor, or it can be a paddle arrangement (disintegrator) in the scrubbing liquid that produces droplets. These units have cut diameters down to 2.0 μm (aerodynamic), or even 1.0 μm for some disintegrator designs (Calvert et al. 1972).

Fluidizied- (Moving) Bed Scrubbers

Fluidized-bed scrubbers are similar to packed-bed units except that the packing material is light enough to float in the gas stream (see Figure 3.5.12). The packing material expands to about twice its original depth. This expansion intimately mixes the particulates and scrubbing liquid and permits effective collection down to cut diameters of 1 μm (aerodynamic) (Calvert 1977).

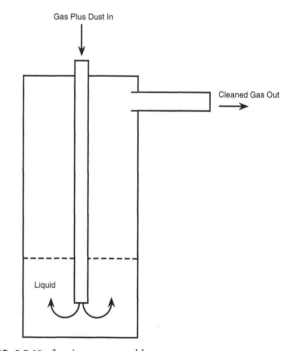

FIG. 3.5.10 Impingment scrubber.

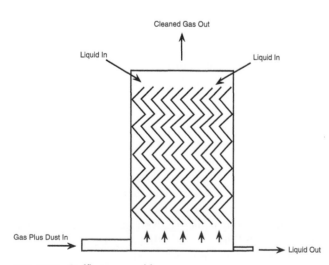

FIG. 3.5.9 Baffle-type scrubber.

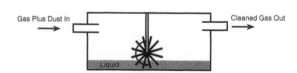

FIG. 3.5.11 Mechanically aided scrubber.

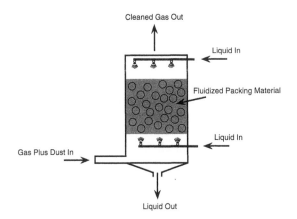

FIG. 3.5.12 Fluidized- or mobile-bed scrubber.

SCRUBBERS USING A COMBINATION OF DESIGNS

Besides the preceding basic scrubber types, a number of scrubber designs use two or more of the collection mechanisms in Table 3.5.1. With a process known as flux force/condensation scrubbing (Calvert and Jhaveri 1974), collection phenomena such as diffusiophoresis or thermophoresis combine with condensation to improve particle removal. This process uses steam to increase the deposition of scrubbing liquid on the dust particles by condensation, making the resulting dust–liquid droplets easier to remove. Temperature gradients or concentration fluxes enhance particle collection. This process is particularly attractive if the incoming dusty gas stream is hot and humid. Otherwise, the process is more costly than more conventional scrubber designs.

Another approach enhances scrubber collection efficiency using electrostatics. The electrostatics can take the form of wet ESPs, charged-dust/grounded-liquid scrubbers, charged-drop scrubbers, or charged-dust/charged-liquid scrubbers (Calvert 1977).

Factors Influencing Collection Efficiency

The collection efficiency of the scrubbing process depends on having high velocities between the dust particles and

TABLE 3.5.1   DEPOSITION PHENOMENA

1.	Interception
2.	Magnetic force
3.	Electrical force
4.	Inertial impaction
5.	Brownian diffusion
6.	Turbulent diffusion
7.	Gravitational force
8.	Photophoretic force
9.	Thermophoretic force
10.	Diffusiophoresis force

the collecting liquid (Overcamp and Bowen 1983). These high velocities enhance the impaction efficiency. Because fine particles rapidly attain the velocity of their surroundings, continually mixing the dirty air stream with the collecting liquid is necessary for good impaction, and short distances between the particles and the collecting surface are required for efficient diffusive collection (Nonhebel 1964).

Considering all collection mechanisms in an ideal scrubber, Figure 3.5.13 shows the shape of a typical efficiency versus particle diameter curve for impaction plus diffusion. In the figure a minimum collection efficiency is apparent that is typical of the low efficiency in the transition from impaction collection processes to diffusion-related processes. Above this minimum, efficiency varies roughly as a function of the dust particle diameter squared, and below it roughly as $(1/d^2)$ (Reist 1993). Efficiency is approximately proportional to particle density. As might be expected, because of the variety of scrubbing configurations available, no one collection mechanism is dominant, and no single approach determines all scrubber performance (Crawford 1976).

CONTACTING POWER RULE

Through experience, investigators have observed that scrubber efficiency for similar aerosols, regardless of scrubber design, is related to the power consumed by the scrubber in making the liquid–particle contact. Thus, scrubber grade efficiency appears to be a function of power input. Scrubbers are high-energy devices. That is, in general, the higher the energy input per unit volume of gas treated, the more efficient the scrubber is for smaller sized particle collection. Also, higher power inputs imply smaller scrubbing liquid drop sizes and more turbulence.

Semrau (1960) shows that the efficiency of any scrubber for similar aerosols is strongly affected by the power

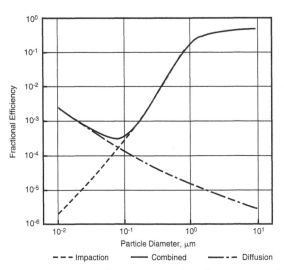

FIG. 3.5.13 Fractional efficiency for combined impaction and diffusion.

dissipated in operating the scrubber (termed *contacting power*) and little affected by scrubber size, geometry, or the manner in which the power is applied.

In this text, contacting power is the power per unit of the gas volumetric flow rate used in contacting the aerosol to be cleaned. This power is ultimately dissipated as heat. The power per unit of the gas volumetric flow rate is considered an effective friction loss and represents friction loss across the scrubbing unit, neglecting losses due to kinetic energy changes in the flowing gas stream or losses due to equipment operating dry (Semrau 1960, 1977).

Contacting power is stated in the following contacting power rule:

$$N_t = \alpha P_T^\gamma \qquad 3.5(3)$$

where N_t is the number of transfer units. Transfer units are related to the fractional collection efficiency as follows:

$$= 1 - e^{-N_t} \qquad 3.5(4)$$

In Equation 3.5(3), the terms α and γ are empirical constants that depend on the properties of the dust being collected. Table 3.5.2 lists the values for α and γ for various dust plus scrubber combinations. The contacting power P_T is calculated from the following equation (Semrau 1960):

$$P_T = P_G + P_L + P_M \qquad 3.5(5)$$

where:

P_G = the power input required to overcome the gas pressure drop across the collector

P_L = the power input required to produce droplets through the spray nozzles

P_M = the power input required to drive any rotor, if present.

Figure 3.5.14 plots N_T as a function of contacting power for various dusts. In the preceding equations, the power terms are expressed in kWhr/1000 m³. Table 3.5.2 gives the factors for converting power to either hp/1000 cfm, or effective friction loss (inches or centimeters of water).

Using the contacting power approach is best when some knowledge of the scrubber's performance for a specific aerosol is available. For example, when pilot plant designs are scaled up to full-size units for a dust control problem,

contacting power is helpful in predicting the overall efficiency of the new, full-size unit. Or, if a scrubbing unit is to be replaced with a new and different design, the contacting power approach is helpful in estimating the performance of the new unit. However, in determining contacting power, a design engineer must insure that only the portion of energy input representing energy dissipated in scrubbing is used (Semrau 1977). Also, little data exists relating gas-phase and mechanical contacting power. Since the contacting power approach is an empirical approach, extrapolating the results into areas with little data requires caution.

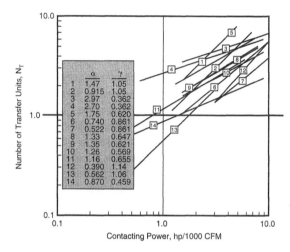

1. Raw lime kiln dust - venturi and cyclonic spray
2. Prewashed lime kiln dust - venturi, pipeline, and cyclonic spray
3. Talc dust - venturi
4. Talc dust - orifice and pipeline
5. Black liquor - venturi and cyclonic spray
6. Black liquor - venturi, pipeline, and cyclonic spray
7. Black liquor - venturi evaporator
8. Phosphoric acid mist - venturi
9. Foundry cupola dust - venturi
10. Open-hearth steel - venturi
11. Talc dust - cyclone
12. Copper sulfate - solivore, and mechanical spray generators
13. Copper sulfate - hydraulic nozzles
14. Ferrosilicon furnace fume - venturi and cyclonic spray

FIG. 3.5.14 Performance curve for scrubbing aerosols. (Adapted from K.T. Semrau, 1960, Correlation of dust scrubber efficiency, *J. APCA* 10:200.)

TABLE 3.5.2 FORMULAS FOR PREDICTING CONTACTING POWER

	Symbol	*Metric Units*[b]	*Dimensional Formula*
Effective friction loss	F_E	cm H_2O	Δp[a]
Gas phase contacting power	P_G	kWh/1000 m³	$0.02724\,F_E$
Liquid phase contacting power	P_L	kWh/1000 m³	$0.02815\,\rho_f\,(Q_L/Q_G)$[c]
Mechnical contacting power	P_M	kWh/1000 m³	$16.67\,(W_s/Q_G)$[c]
Total contacting power	P_T	kWh/1000 m³	$P_G + P_L + P_M$

Notes: [a]The effective friction loss is approximately equal to the scrubber pressure loss Δp.

[b]1.0 kWh/1000 m³ = 2.278 hp/1000 ft³/min

[c]This quantity is actually power input and represents an estimate of contacting power; Q_L is in l/min, Q_G is in m³/min, and W_s is net mechanical power input in kW.

USE OF THE AERODYNAMIC CUT DIAMETER

Because of the variety of parameters involved, accurately designing a scrubber for an installation without prior knowledge of scrubber performance at that or a similar installation is difficult. Factors such as air flow rate, scrubbing liquid flow rate, inlet gas temperature, scrubbing liquid temperature, relative humidity, concentration, size and size distribution of the aerosol in the incoming gas, scrubber droplet size and size distribution, and the type of scrubber can confound the design engineer and cloud the design process. Even so, some equations are available to make initial estimates for scrubber design.

Dust size distributions are often represented by log–normal distributions; that is, the particle frequency for a given diameter is distributed according to a normal curve when plotted against the logarithm of the particle diameter. Then, similar to the standard deviation of a normal distribution, the geometric standard deviation of a log–normal distribution measures the spread of the distribution. With a log–normal distribution having a geometric standard deviation of σ_g, 67% of all particles have diameters between the size ranges d_g/σ_g and $\sigma_g d_g$. Equation 3.5(6) gives the form of the log–normal distribution:

$$\int_0^d f(x)dx = \int_0^d \frac{1}{x \ln \sigma_g (2\pi)^{0.5}} \exp\left[-\frac{(\ln x - \ln d_g)^2}{2 \ln^2 \sigma_g}\right] dx \quad 3.5(6)$$

where:

d_g = the geometric median diameter
σ_g = the geometric standard deviation
x = any particle diameter of interest.

Thus, the right side of Equation 3.5(6) gives on integration the fraction of particles whose diameters are less than or equal to x. Integrating between limits of x = 0 to ∞ gives a result of 1, and between limits of x = 0 to d_g, a result of 0.5.

Two particle size distributions must be considered in scrubber design. Besides the size distribution of the dust to be collected, the grade efficiency must also be considered when the collector operates under specific conditions. As previously discussed, according to Calvert et al. (1972), the grade efficiency can be effectively represented by the *aerodynamic cut diameter* d_{ac}. If $d_{ac} = d_g$, a collection efficiency of 50% is estimated; if $d_{ad} = d_g/\sigma_g$, the estimated collection efficiency is 84%. Hence, for any level of efficiency and a log–normal aerosol, the use of Equation 3.5(6) determines the equivalent d_{ac}.

Figure 3.5.15 represents an integrated form of Equation 3.5(6) which gives any collection efficiency (50% or greater) as a function of the ratio d_{ac}/d_g with σ_g as a parameter. Thus, for example, if a collection efficiency of 99% is needed for a log–normal, unit-density aerosol with $d_g = 5.0 \ \mu m$ and $\sigma_g = 2$, Figure 3.5.15 shows that a multiplication factor of about 0.2 is indicated. The required d_{ac} is then $5 \times 0.2 = 1 \ \mu m$.

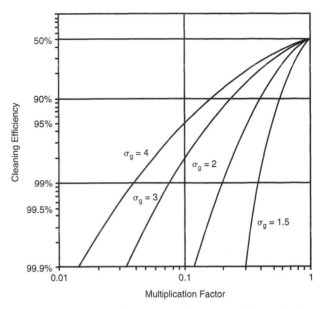

FIG. 3.5.15 Cleaning efficiency as a function of the multiplication factor for log–normal particle distributions of various skewness.

DETERMINATION OF d_{ac} AS A FUNCTION OF SCRUBBER OPERATING PARAMETERS

A number of theoretical equations estimate scrubber performance with reasonable accuracy for a specific scrubber type.

Venturi

Calvert (1970) and Calvert et al. (1972) considered the venturi scrubber and developed the following equation for collection efficiency ε:

$$\varepsilon = \exp\left[-\frac{2Q_L u_G \rho_L d_d}{55 Q_G \mu_G} F(K_{pt}, f)\right] \quad 3.5(7)$$

where:

Q_L/Q_G = the volumetric flow ratio of liquid to gas
u_G = the gas velocity in cm/sec
ρ_L = the density of the liquid in gm/cm³
μ_G = the viscosity of the liquid in poises
d_d = the drop diameter in cm
f = an operational factor which ranges from 0.1 to 0.3, but is often taken as 0.25

The term f includes the influence of factors such as collection by means other than impaction, variation in drop sizes, loss of liquid to venturi walls, particle growth by condensation, and degree of utilization of the scrubbing liquid.

The following equation gives function $F(K_{pt}, f)$:

$$F(K_{pt}, f) =$$
$$\left[0.7 + K_{pt}f - 1.4 \ln\left(\frac{K_{pt}f + 0.7}{0.7}\right) - \frac{0.49}{K_{pt}f + 0.7}\right]\frac{1}{K_{pt}} \quad 3.5(8)$$

where K_p is defined as:

$$K_p = \frac{v_r d_a^2}{9\mu_G d_d} \qquad 3.5(9)$$

The term v_r represents the relative velocity between the dust particle and the droplet, and d_a is the aerodynamic particle diameter as defined in Equation 3.5(1). The average liquid droplet diameter d_d is given by the empirical equation of Nukiyama and Tanasawa (1938) as:

$$d_d = \frac{50}{u_G} + 91.8 \left(\frac{Q_L}{Q_G}\right)^{1.5} \qquad 3.5(10)$$

The drop diameter d_d is expressed in units of centimeters (cm), and the gas velocity u_G in cm/sec.

When $ = 0.5 = 50\%$, Equation 3.5(7) can be solved for $d_a = d_{ac}$ for a variety of Q_L/Q_G ratios. The solutions to these calculations are plotted as shown Figure 3.5.16. For these calculations, $k = 0.25$, $\mu_G = 1.82 \times 10^{-4}$ poises, and $u_G = v_r$ with u_G as a parameter ranging from 5 m/sec to 15 m/sec. Figure 3.5.16 can then be used to estimate d_{ac} for venturi scrubber operation at different Q_L/Q_G ratios and different throat velocities.

Rudnick et al. (1986) evaluated three widely used venturi scrubber equations (Calvert [1970], Yung et al. [1978], and Boll [1973]) and compared the equations with experimental data. They concluded that Yung's model, a modification of the Calvert model, predicts data better than the other two. However, the difference was not that great, and since the Yung model is complicated, the Calvert model remains useful for a first approximation of venturi scrubber performance.

The pressure drop across a venturi scrubber (in centimeters of water) can be estimated from the following equation (Leith, Cooper, and Rudnick 1985):

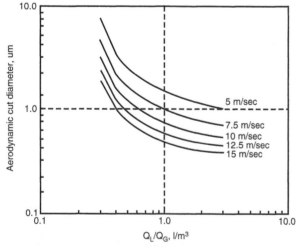

FIG. 3.5.16 Predicted venturi performance. (Adapted from S. Calvert, J. Goldshmid, D. Leith, and D. Mehta, 1972, *Scrubber handbook*, Riverside, Calif.: Ambient Purification Technology, Inc.)

$$\Delta p = 1.02\rho_L \frac{Q_L}{Q_G} u_{Gt}^2 \left[\beta\left(1 - \frac{u_{Gf}}{u_{Gt}}\right) + \left(\frac{u_{Gf}}{u_{Gt}}\right)^2\right] \qquad 3.5(11)$$

The term u_{Gf} is the gas velocity at the exit of the venturi, and u_{Gt} is the gas velocity in the throat. The term β can be estimated from the following equation:

$$\beta = 2(1 - X^2 + \sqrt{X^4 - X^2}) \qquad 3.5(12)$$

and X is defined as follows:

$$X = 1 + 3L_t C_D \rho_G/(16 d_d \rho_L) \qquad 3.5(13)$$

where L_t is the throat length, and C_D is the drag coefficient, which can be estimated up to a flow Reynolds number of about 1000 by the following equation:

$$C_D = \frac{24}{Re}(1 + 0.15 Re^{0.687}) \qquad 3.5(14)$$

Above this Reynolds number, the drag coefficient is estimated by $C_D = 0.44$. The Reynolds number is given by the following equation:

$$Re = \frac{u_G d \rho_g}{\mu_G} \qquad 3.5(15)$$

Countercurrent Spray Tower

For a countercurrent spray tower as shown in Figure 3.5.4, the following equation estimates the collection efficiency (Calvert 1968):

$$ = 1 - \exp\left[-\frac{3Q_L v_{d/g} Z \eta_d}{2Q_G v_{d/w} d_d}\right] \qquad 3.5(16)$$

where:

$v_{d/g}$ = the velocity of the droplets relative to the gas
$v_{d/w}$ = the velocity of the droplets relative to the walls
$\qquad (v_{d/w} = v_{d/g} + v_{g/w})$
Z   = the height or length of the scrubber in cm

The following equation gives η_d:

$$\eta_d = \left(\frac{Stk}{Stk + 0.7}\right)^2 \qquad 3.5(17)$$

The following equation gives the Stokes number Stk:

$$Stk = \frac{d_a^2 \rho_p v_{a/g}}{9\mu_G d_d} \qquad 3.5(18)$$

Impingement and Entrainment-Type Scrubber

For this type of unit (see Figure 3.5.10), the efficiency can be estimated from the following equation (Calvert et al. 1972):

$$ = 1 - \exp\left[\frac{2\Delta p d_d}{55 u_G \mu_G} F(K_{pt}, f)\right] \qquad 3.5(19)$$

Usually a value of 0.25 is chosen for f. According to Schiffter and Hesketh (1983), the drop diameter d_d is about one-third that calculated from Equation 3.5(10). The pressure drop in these units is given by the following equation:

$$\Delta p = \frac{\rho_L Q_L u_G}{A} \qquad 3.5(20)$$

where A is the collecting area, defined as the total surface area of the collecting elements perpendicular to the gas flow direction.

—Parker C. Reist

References

Batel, W. 1976. *Dust extraction technology.* Stonehouse, England: Technicopy Ltd.

Boll, R.H. 1973. *Ind. Eng. Chem. Fundam.* 12:40–50.

Calvert, S. 1968. Vol. 3 in *Air pollution.* 2d ed. Edited by A.C. Stern. New York: Academic Press.

———. 1970. Venturi and other atomizing scrubbers: Efficiency and pressure drop. *AIChE J.* 16:392–396.

———. 1974. Engineering design of fine particle scrubbers. *J. APCA* 24:929–934.

———. 1977. How to choose a particulate scrubber. *Chemical Eng.* (29 August):55.

Calvert, S., J. Goldshmid, D. Leith, and D. Mehta. 1972. *Scrubber handbook.* Riverside, Calif.: Ambient Purification Technology, Inc.

Calvert, S. and N.C Jhaveri. 1974. Flux force/condensation scrubbing. *J. APCA* 24:946–951.

Crawford, M. 1976. *Air pollution control theory.* New York: McGraw-Hill, Inc.

Leith, D., D.W. Cooper, and S.N. Rudnick. 1985. Venturi scrubbers: Pressure loss and regain. *Aerosol Sci. and Tech.* 4:239–243.

Nonhebel, G. 1964. *Gas purification processes.* London: George Newnes, Ltd.

Nukiyama, S. and Y. Tanasawa. 1938. Experiment on atomization of liquids. *Trans. Soc. Mech. Eng. (Japan)* 4:86.

Overcamp, T.J., and S.R. Bowen. 1983. Effect of throat length and diffuser angle on pressure loss across a venturi scrubber. *J. APCA* 33:600–604.

Reist, P.C. 1993. *Aerosol science and technology.* New York: McGraw-Hill.

Rudnick, S.N, J.L.M. Koelher, K.P. Matrin, D. Leith, and D.W. Cooper. 1986. Particle collection efficiency in a venturi scrubber: Comparison of experiments with theory. *Env. Sci. and Tech.* 20:237–242.

Schifftner, K.C. and H. Hesketh. 1983. *Wet scrubbers.* Ann Arbor, Mich.: Ann Arbor Science.

Semrau, K.T. 1960. Correlation of dust scrubber efficiency. *J. APCA* 10:200–207.

———. 1977. Practical process design of particulate scrubbers. *Chemical Eng.* (26 September):87–91.

Yung, S., S. Calvert, H.F. Barbarika, and L.E. Sparks. 1978. *Env. Sci. and Tech.* 12:456–459.

3.6
GASEOUS EMISSION CONTROL

This section presents some options for reducing the emission of pollutants. It focuses on the complex series of processes and phenomena generally grouped under acid rain. Acid rain is associated with the release of sulphur and nitrogen oxides into atmosphere via the burning of fossil fuels. The deposition of sulfur and nitrogen compounds is one of the most pressing large-scale air pollution problems.

The options presented are based on general pollution prevention techniques. This section focuses on source control via technology changes. Source reduction measures aim at long-term reduction, while other options can limit pollution during unfavorable conditions.

Energy Source Substitution

This option includes the transition to solar energy, nuclear energy, and other methods of obtaining energy. Of the pollutants generated by burning fuels, the emission of sulfur oxides and ash are directly attributable to fuel composition. Using the right fuel can cut sulfur dioxide and ash emissions. This option means switching to coal with lower sulfur content or replacing coal with gaseous fuel or desulfurized fuel oils. The sulfur in heavy fuel oils can be removed in a high-pressure catalytic reactor, in which hydrogen combines with the sulfur to form hydrogen sulfide which is then recovered.

An alternative to using lower-sulfur coal is to remove some of the sulfur in the coal (see Figure 3.6.1). Sulfur in the form of discrete iron pyrite crystals can be removed because they have different properties from the organic coal matrix in which they are embedded. This difference makes separation by physical processing possible. The key economic factor is the cost of the losses of combustible material which occur during the cleaning process. To date, none of the coal cleaning processes have been proven commercially successful. Their attractiveness is also diminished by their inability to recover, on average, more than half the sulfur content (e.g., the pyritic fraction) of the fuel (Bradshaw, Southward, and Warner 1992).

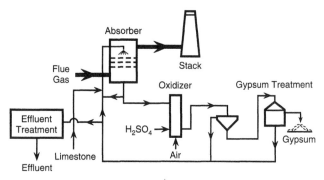

FIG. 3.6.1 Schematic diagram of a limestone–gypsum FGD plant.

Nitrogen oxides form at firebox temperatures by the reaction of the oxygen and nitrogen in the air and fuel. The thermal fixation of atmospheric nitrogen and oxygen in the combustion air produces *thermal NO_x*, while the conversion of chemically bound nitrogen in the fuel produces *fuel NO_x*.

For natural gas and light distillate oil firing, nearly all NO emissions result from thermal fixation. With residual fuel oil, the contribution of fuel-bound nitrogen can be significant and in some cases predominant. This contribution is because the nitrogen content of most U.S. coals ranges from 0.5 to 2% whereas that of fuel oil ranges from 0.1 to 0.5%. The conversion efficiencies of fuel mixture to NO_x for coals and residual oils have been observed between 10 and 60% (U.S. EPA 1983). Figure 3.6.2 shows the possible fates of fuel nitrogen. One option of reducing NO_x is to use low-nitrogen fuel.

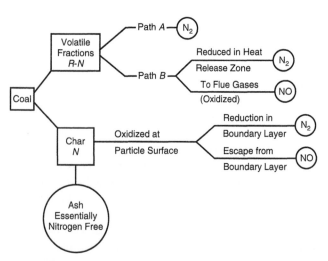

FIG. 3.6.2 Possible fates of nitrogen contained in coal. (Reprinted, with permission, from M.P. Heap et al., 1976, The optimization of burner design parameters to control NO formation in pulverized coal and heavy oil flames, *Proceedings of the Stationary Source Combustion Symposium—Vol. II: Fuels and Process Research Development*, EPA-600/2-76-152b [Washington, D.C.: U.S. EPA].

Process Modifications

The formation rates of both thermal NO_x and fuel NO_x are kinetically or aerodynamically limited, with the amount of NO_x formed being less than the equilibrium value (Wark and Warner 1981). Combustion conditions dominate the formation rate of NO_x, and modifying the combustion process can suppress it. Rapidly mixing oxygen with the fuel promotes both thermal and fuel NO_x formation.

MacKinnon (1974) developed a kinetic model from experimental studies of a heated mixture of N_2, O_2, and air as well as air. His kinetic equations provide insight to the strategies for controlling the formation of thermal NO_x as follows:

The peak temperature should be reduced.
The gas residence time at the peak temperature should be reduced.
The oxygen concentration in the highest temperature zone should be reduced.

Regardless of the mechanisms, several general statements can be made about fuel NO_x. It depends highly on the air/fuel ratio. The percent conversion to fuel NO_x declines rapidly with an increasing fuel equivalent ratio. (The *fuel equivalent ratio* is a multiple of the theoretical fuel/air ratio and is the inverse of the stoichiometric ratio. The stoichiometric ratio is unity when the actual air/fuel ratio equals the theoretical air to fuel needed for complete combustion.) The fuel equivalent ratio primarily affects the oxidation of the volatile R–N fraction (where R represents an organic fragment) rather than the nitrogen remaining in the char. The degree of fuel–air mixing also strongly affects the percent conversion of fuel nitrogen to NO_x, with greater mixing resulting in a greater percent conversion.

COMBUSTION CONTROL

Several process modifications can reduce NO_x formation. Table 3.6.1 summarizes the commercially available NO_x control technologies as well as their relative efficiencies, advantages and disadvantages, applicability, and impacts.

The simplest combustion control technology is the *low-excess-air* (LEA) operation. This technology reduces the excess air level to the point of some constraint, such as carbon monoxide formation, flame length, flame stability, and smoke. Unfortunately, the LEA operation only moderately reduces NO_x.

Off-Stoichiometric Combustion (OSC)

OSC, or staged combustion, combusts fuel in two or more stages. The primary flame zone is fuel-rich, and the secondary zones are fuel-lean. This combustion is achieved through the following techniques. These techniques are generally applicable only to large, multiple-burner, combustion devices.

TABLE 3.6.1 SCREENING POTENTIAL NO$_x$ CONTROL TECHNOLOGIES

Technique	Description	Advantages	Disadvantages	Impacts To Consider	Applicability	NO$_x$ Reduction
LEA	Reduces oxygen availability	Easy operational modification	Low NO$_x$ reduction potential	High carbon monoxide emissions, flame length, flame stability	All fuels	1–15%
OSC a. BOOS b. OFA c. Air Lances	Staged combustion, creating fuel-rich and fuel-lean zones	Low operating cost, no capital requirement required for BOOS	a. Typically requires higher air flow to control carbon monoxide b. Relatively high capital cost c. Moderate capital cost	Flame length, forced draft fan capacity, burner header pressure	All fuels; Multiple-burner devices	30–60%
LNB	Provides internal staged combustion, thus reducing peak flame temperatures and oxygen availability	Low operating cost, compatible with FGR as a combination technology to maximize NO$_x$ reduction	Moderately high capital cost; applicability depends on combustion device and fuels, design characteristics, and waste streams	Forced-draft fan capacity, flame length, design compatibility, turndown flame stability	All fuels	30–50%
FGR	Up to 20–30% of the flue gas recirculated and mixed with the combustion air, thus decreasing peak flame temperatures	High NO$_x$ reduction potential for natural gas and low-nitrogen fuels	Moderately high capital cost, moderately high operating cost, affects heat transfer and system pressures	Forced-draft fan capacity, furnace pressure, burner pressure drop, turndown flame stability	Gas fuels and low-nitrogen fuels	40–80%
W/SI	Injection of steam or water at the burner, which decreases flame temperature	Moderate capital cost, NO$_x$ reductions similar to FGR.	Efficiency penalty due to additional water vapor loss and fan power requirements for increased mass flow	Flame stability, efficiency penalty	Gas fuels and low-nitrogen fuels	40–70%
RAPH	Air preheater modification to reduce preheat, thereby reducing flame temperature	High NO$_x$ reduction potential	Significant efficiency loss (1% per 40°F)	Forced-draft fan capacity, efficiency penalty	Gas fuels and low-nitrogen fuels	25–65%
SCRI	Catalyst located in flue gas stream (usually upstream of air heater) promotes reaction of ammonia with NO$_x$	High NO$_x$ removal	Very high capital cost, high operating cost, extensive ductwork to and from reactor required, large volume reactor must be sited, increased pressure drop may require induced-draft fan or larger forced-draft fan, reduced efficiency, ammonia sulfate removal equipment for air heater required, water treatment of air heater wash required	Space requirements, ammonia slip, hazardous waste disposal	Gas fuels and low-sulfur liquid and solid fuels	70–90%
SNCR—Urea Injection	Injection of urea into furnace to react with NO$_x$ to form nitrogen and water	Low capital cost, relatively simple system, moderate NO$_x$ removal, nontoxic chemical, typically low energy injection sufficient	Temperature dependent, design must consider boiler operating conditions and design, NO$_x$ reduction may decrease at lower loads	Furnace geometry and residence time, temperature profile	All fuels	25–50%
SNCR—Ammonia Injection	Injection of ammonia into furnace to react with NO$_x$ to form nitrogen	Low operating cost, moderate NO$_x$ removal	Moderately high capital cost; ammonia handling, storage, vaporization	Furnace geometry and residence time, temperature profile	All fuels	25–50%

149

The *burner-out-of-service* (BOOS) technique terminates the fuel flow to selected burners while leaving the air registers open. The remaining burners operate fuel rich, thereby limiting oxygen availability, lowering peak flame temperatures, and reducing NO_x formation. The unreacted products combine with the air from the terminated fuel burners to complete burnout before exiting the furnace.

Installing *air-only (OFA) ports* above the burner zone also achieves staged combustion. This technique redirects a portion of the air from the burners to the OFA ports. A variation of this concept, *lance air,* has air tubes installed around the periphery of each burner to supply staged air.

Combustion Temperature Reduction

Reducing the combustion temperature effectively reduces thermal NO_x but not fuel NO_x. One way to reduce the temperature further is to introduce a diluent, as in *flue gas recirculation* (FGR). FGR recirculates a portion of the flue gas back into the windbox. The recirculated flue gas, usually 10–20% of the combustion air, provides sufficient dilution to decrease NO_x emissions.

An advantage of FGR is that it can be used with most other combustion control methods. However, in retrofit applications, FGR can be expensive. In addition to requiring new large ducts, FGR may require major modifications to fans, dampers, and controls.

Water or steam injection (W/SI) is another method that works on the principle of combustion dilution, similar to FGR. In addition, W/SI reduces the combustion air temperature. In some cases, W/SI is a viable option when moderate NO_x reductions are required for compliance.

OTHER MODIFICATIONS

Reduction of the air preheat temperature (RAPH) is another viable technique for cutting NO_x emissions. This technique lowers the peak flame temperatures, thereby reducing NO_x formation. The thermal efficiency penalty, however, can be substantial.

Post-combustion control techniques such as SCR and SNCR by ammonia or urea injection are described in Section 4.2.

The techniques of OSC, FGR, and RAPH can be effectively combined. The OSC techniques, namely LEA and two-stage combustion, reduce the quantities of combustion gases reacting at maximum temperature, while FGR and RAPH directly influence the maximum level of combustion temperatures. The reductions obtained by combining individual techniques are not additive but multiplicative. However, the combined conditions necessary to achieve such low levels of oxides are not compatible with operational procedures.

Figure 3.6.3 summarizes the NO_x control technology choices. An environmental engineer can use this figure and Table 3.6.1 to identify the potential control technologies for boilers and process heaters. After identifying the applicable technologies, the engineer must conduct an economic analysis to rank the technologies according to their cost effectiveness. Management can then select the optimum NO_x control technology for a specific unit.

Design Feature Modifications

Several design feature modifications also reduce NO_x emissions including modified burners, burner location and spacing, tangential firing, steam temperature control, air and fuel flow patterns, and pressurized fluidized-bed combustion.

MODIFIED BURNERS

While SCR and SNCR control NO_x emissions by treating the NO_x after it has been formed in the combustion reaction, modifications to the combustion equipment or burners also significantly reduces NO_x formation. Using such modified burners has a number of advantages, the major ones being simplicity and low cost. At the same time, since burners are the primary component of a furnace, implementing new ones should be tried cautiously.

Stage-Air Burners

Staged-air burner systems divide incoming combustion air into primary and secondary paths. All fuel is injected into the throat of the burner and is combined with the primary

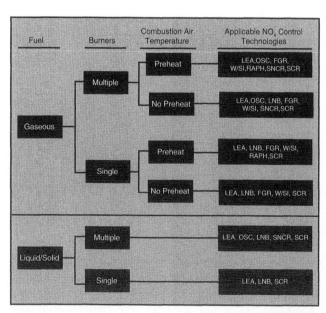

FIG. 3.6.3 Guidelines to identify potential NO_x control technologies. (Reprinted, with permission, from A. Garg, 1994, Specify better low-NO burners for furnaces, *Chem. Eng. Prog.* [November].)

air, which flows through the venturi and burns (see the right side of Figure 3.6.4).

Staged-air burners are simple and inexpensive, and NO_x reductions as high as 20 to 35% have been demonstrated (Garg 1992). These burners are most suitable for forced-draft, liquid-fuel applications and lend themselves to external, flue gas circulations. The main disadvantage of these burners is the long flames, which must be controlled.

Staged-Fuel Burners

In staged-fuel burners, fuel is injected into the combustion zone in two stages, thus creating a fuel-lean zone and delaying completion of the combustion process. This lean combustion reduces peak flame temperatures and reduces thermal NO_x. The remainder of the fuel–gas is injected into the secondary zone through secondary combustion nozzles (see the left side of Figure 3.6.4).

The combustion products and inert gases from the primary zone reduce the peak temperatures and oxygen concentration in the secondary zone, further inhibiting thermal NO_x formation. Some of the NO_x formed in the first stage combustion zone is reduced by the hydrogen and carbon monoxide formed in the staged combustion.

Staged-fuel burners can reduce NO_x emissions by 40–50% (Garg 1994). The flame length of this type of burner is about 50% longer than that of a standard gas burner. Staged-fuel burners are ideal for gas-fired, natural-draft applications.

Ultra-Low-NO$_x$ Burners

Several designs combine two NO_x reduction steps into one burner without any external equipment. These burners typically incorporate staged air with internal FGR or staged fuel with external FGR. In the staged-

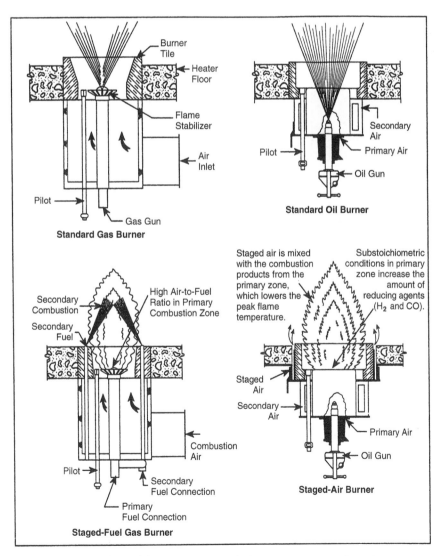

FIG. 3.6.4 OFC achieved by air staging or fuel staging. (Reprinted, with permission, from Garg, 1994.)

fuel burners with internal FGR, the fuel–gas pressure induces recirculation of the flue gas (see Figure 3.6.5), creating a fuel-lean zone and reducing oxygen partial pressure.

In staged-air burners with internal FGR, fuel mixes with part of the combustion air, creating a fuel-rich zone. The high-pressure atomization of liquid fuel or fuel gas creates FGR. Pipes or parts route the secondary air in the burner block to complete combustion and optimize the flame profile.

Air or fuel–gas staging with internal FGR can reduce NO_x emissions by 55 to 75%. The latter design can be used with liquid fuels, whereas the former is used mostly for fuel–gas applications.

Low-NO_x burners have been installed in a variety of applications in both new and revamped plants. Table 3.6.2 summarizes the performance of several of these installations.

BURNER LOCATIONS AND SPACING

NO_x concentrations vary with burner type, spacing, and location in coal-fired generation plants. Cyclone burners are known for their highly turbulent operations and result in high-level emissions of NO_x in these plants.

The amount of heat released in the burner zone seems to have a direct effect on NO_x concentrations (see Figure 3.6.6). The OFC reduction techniques are the most effective for larger generating units with larger burners. These techniques are effective since essentially all NO is formed in the primary combustion zone, and as the burner size increases, the primary zone becomes less efficient.

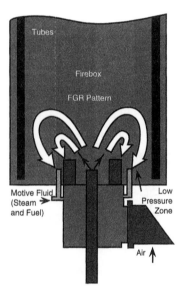

FIG. 3.6.5 Combining staged fuel burners with internal FGR. (Reprinted, with permission, from A. Garg, 1992, Trimming NO_x from furnaces, *Chem. Eng. Prog.* [November].)

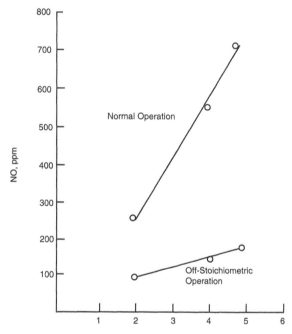

FIG. 3.6.6 Effect of nitric oxide concentrations with increasing burner zone heat release rate: 480 MW unit, natural gas fuel.

TABLE 3.6.2   APPLICATIONS OF LNB BURNERS

Heater Application	Burner Type	NO_x Emission Level, ppm
Crude, vacuum, and coker heaters (cabin)	Forced-draft, staged fuel, preheated air	60
Vertical cylindrical refinery heaters	Natural-draft, staged-fuel, with internal FGR	25
Down-fired hydrogen reformer	Induced-draft, staged-fuel-gas	60*
Vertical cylindrical refinery heaters	Forced-draft, staged-fuel, with internal FGR, preheated air	60
Upfired ethane cracker	Natural-draft, staged-fuel	85

Source: A. Garg, 1994, Specify better low-NO_x burners for furnaces, *Chem. Eng. Prog.* (January).
Notes: *With steam injection.

Different burner spacing has essentially no effect on the NO_x concentration in boiler emissions when the boilers are at full-load operations. However, at a reduced load, closer burner spacing results in higher NO_x releases because close burner spacing inhibits effective bulk recirculation into the primary combustion zone.

The distribution of air, through the primary air ducts located immediately above and below the fuel nozzles and through the secondary air ducts located immediately above and below the primary ducts, can be used as a means to reduce NO_x emissions.

TANGENTIAL FIRING

In tangential firing, the furnace is used as a burner, resulting in a lower flame temperature and in a simultaneous reduction (as much as 50 to 60% compared with conventional firing methods) in NO_x emissions.

STEAM TEMPERATURE CONTROL

Product gas recirculation, burner tilt, and high excess air are techniques for steam temperature control. The burner tilt and high excess air increase the NO_x formation. Changing the burner tilt from an angle inclined 30° downstream from horizontal to 10° upstream increases the NO_x concentration from 225 ppm to 335 ppm (Shah 1974).

AIR AND FUEL FLOW PATTERNS

For effective use of the OFA technique, uniform air and fuel flow to the burners is essential. High CO concentrations, localized in a small zone of the furnace exhaust gas, occur when one burner operates at a rich fuel–air mode. Reducing the high CO concentration involves increasing excess air or decreasing the degree of OFA, both of which increase the NO_x emissions. Therefore, the effective solution is to reduce the fuel flow to the burner.

PRESSURIZED FLUIDIZED-BED COMBUSTION

Mixing an adsorbent (e.g, limestone) directly into the fluidized bed in which coal is burned achieves direct sulfur dioxide control (see Figure 3.6.7). Fluidization is achieved via the combustion air that enters the base of the bed (Halstead 1992). Environmental engineers can apply this basic concept in several ways to meet the needs of chemical industries as well as power generation plants.

The potential advantages of fluidized-bed combustion are as follows:

Lower combustion temperature resulting in less fouling and corrosion and reduced NO_x formations
Fuel versatility including range of low-grade fuels, such as char from synthetic fuel processing

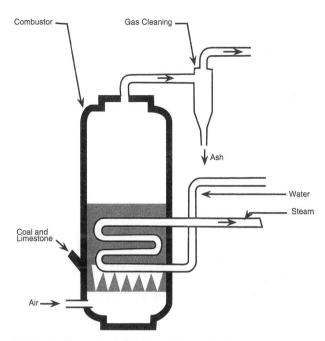

FIG. 3.6.7 Pressurized fluidized-bed combustion system.

Higher thermal efficiency including high heat release and heat transfer
Waste solids in dry form

The potential disadvantages are as follows:

Large particulate loading in the flue gas
Potentially large amounts of solid waste, which are SO_2 absorbent.

Pollution Monitoring

Continuous stack and ambient monitoring provides the information needed to formulate strategies for pollutant control (see Sections 2.1, 2.3, and 2.4). In Osaka, Japan, telemeters monitoring pollutant emissions from factories and stations sampling ambient quality perform environmental monitoring. An environmental monitoring system uses remote-sensing by earth-observing satellites and an environmental information system to forecast the environmental impact of pollution.

A procedure change can ease the effects of sulfur or nitrogen oxide pollution. Automatic changes can be made during high levels of ambient pollution. For example, the Tennessee Valley Authority (TVA) has been switching to low-sulfur fuels during adverse meteorological conditions.

—David H.F. Liu

References

Bradshaw, A.D., Sir Richard Southward, and Sir Frederick Warner, eds. 1992. *The treatment and handling of wastes.* London: Chapman & Hall.

Garg, A. 1992. Trimming NO$_x$ from furnaces. *Chem. Eng. Prog.* (November).

———. 1994. Specify better low-NO$_x$ burners for furnaces. *Chem. Eng. Prog.* (January).

MacKinnon, D.J. 1974. Nitric oxide formation at high temperatures. *Journal of the Air Pollution Control Association* 24, no. 3 (March).

Shah, I. 1974. Furnace modifications. In Vol. 2 of *Environmental engi-neers' handbook,* edited by B.G. Liptak. Radnor, Pa.: Chilton Book Company.

U.S. Environmental Protection Agency (EPA). 1983. *Control techniques for nitrogen oxide emissions from stationary sources.* Revised 2d ed. EPA-450/3-83-002. Research Triangle Park, N.C.: U.S. EPA.

Wark, K., and C.F. Warner. 1981. Air pollution, its origin and control. New York: Harper & Row Publishers.

3.7
GASEOUS EMISSION CONTROL: PHYSICAL AND CHEMICAL SEPARATION

Major air pollutants are gases such as carbon monoxide, nitrogen oxides, sulfur oxides, and VOCs. Generally, the pollutant concentrations in waste air streams are relatively low, but emissions can still exceed the regulatory limits. Removing air pollutants is achieved by the following methods:

Absorption by a liquid solution
Condensation of pollutants by cooling the gas stream
Adsorption on a porous adsorbent
Chemical conversion of pollutants into harmless compounds

Sometimes methods are combined to treat a feed stream. For example, the absorption of SO$_2$ can be performed in an absorber using an aqueous lime solution. The key step in the separation is absorption although reactions occur between the SO$_2$ and lime in the absorber. Therefore, this SO$_2$ separation is considered *absorption.*

Choosing the air pollutant removal method depends mostly on the physical and chemical properties of the pollutant and the conditions (i.e., temperature, pressure, volume, and concentration) under which the pollutant is treated. The methods chosen for reducing air pollution must not increase pollution in other sectors of the environment. For example, transferring the air pollutant into liquid or solid absorption agents that subsequently contaminate the environment is not a solution to the problem.

Absorption

Absorption is a basic chemical engineering operation and is probably the most well-established gas control technique. It is used extensively in the separation of corrosive, hazardous, or noxious pollutants from waste gases. The major advantage of absorption is its flexibility; an absorber can handle a range of feed rates.

Absorption, also called scrubbing, involves transferring pollutants from a gas phase to a contacting solvent. The transfer occurs when the pollutant partial pressure in the gas phase is higher than its vapor pressure in equilibrium with the solvent. To maximize the mass-transfer driving force (i.e., the difference in pollutant concentration between the gas and liquid phase), the absorber generally operates in a countercurrent fashion.

Absorption systems can be classified as physical absorption and absorption with a chemical reaction. In physical absorption, the pollutants are dissolved in a solvent and can be desorbed for recovery. The absorption of ammonia by water or the absorption of hydrocarbon by oil are typical examples.

If a solvent to absorb a significant quantity of the pollutant cannot be found, a reactant mixed with the solvent can be used. The pollutants must first be absorbed into the liquid phase for the reaction to occur. In this case, the pollutant concentration in the liquid is reduced to a low level. As a result, high absorption capacity is achieved.

The reaction can be reversible or irreversible. Typical reversible reactions are H$_2$S/ethanolamines, CO$_2$/alkali, carbonates, and some flue gas desulfurization (FGD) systems. The reversible reactions allow the pollutants to be recovered in a concentrated form and the solvent to be recycled to the absorber. If the reactions are irreversible, the reaction products must be disposed or marketed (e.g. ammonium sulfate). A few FGD systems are irreversible (vide infra).

ABSORPTION OPERATIONS

Absorption systems design involves selecting a solvent and the design of the absorber.

Solvent Selection

Solubility is the most important consideration in the selection of a solvent for absorption. The higher the solubility, the lower the amount of solvent required to remove a given amount of pollutants. The solvent should also be relatively nonvolatile to prevent an excessive carryover in the gas effluent. Other favorable properties include low flammability and viscosity, high chemical stability, acceptable corrosiveness, and low toxicity and pollution potential. The final selection criterion is an economic comparison with other control technologies.

Absorber Design

Any gas–liquid contactors that promote the mass transfer across the phase boundary can be used in absorption operations. The most popular devices are spray, packed, and tray columns as well as venturi scrubbers. For gas pollution control, the combination of high gas flowrate and low pollutant concentration suggests that the absorber should exhibit a low pressure drop. The mass-transfer efficiency of the absorber determines the height of the column, but it is not as important a consideration as the pressure drop. Clearly, spray and packed columns are the best devices to satisfy the preceding criteria. Spray columns are used where fouling and low pressure drops are encountered. The design of a spray column is straightforward and is detailed in other publications (Kohl 1987). The packed column is the device used most often.

Two types of packings are used for absorption: random packing and structured packing. Table 3.7.1 shows a list of packing suppliers. Structured packing tends to be proprietary, and the design procedure is normally obtained from suppliers. Random packings (e.g., Pall rings and Intalox saddles) can be purchased from several suppliers,

and the system design procedure is published. Table 3.7.2 shows the properties of typical random packings.

As a general rule, randomly packed columns have the following characteristics (Strigle 1987):

A pressure drop between 2 and 5 cm H_2O/m of packed depth

Air velocity between 1.7 and 2.4 m/sec for modern, high-capacity plastic packings

An inlet concentration of pollutants below 0.5% by volume

A superficial liquid rate in the range of 1.35 to 5.5 $L/m^2 \cdot sec$

Figure 3.7.1 is a schematic diagram of a typical packed column installation.

Design Procedure

In designing a packed column, the design engineer must consider the solvent rate, column diameter, and column height.

SOLVENT RATE

Once the feed and effluent specifications are established, the design engineer can calculate the minimum solvent rate. The minimum rate is the rate below which separation is impossible even with a column of infinite height. The engineer calculates flowrates as illustrated in Figure 3.7.2. The mole fraction of the pollutant in the solvent (X_T) and effluent gas (Y_T) at the top of the column as well as that in the feed gas (Y_B) are known. A straight line is drawn from (X_T, Y_T) to intercept the equilibrium line at Y_B. The slope of this line represents the ratio of the minimum solvent rate to the feed gas rate.

The normal operating solvent rate should be 30 to 70% above the minimum rate.

TABLE 3.7.1   ABSORPTION PACKING SUPPLIERS

Supplier	Address	Random Packing	Structured Packing
ACS Industries	Woonsocket, RI		x
Ceilcote Air Pollution Control	Berea, OH	x	
Chem-Pro Corp.	Fairfield, NJ	x	
Clean Gas Systems, Inc.	Farmingdale, NY	x	
Glitsch, Inc.	Dallas, TX	x	x
Jaeger Products, Inc.	Houston, TX	x	
Julius Montz Co.	Hilden, Germany		x
Koch Engineering Co., Inc.	Wichita, KS	x	x
Kuhni, Ltd.	Basel, Switzerland		x
Lantec Products, Inc.	Agoura Hills, CA	x	
Munters Corp.	Fort Myers, FL		x
Norton Chemical Process Product Corp.	Akron, OH	x	x
Nutter Engineering	Tulsa, OK	x	x
Sulzer Brothers Ltd.	Winterthur, Switzerland		x

TABLE 3.7.2   PROPERTIES OF RANDOM PACKINGS

Packing Type	Nominal Size (mm)	Elements (per m³)	Bed Weight (kg/m³)	Surface Area (m²/m³)	ε Void Fraction	F_p Packing Factor (m⁻¹)
Intalox saddles (ceramic)	13	730,000	720	625	0.78	660
	25	84,000	705	255	0.77	320
	38	25,000	670	195	0.80	170
	50	9400	670	118	0.79	130
	75	1870	590	92	0.80	70
Intalox saddles (metal)	25	168,400	350	n.a.	0.97	135
	40	50,100	230	n.a.	0.97	82
	50	14,700	181	n.a.	0.98	52
	70	4630	149	n.a.	0.98	43
Pall rings (metal)	16				0.92	230
	25	49,600	480	205	0.94	157
	38	13,000	415	130	0.95	92
	50	6040	385	115	0.96	66
	90	1170	270	92	0.97	53
Raschig rings (ceramic)	13	378,000	880	370	0.64	2000
	25	47,700	670	190	0.74	510
	38	13,500	740	120	0.68	310
	50	5800	660	92	0.74	215
	75	1700	590	62	0.75	120
Berl saddles (ceramic)	13	590,000	865	465	0.62	790
	25	77,000	720	250	0.68	360
	38	22,800	640	150	0.71	215
	50	8800	625	105	0.72	150
Intalox saddles (plastic)	25	55,800	76	206	0.91	105
	50	7760	64	108	0.93	69
	75	1520	60	88	0.94	50
Pall rings (plastic)	16	213,700	116	341	0.87	310
	25	50,150	88	207	0.90	170
	50	6360	72	100	0.92	82

Source: A.L. Kohl, 1987, Absorption and stripping, in *Handbook of separation process technology* (New York: John Wiley).
Note: n.a. = not applicable.

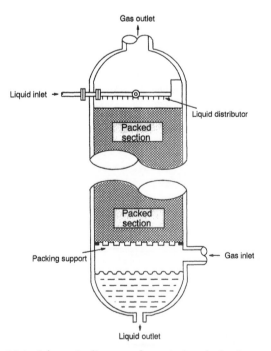

FIG. 3.7.1 Schematic diagram of a typical packed column.

COLUMN DIAMETER

Since the pressure drop directly impacts the operating cost for the blower to deliver the feed gas, the design engineer should choose the tower diameter to meet the pressure drop specification. Figure 3.7.3 shows the value for a widely used correlation for the pressure drop calculation. The correlation includes a parameter, packing factor F_P, which is a constant for a given packing size and shape (Fair 1987). The value of F_p can be obtained from Table 3.7.2. With known physical properties and flowrates for the liquid and gas streams, the engineer first determines the specific gas rate G, kg/m² · sec in the column and then calculates the column cross-sectional area by dividing the total gas flowrate (G_T,kg/s) by G.

In cases where the pressure drop is not a prime consideration (e.g., treatment of a pressurized process stream), the column can be designed to operate at 70–80% of flooding rate. For packings below 2.5 cm size, the design engineer can use the flooding line shown in Figure 3.7.3 to calculate the flooding rate. If the packing is larger than 3.7 cm, the following empirical equation determines the pressure drop at flooding (McCabe 1993):

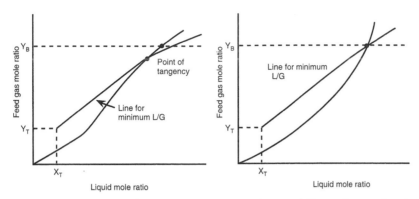

FIG. 3.7.2 Calculation of minimum solvent rates for two different shapes of equilibrium line.

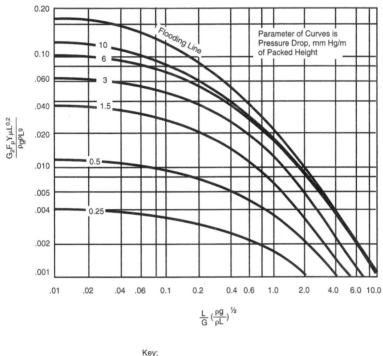

Key:
L = Liquid rate, kg/s m²
G = gas rate, kg/s m²
ρL = Liquid density, kg/m³
ρg = Gas density, kg/m³
Fp = Packing factor, m⁻¹
μL = Viscosity of Liquid, mPa s
Y = Ratio, (density of water)/(density of liquid)
g = Gravitational constant, 9.81 m/s²

FIG. 3.7.3 Generalized correlation for the calculation of pressure drops in packed columns.

$$\Delta P_{flood} = 0.307 F_P^{0.7} \qquad 3.7(1)$$

Using Figure 3.7.3, the engineer can then calculate the G value and hence the required column diameter.

COLUMN HEIGHT

The required height of an absorption column is determined primarily by the degree of pollutant removal and the mass-transfer characteristics of the system. The degree of pollu-

tant removal can be represented by the number of transfer units (NTU) which is a function of the average driving force for mass transfer and the degree of separation as follows:

$$NTU = \int_{y_T}^{y_B} \frac{dy}{y - y^*} \qquad 3.7(2)$$

where y and y^* are the mole fraction of the pollutant in the gas phase and in equilibrium with the contacting solvent, respectively. The mass-transfer characteristics include the fluid properties, operating conditions, and column internal parameters. The following equation expresses these effects:

$$\text{HTU} = \text{Height of a Transfer Unit} = \frac{G}{K_y a} \qquad 3.7(3)$$

where K_y is the overall mass-transfer coefficient and a is the effective interfacial area available for mass transfer. The required column height H can then be calculated from the following expression:

$$H = \text{HTU} \cdot \text{NTU} = \frac{G}{K_y a} \int_{y_T}^{y_B} \frac{dy}{y - y^*} \qquad 3.7(4)$$

McCabe (1993) provide the derivation of the preceding equations. To solve Equation 3.7(4), the design engineer must specify y_B and y_T and obtain G from the column diameter calculations, y^* from the equilibrium relationship, and $K_y a$ from engineering correlations. The values for y^* for physical absorption are available from Henry's law equation as follows:

$$y^* = \frac{H}{P_T} x \qquad 3.7(5)$$

where H is the Henry's law constant and P_T is the total pressure. The accuracy of estimating column height depends mostly on the $K_y a$ correlation chosen for the calculation. In mass-transfer operations, K_y can be broken down to the mass-transfer coefficient in the gas phase k_y and in the liquid phase k_x. This two-resistance theory is widely accepted (McCabe 1993) and leads to the following relationships:

$$\frac{1}{K_y} = \frac{1}{k_y} + \frac{m}{k_x} \qquad \text{and} \quad y^* = mx \qquad 3.7(6)$$

These relationships assume that the resistances to mass transfer in the gas and liquid phases occur in series, and the gas–liquid interface does not contribute significant resistance to the mass transfer. If k_y is much greater than k_x/m, then $K_y \cong m/k_x$, and the rate of absorption is controlled by the liquid-phase transfer. On the other hand, if $k_y \ll k_x/m$, then the system is gas-phase controlled. To some extent, the value of m or H controls the phase control mechanism. For a soluble system (e.g., absorption of NH_3 by water), the value of m is small, and thus the mass transfer is gas-phase controlled. For a CO_2/water system, the value of m is large, and the absorption is liquid-phase controlled.

The calculations also require estimating the effective interfacial area a. The area a is generally smaller than the geometric area of the packing because some packing area is not wetted and some area is covered by a stagnant, liquid film that is already saturated with the pollutant (i.e., inactive for mass transfer).

Onda, Takeuchi, and Okumoto (1968) have expressions for the individual mass-transfer coefficients k_x and k_y as well as the effective interfacial area a, and the de-

rived values can be used in the absence of experimental data. The equations are as follows:

$$k_y \left(\frac{RT}{a_p D_G} \right) = 5.23 \left(\frac{G}{a_p \mu_G} \right)^{0.7} (Sc_G)^{1/3} (a_p d_p)^{-2} \qquad 3.7(7)$$

$$k_x \left(\frac{\rho L}{g \mu_L} \right)^{1/3} = 0.0051 \left(\frac{L}{a_p \mu_L} \right)^{2/3} (Sc_L)^{-1/2} (a_p d_p)^{0.4} \qquad 3.7(8)$$

where:

   L = mass flowrate of liquid (kg/sec · m^2)
   G = mass flowrate of gas (kg/sec · m^2)
   D_G = diffusion coefficient (m^2/sec)
   S_{cL} = Schmidt number, $(\mu_L/\rho_L D_L)$
   k_y = gas-phase mass-transfer coefficient (kmol/N · s)
   k_x = liquid-phase mass-transfer coefficient (m/s)
   ρ_L = density of liquid (kg/m^3)
   μ_L = viscosity of liquid (N · s/m^2)
   d_p = effective packing diameter (the diameter of a
       sphere with equal surface area)
   a_p = dry outside surface area of packing

In this model, a is assumed to equal a_w, the wetted area of the packing, and is calculated by the following equation:

$$a = a_w$$
$$= a_p \left\{ 1 - \exp \left[-1.45 (Re_L)^{-0.1} (Fr_L)^{-0.05} (We_L)^{0.2} \left(\frac{\sigma}{\sigma_c} \right)^{-0.75} \right] \right\}$$
$$3.7(9)$$

where:

   a   = the effective interfacial area for mass transfer
       (m^2/m^3)
   a_w = wetted area of packing
   σ_c = a critical surface tension = 61 dyn/cm for ceramic packing, 75 dyn/cm for carbon steel packing, and 33 dyn/cm for polyethylene packing
   Re_L = $L/a_p \mu_L$
   Fr_L = $a_p L^2/(g \rho_L)^2$
   We_L = $L^2/a_p \rho_L$
   Sc_L = $\dfrac{\mu_L}{\rho_L D_L}$
   Sc_G = $\dfrac{\mu_G}{\rho_G D_G}$

Absorption with a chemical reaction in the liquid phase is involved in most applications for gas control. Reaction in the liquid phase reduces the equilibrium partial pressure of the pollutant over the solution, which increases the driving force for mass transfer. If the reaction is irreversible, then $y^* = 0$, and the NTU is calculated as follows:

$$\text{NTU} = \int_{y_T}^{y_B} \frac{dy}{y - y^*} = \ln \frac{y_B}{y_T} \qquad 3.7(10)$$

A further advantage of the reaction is the possible increase in the liquid-phase mass-transfer coefficient k_x and the effective interfacial area a. Perry and Green (1984) describe the design methods in detail.

COMMERCIAL APPLICATIONS

Most commercial applications of absorption for gas control fall into two categories: regenerative and nonregenerative systems. For regenerative systems, the solvent leaving the absorber is enriched in pollutant. This mixture is normally separated by distillation. The solvent is then recycled back to the absorber, and the pollutant can be recovered in a concentrated form. Regeneration is possible when the absorber uses a physical solvent or a solvent containing compounds that react reversibly with the pollutant. The recovery of H_2S from hydrocarbon processing using an amine as the solvent is a major application of regenerative absorption. Figure 3.7.4 shows a typical flowsheet.

For the capture of H_2S and volatile sulfur compounds (e.g., thiols, COS, and CS_2) released during the transportation of sour liquids, a portable system is required. Am-gas Scrubbing Systems Ltd. (Calgary, Canada) manufactures both portable and stationary units that use 26% aqueous ammonia. The spent solution oxidizes to form ammonium sulfate, which is recovered from the spent absorbent and can be marketed as fertilizer. Alternate absorbents for portable units include basic hypochlorite (bleach) solutions, but the stability of the solution decreases with consumption of the base as the H_2S is absorbed. A complicating factor in the use of basic fluid absorbent systems is that CO_2 is also a weak acid and is absorbed to sequentially form bicarbonate (HCO_3^-) and carbonate (CO_3^{2-}) ions, thereby consuming the absorbent.

The absorption of SO_2 by dimethylaniline from off-gases at copper, zinc, lead, and nickel smelters is another application. The desorbed SO_2 produces H_2SO_4. The SO_2 concentration in the off-gases is approximately 10%.

For nonregenerative or throwaway systems, irreversible reactions occur in the liquid phase. The reagents in the solvent are consumed and must be replenished for absorption efficiency. The resulting products are discarded. Typical examples are the absorption of acid gases by caustic solutions. Nonregenerative systems are economical only when the reagents are inexpensive or the volume involved is small.

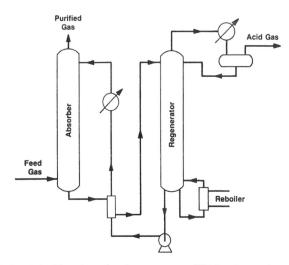

FIG. 3.7.4 Flowsheet for the recovery of H_2S using amine as a solvent.

TABLE 3.7.3 FLUE GAS DESULFURIZATION PROCESSES

Process Generics	Process Operations	Active Material	Key Sulfur Product
Throwaway processes			
1. Lime or limestone	Slurry scrubbing	CaO, $CaCO_3$	$CaSO_3$/$CaSO_4$
2. Sodium	Na_2SO_3 solution	Na_2CO_3	Na_2SO_4
3. Dual alkali	Na_2SO_3 solution, regenerated by CaO or $CaCO_3$	$CaCO_3$/Na_2SO_3 or CaO/NaOH	$CaSO_3$/$CaSO_4$
4. Magnesium promoted-lime or limestone	$MgSO_3$ solution, regenerated by CaO or $CaCO_3$	MgO/$MgSO_4$	$CaSO_3$/$CaSO_4$
Regenerative processes			
1. Magnesium oxide	$Mg(OH)_2$ slurry	MgO	15% SO_2
2. Sodium (Wellman–Lord)	Na_2SO_3 solution	Na_2SO_3	90% SO_2
3. Citrate	Sodium citrate solution	H_2S	Sulfur
4. Ammonia	Ammonia solution, conversion to SO_2	NH_4OH	Sulfur (99.9%)
Dry processes			
1. Carbon adsorption	Adsorption at 400°K, reaction with H_2S to S, reaction with H_2 to H_2S	Activated carbon/H_2	Sulfur
2. Spray dryer	Absorption by sodium carbonate or slaked lime solutions	Na_2CO_3/$Ca(OH)_2$	Na_2SO_3/Na_2SO_4 or $CaSO_3$/$CaSO_4$

The most important application of absorption for gas control technology is FGD. Because flue gas from power plants contributes about two-thirds of the U.S. emissions, large efforts have been spent on developing an effective control technology. Currently, several hundred FGD systems are in commercial operation. The absorption can be regenerative or nonregenerative as well as wet or dry; in effect, four categories of FGD processes exist. Table 3.7.3 lists the processes in use. The following sections describe the most important FGD processes.

Nonregenerative Systems

Most FGD systems are nonregenerative. The reagents used in the absorber are alkaline compounds that react with SO_2. Most well-established technology (see Figure 3.7.5) is based on limestone and lime. Other processes are based on $NaOH$, Na_2CO_3, and NH_4OH. Since lime and limestone are the most inexpensive and abundant reagents, they are used in about 75% of the installed FGD systems. In this process, the reaction products are $CaSO_3$ and $CaSO_4$ in a sludge form that must be disposed. These systems can remove 90% of the SO_2 in the flue gases. Although lime is more reactive than limestone, it is more expensive; therefore, lime is not used as widely as limestone.

The main drawback of lime- and limestone-based FGD systems is scaling and plugging of the column internals. This problem is eliminated in the dual alkali system (see Figure 3.7.6) by absorption with a Na_2SO_3/Na_2SO_4 solution which is then sent to a separate vessel where lime or limestone and some $NaOH$ are introduced. The lime or limestone precipitates the sulfite and sulfate and regenerates $NaOH$ for reuse in the absorption column. The major disadvantage of this process is the loss of soluble sodium salts into the sludge which may require further treatment. As a result, this process is not used as widely as the lime and limestone process.

Regenerative Systems

Regenerative processes have higher costs than throwaway processes. However, regenerative processes are chosen when disposal options are limited. Regenerative processes produce a reusable sulfur product. In Japan, where the government mandates FGD, regenerative processes are used almost exclusively.

The Wellman–Lord process is the most well-established regenerative process which uses an aqueous sodium sulfite solution as the solvent. Figure 3.7.7 is a schematic diagram of the Wellman–Lord process. This process consists of the four following subprocesses:

Flue gas pretreatment. In this subprocess, flue gas from an ESP is blown through a venturi prescrubber. The prescrubber removes most of the remaining particles and any existing SO_3 and HCl, which upset the SO_2 absorption chemistry. The prescrubber also cools and humidifies the flue gas. A liquid purge stream from the prescrubber removes the solids and chlorides.

SO_2 absorption by sodium sulfite solution. In the SO_2 absorber tower, the flue gas from the prescrubber contacts the aqueous sodium sulfite, and the SO_2 is absorbed and reacted with Na_2SO_3 in the liquid to form sodium bisulfite. Since excess O_2 is always present, some of the Na_2SO_3 oxidizes to Na_2SO_4; some of the Na_2SO_3 reacts with the residual SO_3 to form Na_2SO_4 and sodium bisulfite. The sodium sulfate does not further SO_2 absorption. A continuous purge from the bottom of the absorber prevents excessive sulfate buildup. Since the absorber bottom is rich in bisulfite, most of the stream is routed for further processing.

Purge treatment. Part of the liquid stream leaving the absorber is sent to the chiller and crystallizer, where the less soluble, sodium sulfate crystals are formed. The slurry is centrifuged, and the solids are dried and discarded. The bisulfite-rich, centrifugal material is returned to the process.

Sodium bisulfite regeneration. The remaining part of the liquid stream from the absorber is sent to a heated evap-

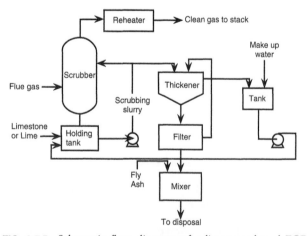

FIG. 3.7.5 Schematic flow diagram of a limestone-based FGD system.

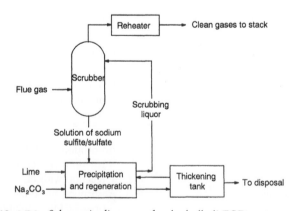

FIG. 3.7.6 Schematic diagram of a dual alkali FGD system.

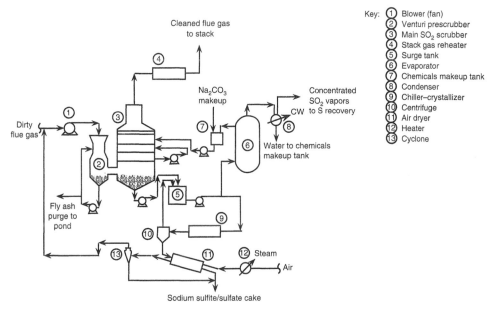

Key:
1. Blower (fan)
2. Venturi prescrubber
3. Main SO$_2$ scrubber
4. Stack gas reheater
5. Surge tank
6. Evaporator
7. Chemicals makeup tank
8. Condenser
9. Chiller–crystallizer
10. Centrifuge
11. Air dryer
12. Heater
13. Cyclone

FIG. 3.7.7 Schematic process flow diagram of the Wellman–Lord SO$_2$ scrubbing and recovery system. (Reprinted from U.S. Environmental Protection Agency (EPA), 1979.)

orator and crystallizer, where sodium bisulfite is decomposed to Na$_2$SO$_3$ and SO$_2$. The gas stream contains 85% SO$_2$ and 15% H$_2$O, thus the SO$_2$ can be used as feed stock for producing S or sulfuric acid. To replace the lost Na$_2$SO$_3$, this subprocess adds soda ash (Na$_2$CO$_3$) to make up sodium. The Na$_2$CO$_3$ reacts readily with SO$_2$ in the absorber tower to give sodium sulfite.

Condensation

In cases where pollutants have low vapor pressures, condensation is effective for removing a significant part of the vapor. The condenser works by cooling the feed gas to a temperature below the dew point of the feed gas. Although condensation can also occur by increasing the pressure without changing the temperature, this method is seldom used for the control of air pollution.

Because condensation is a simple process and the design techniques for condensers are well-established, condensation is also used as a pretreatment process for reducing the load and operating problems of other pollution control devices.

Cooling water is the most commonly used coolant although brine solutions are also used when low temperatures are required.

Two types of condensers are surface and contact condensers. Surface condensers are used when the coolant and pollutant form a miscible mixture. This mixture must be further separated (e.g., by distillation) for recovery of the pollutants. If the coolant is immiscible with the pollutants, contact condensers are used because phase separation is a relatively easy operation. However, the coolant is not contaminated to the extent of causing water pollution problems.

Surface condensers are normally shell and tube type and should be set vertically. Vapor should only condense inside the tubes. This arrangement prevents a stagnant zone of inert gas (air) that might blanket the heat transfer surfaces. Figure 3.7.8 shows a typical surface condenser. The feed gas enters the top of the condenser and flows concurrently downward with the condensate.

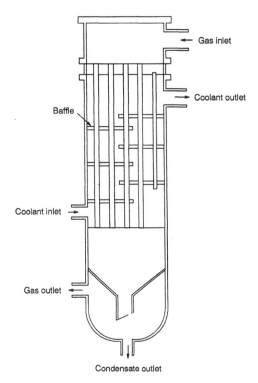

FIG. 3.7.8 Surface condenser.

Contact condensers are much smaller and less expensive than surface condensers. In the design shown in Figure 3.7.9, part of the coolant is sprayed into the gas space near the top of the condenser. This design is similar to the spray column used in absorption. The remainder of the coolant is directed into a discharge throat to complete the condensation. The pressure regain in the downstream cone of the venturi is often sufficient to eliminate the need for a pump or a barometric leg.

Because the coolant required to operate contact condensers is ten to twenty times that for surface condensers, the latter is the predominate device used in air pollution control applications. The design of a surface condenser is based on complete resistance to heat transfer on the condensing side in the layer of condensate. A mean condensing coefficient is calculated from appropriate correlations. Calculating the overall heat transfer coefficient and the log–mean temperature driving force estimates the required heat transfer area. Many engineering textbooks (Perry and Green 1984) provide additional design details.

Adsorption

Adsorption is a separation process based on the ability of adsorbents to remove gas or vapor pollutants preferentially from a waste gas stream (Yang 1987; Kohl and Riesenfeld 1985). The process is particularly suitable when pollutants are (1) noncombustible, (2) insoluble in liquids, or (3) present in dilute concentration.

The mechanism of adsorption can be classified as either physical adsorption or chemisorption. In physical adsorption, gas molecules adhere to a solid surface via van der Waals forces. The process is similar to the condensation of a vapor. It is a reversible process. Desorption occurs by lowering the pressure, increasing the temperature, or purging the adsorbent with an inert gas. In chemisorp-

tion, gas molecules are adsorbed by forming chemical bonds with solid surfaces. This process is sometimes irreversible. For example, oxygen chemisorbed on activated carbon can only be desorbed as CO or CO_2. Chemisorbed spent adsorbents cannot normally be regenerated under mild temperatures or vacuum.

ADSORPTION ISOTHERM

For a particular single gas–solid combination, one of five types of adsorption isotherm is found (see Figure 3.7.10). In separation processes, the favorable isotherms, types 1 and 2 are most frequently encountered.

From considering kinetics for a single gas impacting a uniform solid surface and adsorbing without chemical change, Langmuir (1921) deduced that the fraction of a surface covered by a monolayer varies with the partial pressure of the adsorbate P_a. The following equation gives the mass of adsorbate adsorbed per unit mass of adsorbent m_a:

$$m_a = \frac{k_1 P_a}{k_2 P_a + 1} \qquad 3.7(11)$$

where k_1 and k_2 are constants. Equation 3.7(11) is known as the Langmuir isotherm, and the shapes are types 1 and 2 in Figure 3.7.10.

Assuming that the number of sites of energy Q (N_Q) is related to a base value for Q, Q_0, as follows:

$$N_Q = ae^{-Q/Q_0} \qquad 3.7(12)$$

then Equation 3.8(11) reduces to the following approximation:

$$\Theta = const.P^{1/n} \qquad 3.7(13)$$

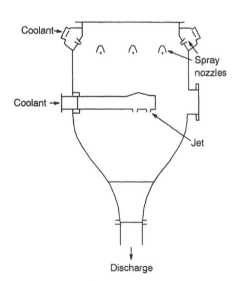

FIG. 3.7.9 Contact condenser.

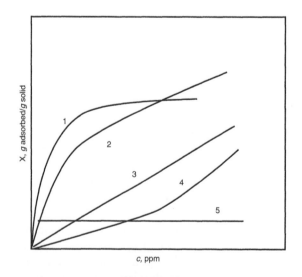

Key: 1 = High loading (highly favorable)
2 = Favorable
3 = Linear
4 = Low loading (unfavorable)
5 = Irreversible

FIG. 3.7.10 Adsorption loading profiles.

When expressed as the volume of adsorbate adsorbed per unit mass of adsorbent ν, as follows:

$$\nu = kP^{1/n} \qquad 3.7(14)$$

or expressed in terms of concentration in the gas phase c and at the absorbent surface w, as follows:

$$c = \alpha w^{\beta} \qquad 3.7(15)$$

the expression is known as the Freundich isotherm. For a binary, or higher, mixture in which the components compete for surface sites, the expression is more complex (Yang 1987). Table 3.7.4 gives the values for parameters k and n in the Freundich isotherm for selected adsorbates.

Ideally, an adsorbent adsorbs the bulk of a pollutant from air even when that material is in low concentration. The adsorption profile for such behavior is called *favorable* and is shown as curves 1 and 2 in Figure 3.7.10. If high concentrations must be present before significant quantities are adsorbed (curve 4), the profile is called *unfavorable adsorption.*

As a pollutant is adsorbed onto the adsorbent, the concentration in the air stream falls. The adsorbent continues to absorb the pollutant from the air stream until it is close to saturation. Thus, as the air stream passes through a bed of adsorbate, the pollutant concentration varies along the bed length. This adsorption wave progresses through the bed with time (see Figure 3.7.11). At time t_1, the bed is fresh, and essentially all pollutant is adsorbed close to the entrance of the bed. At time t_2, the early part of the bed is saturated, but the pollutant is still effectively adsorbed. At time t_3, a small concentration of pollutant remains in the air stream at the exit. When the pollutant concentration at the exit meets or exceeds the limiting value, the bed is spent and must be replaced or regenerated. For some pollutants, the breakthrough of detectable amounts of pollutants is unacceptable. To avoid exceeding acceptable or regulated limits for pollutant concentrations in exit gases, replacing or regenerating adsorbent beds well before the end of the bed's lifetime is essential.

The profiles shown in Figure 3.7.11 are for a single pollutant in an air stream. An adsorber normally operates in a vertical arrangement to avoid bypassing gases. For multiple pollutants, the process is more complex as each pollutant can exhibit different behavior or competition can occur between pollutants for adsorption sites (Yang 1987; Kast 1981). The bed design must be able to adsorb and retain each pollutant.

ADSORPTION EQUIPMENT

The major industrial applications for adsorption processes have both environmental and economic objectives. The abatement of air pollution includes removing noxious and odorous components. Additionally, adsorption is used for the recovery of solvents and reagents, such as carbon disulfide, acetone, alcohol, aliphatic and aromatic hydrocarbons, and other valuable materials, from effluent streams (Kast 1981).

Typically, a process in which the adsorbate is recovered uses temperature swing adsorption (TSA), pressure swing adsorption (PSA), or continuous or periodic removal of the adsorbent from the system (Kast 1981). The TSA and PSA technologies are normally used only for the removal of high concentrations of adsorbate. For the majority of environmental applications, small concentrations must be removed from air, and continuous or periodic removal technologies are more appropriate.

TABLE 3.7.4 FREUNDICH ISOTHERM PARAMETERS FOR SOME ADSORBATES

Adsorbate	Temperature (K)	k × 100	n	Partial Pressure (Pa)
Acetone	311	1.324	0.389	0.69–345
Acrylonitrile	311	2.205	0.424	0.69–103
Benzene	298	12.602	0.176	0.69–345
Chlorobenzene	298	19.934	0.188	0.69–69
Cyclohexane	311	7.940	0.210	0.69–345
Dichloroethane	298	8.145	0.281	0.69–276
Phenol	313	22.116	0.153	0.69–207
Toluene	298	20.842	0.110	0.69–345
TCA	298	25.547	0.161	0.69–276
m-Xylene	298	26.080	0.113	0.69–6.9
p-Xylene	298	28.313	0.0703	6.9–345

Source: U.S. Environmental Protection Agency (EPA), 1987, *EAB control cost manual,* 3d ed. (Research Triangle Park, N.C. U.S. EPA).
Notes: The amount adsorbed is expressed in kg adsorbate/kg adsorbent.
The equilibrium partial pressure is expressed in Pa.
Data are for the adsorption on Calgon-type BPL activated carbon (4 × 10 mesh).
Data should not be extrapolated outside of the partial pressure ranges shown.

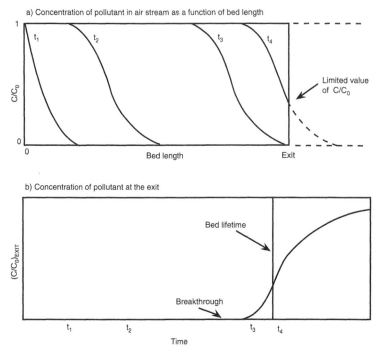

FIG. 3.7.11 Adsorption profiles for a single contaminant in an air stream.

Periodic Removal

This system is the simplest to design or build but is not necessarily the most efficient or cost-effective to operate. Essentially, the adsorbate is left in a fixed-bed or fluidized-bed adsorber until the adsorptive capacity of the bed is approached. Then, the adsorbate is removed for destruction, disposal, or regeneration. A new adsorbent is then charged to the system. For systems in which either the adsorbent or adsorbate is hazardous or contact with air is deleterious, this method presents issues requiring expensive solutions. Further, during changeover of the spent adsorbent, the adsorber is not used. Therefore, a minimum of two adsorber units is necessary for any plant in continuous operation.

Continuous Removal

By simultaneously introducing and removing adsorbent, this system achieves continuous operation. This system requires the bed, or the material in the bed, to be in continuous motion. A fixed-bed can rotate so that the fraction of the bed exposed to the feed stream is continuously charged. The spent adsorbent then rotates to a separate zone where the system regenerates it either by flushing with a carrier gas, contact with a reagent, or desorption of the adsorbate by sweeping with a hot gas. Because the adsorbent is necessarily porous, no significant reverse pressure differential exists between the feed and flushing systems.

A solid adsorbent cascading through the feed gas can be continuously removed from the base of the adsorber unit. Regenerated adsorbent can then be recycled to the unit.

TSA

When the bond between the adsorbent and adsorbate is strong (i.e., chemisorption), regenerating the adsorbent requires elevated temperatures and purging with a nonadsorbing gas. If this process occurs in the adsorber unit, a TSA process is used. For continuous processing, a minimum of three adsorber units are required for economic operation: at least one in operation, one undergoing regeneration, and one cooling down following regeneration. Typically, the gas flushing the unit that is cooling down is then passed through the bed undergoing regeneration. A heater ensures that the flush gas is at a temperature to effect desorption. The effluent stream is then cooled, and if the adsorbate is to be recovered, the flush gas and adsorbate are separated by physical methods (distillation, condensation, or decanting). Regeneration can also require additional steps, including physical or chemical treatment, washing, and drying.

PSA

When the adsorption capacity of an adsorbent is a function of pressure or one component in a gaseous mixture is preferentially adsorbed at high pressures, PSA can be used for gas separation. Essentially, a bed of adsorbent in one unit is initially exposed to the gas mixture at a high pressure. The feed stream is then diverted to another unit, while the initial unit is vented or flushed at a lower pressure. A significant fraction of the adsorbate desorbs, and the bed in this unit is regenerated and reused following repressurization.

For each of these systems, more than one bed is required for continuous operations to accommodate regeneration of spent material and servicing. The setup required is similar to that shown in Figure 3.7.12.

ADSORBER DESIGN

In adsorber design, a design engineer must have knowledge about adsorbents, the adsorbent operation, and the regeneration operation.

Adsorbents

Favorable properties of an adsorbent include good thermal stability; good mechanical integrity, especially for use in a continuous removal or fluidized-bed system; high surface area and activity; a large capacity for adsorption; and facile regeneration. In some cases, the spent adsorbent must be disposed of, in which case it must be inexpensive and environmentally benign or of commercial value following minor processing.

Activated carbon and the series of microporous aluminosilicates know as zeolites are important adsorbents. Table 3.7.5 presents the properties of the widely used zeolites. Because water vapor is not considered an air pollutant, the adsorption capacity of a zeolite for water is not a benefit.

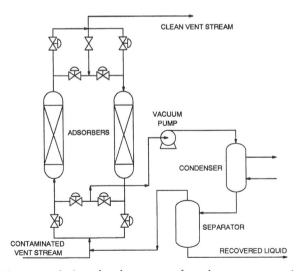

FIG. 3.7.12 Activated carbon system for solvent recovery using modified PSA. (Data from AWD Technologies, Inc., 1993, *Hydrocarbon processing* [August]:76.)

Activated carbon is a highly porous adsorbent with a high capacity for a range of adsorbates (see Table 3.7.6).

Thermal or vacuum desorption permits the recovery of several adsorbates from both zeolites and activated carbon. Thus, these materials are valuable for the recovery of fugitive reagents or solvents. Flushing with water removes inorganic adsorbates such as SO_2 adsorbed on carbon but the carbon must be dried before reuse. Figure 3.7.13 shows a zeolite-based adsorption system for solvent vapors.

Adsorbent Selection

In selecting an adsorbent, the design engineer must ensure that the adsorption isotherm is favorable. If the adsorption isotherm is unfavorable, the loading is low, and the process is rarely economic. Irreversible adsorption should be avoided unless regeneration is not required.

If the feed gas is at an elevated temperature, cooling the gas stream may be required. The cost of cooling is significant and should be included in cost estimates.

Some adsorbates decompose in adsorption or regeneration. Many decomposition products (e.g., from chlorinated hydrocarbons, ketones, and acetates) are acidic and can corrode the system. They also need an additional separation step if the recovered adsorbate is valuable.

Using an adsorbent system for continuous solvent recovery from a vent stream requires using multiple beds. Regeneration of the adsorbent and recovery of the solvent can be accomplished by several means, including PSA (see Figure 3.7.12). Alternatively, a system can recover the solvent by sequentially steam treating the spent adsorbate to remove the adsorbed solvent, drying the carbon with hot air, and cooling the bed to the operating temperature. The number of beds required depends on the effective time-on-stream of the adsorbent bed and the time required to regenerate the carbon. The determination of these parameters is described next. Table 3.7.7 shows typical operating parameters for adsorbers.

Adsorption Operation

The design and operation of an adsorber requires knowledge of the adsorption isotherm, mass transfer, and axial dispersion of adsorbates. Heat transfer is also significant at high pollutant concentrations. Generally, if the feed rate is less than 12 m^3/sec, purchasing an adsorber package from a reputable supplier is more economical than custom building a unit.

Figure 3.7.14 shows the operation of a typical adsorber. For an ideal breakthrough curve, all the pollutant fed in time t_s is adsorbed. During this time period, the concentration on the adsorbent surface increases from the initial value w_o to the saturation (or equilibrium) value w_s. Thus, the following equation applies for a feed gas with linear velocity u_o and concentration c_o:

TABLE 3.7.5   TYPICAL PROPERTIES OF UNION CARBIDE TYPE X MOLECULAR SIEVES

Basic Type	Nominal pore diameter Å	Available from	Bulk density, lb/ft^3	Heat of adsorption (maximum), Btu/lb H_2O capacity, % wt	Equilibrium H_2O capacity % wt	Molecules Adsorbed	Molecules Excluded	Applications
3A	3	Powder $\frac{1}{16}$-in pellets $\frac{1}{8}$-in pellets	30 44 44	1800	23 20 20	Molecules with an effective diameter <3 Å, e.g., including H_2O and NH_3	Molecules with an effective diameter > 3 Å, e.g., ethane	The preferred molecular sieve adsorbent for the commercial dehydration of unsaturated hydrocarbon streams such as cracked gas, propylene, butadiene, and acetylene. Also used for drying polar liquids such as methanol and ethanol.
4A	4	Powder $\frac{1}{16}$-in pellets $\frac{1}{8}$-in pellets 8 × 12 beads 4 × 8 beads 14 × 30 mesh	30 45 45 45 45 44	1800	28.5 22 22 22 22 22	Molecules with an effective diameter < 4 Å, including ethanol, H_2S, CO_2, SO_2, C_2H_4, C_2H_6, and C_3H_6	Molecules with an effective diameter > 4 Å, e.g., propane	The preferred molecular sieve adsorbent for static dehydration in a closed-gas or liquid system. Used as a static desiccant in household refrigeration systems; in packaging of drugs, electronic components, and perishable chemicals; and as a water scavenger in paint and plastic systems. Also used commercially in drying saturated hydrocarbon streams.
5A	5	Powder $\frac{1}{16}$-in pellets $\frac{1}{8}$-in pellets	30 43 43	1800	28 21.5 21.5	Molecules with an effective diameter < 5 Å, including n-C_4H_9OH, n-C_4H_{10}, C_3H_7 to $C_{22}H_{46}$, R-12	Molecules with an effective diameter < 5 Å, e.g., iso compounds and all 4-carbon rings	Separates normal paraffins from branched-chain and cyclic hydrocarbons through a selective adsorption process.
10X	8	Powder $\frac{1}{16}$-in pellets $\frac{1}{8}$-in pellets	30 36 36	1800	36 28 28	Iso paraffins and olefins, C_6H_6, molecules with an effective diameter < 8 Å	Di-n-butylamine and larger	Aromatic hydrocarbon separation.
13X	10	Powder $\frac{1}{16}$-in pellets $\frac{1}{8}$-in pellets 8 × 12 beads 4 × 8 beads 14 × 30 mesh	30 38 38 42 42 38	1800	36 28.5 28.5 28.5 28.5 28.5	Molecules with an effective diameter > 10 Å	Molecules with an effective diameter > 10 Å, e.g., $(C_4F_9)_3N$	Used commercialy for general gas drying, air plant feed purification (simultaneous removal of H_2O and CO_2), and liquid hydrocarbon and natural gas sweetening (H_4S and mercaptan removal)

Source: J.L. Kovach, 1988, Gas-phase adsorption, in *Handbook of separation techniques for chemical engineers* 2d ed. (New York: McGraw-Hill).

TABLE 3.7.6   MAXIMUM CAPACITY OF ACTIVATED CARBON FOR VARIOUS SOLVENTS FROM AIR AT 20°C AND 1 ATM

Adsorbate	Maximum Capacity, kg/kg Carbon
Carbon tetrachloride, CCl_4	0.45
Butyric acid, $C_4H_8O_2$	0.35
Amyl acetate, $C_7H_{14}O_2$	0.34
Toluene, C_7H_8	0.29
Putrescene, $C_4H_{12}N_2$	0.25
Skatole, C_9H_9N	0.25
Ethyl mercaptan, C_2H_6S	0.23
Eucalyptole, $C_{10}H_{18}O$	0.23
Ethyl acetate, $C_4H_5O_2$	0.19
Sulfur dioxide, SO_2	0.10
Acetaldehyde, C_2H_4O	0.07
Methyl chloride, CH_3Cl	0.05
Formaldehyde, $HCHO$	0.03
Chlorine, Cl_2	0.022
Hydrogen sulfide, H_2S	0.014
Ammonia, NH_3	0.013
Ozone, O_3	decomposes to O_2

Source: P.C. Wankat, 1990, *Rate-controlled separations* (London: Elsevier).

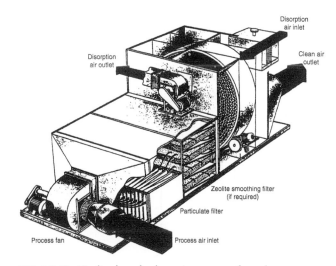

FIG. 3.7.13 Zeolite-based adsorption system for solvent vapors. The large wheel in the middle of the unit adsorbs vapors and is simultaneously regenerated.

In a fixed, cylindrical bed adsorber, the adsorption zone shown in Figure 3.7.14 moves down the column at a nearly constant pattern. The shape and velocity of the adsorption zone are influenced by the adsorption isotherm. For Freundich-type adsorption, in which $c_o = \alpha w_s^\beta$ as expressed in Equation 3.7(15), the following equation gives velocity of the adsorption zone:

$$u_{ad} = \frac{u_o}{\rho_s} (\alpha)^{1/\beta} (c_o)^{\beta - 1/\beta} \qquad 3.7(17)$$

and the following equation expresses the height of the adsorption zone:

$$z = \frac{u_o}{K} \int_0^1 \frac{d\gamma}{\gamma - \gamma^\beta} ; \qquad \gamma = c/c_o \qquad 3.7(18)$$

where K is the overall gas–solid mass-transfer coefficient (s^{-1}) and c is the breakthrough concentration set by the

$$t_s = \frac{LB_s(w_s - w_o)}{u_o c_o} \qquad 3.7(16)$$

where L and P_s are the bed length and the bulk density of the adsorber, respectively. For new and completely regenerated adsorbents, $w_o = 0$.

The actual breakpoint time t is always less than t_s. If the adsorption zone is small compared to the bed length L, most of the adsorbent is used.

In an ideal case of no mass-transfer resistance and no axial dispersion, the adsorption zone becomes zero, i.e., $t = t_s$. However, since mass-transfer resistance always exists, the practical operating capacity is normally 25–50% of the theoretical isotherm value.

TABLE 3.7.7   TYPICAL OPERATING PARAMETERS FOR ADSORBERS

Parameter	Range	Design
Superficial gas velocity	20 to 50 cm/sec (40 to 100 ft/min)	40 cm/sec (80 ft/min)
Adsorbent bed depth*	3 to 10 HAZ	5 HAZ
Adsorption time	0.5 to 8 hr	4 hr
Temperature	−200 to 50°C	
Inlet concentration	100 to 5000 vppm	
Adsorbent particle size	0.5 to 10 mm	4 to 8 mm
Adsorbent void volume	38 to 50%	45%
Steam regeneration temperature	105 to 110°C	
Inert gas regenerant temperature†	100 to 300°C	
Regeneration time	½ adsorption time	
Number of adsorbers	2 to 6	2 to 3

*HAZ is the height of the adsorption zone (see Figure 3.7.1).
†Maximum temperature is 900°C for carbon in nonoxidizing atmosphere and 475°C for molecular sieves.

a) Schematic of adsorption bed in operation.

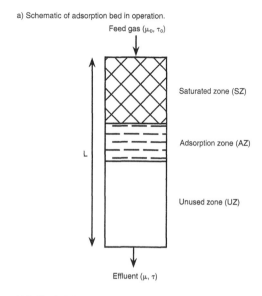

b) Profile of relative concentration of a pollutant through the bed.

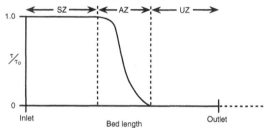

FIG. 3.7.14 Adsorber operation.

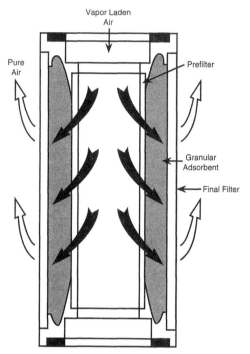

FIG. 3.7.15 Canister adsorption system used to remove trace contaminants from air. (Reprinted, with permission, from Balston, Inc.)

design engineer (kg/m³) to meet the environmental standards. The value of the integral on the right side of Equation 3.7(18) is undefined for limits of 0 and 1. However, taking limits close to these values (e.g., 0.01 and 0.99) defines the integral and solves the equation. The breakthrough time t_β is then as follows:

$$t_\beta = L - \frac{z}{u_{ad}} \qquad 3.7(19)$$

These design equations estimate operating time before breakthrough occurs. The accuracy of the calculations depends on the estimation of K, which is determined by the combination effect of diffusion external to the adsorbent, in the pores and on the pore surface, as well as axial dispersion. Predictions of K values from existing mass-transfer correlations are available but often are unreliable. Thus, adsorber designs are generally based on laboratory data. The scale-up then involves using the same adsorbent size and superficial gas velocity.

For longer beds, the column height of the adsorption zone is a small fraction of the bed length. A longer bed leads to better utilization of the adsorbent. The longer bed also results in a lower degree of backmixing (i.e., higher mass-transfer coefficient). However, proper design is required to avoid an excessive pressure drop. The energy consumption for a blower to overcome a pressure drop is a significant part of the overall operating cost. In the absence of experimental data, the Ergun equation or

equations provided by adsorbent suppliers (e.g., UOP) can provide an estimate. The Ergun equation is as follows:

$$\frac{\Delta P}{L} = \frac{(1 - \varepsilon)G^2}{g_c\varepsilon^3 d_p\rho_G}\left[1.75 + \frac{150(1 - \varepsilon)\mu_G}{d_pG}\right] \qquad 3.7(20)$$

where:

ΔP = pressure drop, lb_f/ft^2
L = bed depth, ft
g_c = gravitational constant, $4.17 \times 10^8 \, lb_m$ = ft/lb_f – h^2
ε = void fraction
ρ_G = gas density, lb_m/ft^3
G = superficial gas velocity, lb_m/ft^2 – h
μ_G = gas viscosity, lb_m/ft^2 – h
d_p = particle diameter, ft

For nonspherical particles, d_p is defined as the equivalent diameter of a sphere having the same specific surface (external area of the particle/bed volume) as the particle.

When the pressure drop across the adsorbent bed must be small, the shape of both the bed and, consequently, the adsorption wave are different from those for a cylindrical packed adsorber. Figure 3.7.15 shows an example of using a canister-type bed for low pressure drop operations.

Regeneration Operation

For continuous operation, when breakthrough occurs, the adsorbent has reached its operating capacity. The adsorption operation must then be performed in another, unsaturated bed. The spent adsorbent is regenerated unless it is used in the treatment of very dilute gases or in specialty applications such as medical uses or gas masks. The most common fluids used in regeneration are hot air and steam. During regeneration, the equations for calculating the velocity and height of the desorption zone can be derived from the Freundich isotherm ($c_R = \alpha_R w_s^\beta R$) similar to Equations 3.7(17), 3.7(18), and 3.7(19) as follows:

$$c_R = \alpha_R \left(\frac{c_o}{\alpha}\right)^{\beta_R/\beta} \qquad 3.7(21)$$

where subscript R refers to regeneration.

Theoretical predictions are not reliable and should be treated as estimates only. Laboratory data may not be available for scale-up because simulating the conditions for a large adiabatic bed is difficult using a small column. For the regeneration of carbon, the reverse flow of steam at 105 to 110°C is often used. The regeneration stops soon after the temperature front reaches the exit, and the typical steam consumption is 0.2 to 0.4 kg/kg carbon. Under these conditions, the carbon is not free of adsorbate. However, prolonging the desorption period is not economic because an excessive amount of steam is required to remove the remaining adsorbate due to the unfavorable shape of the tail of the desorption wave.

If the regeneration time is longer than the adsorption time, three or more beds are required to provide continuous operation.

Chemical Conversion

Removing air pollutants by chemical conversion is used for VOCs, hydrogen sulfide, and nitrogen oxides.

VOCs

When the concentration of VOCs in an air stream must be eliminated or severely reduced, the VOC can be destroyed thermally or catalytically by oxidation. For thermal destruction technologies, see Section 3.8. The VOCs include hydrocarbons (gasoline vapor, solvents, and aromatics), halogenated organics (solvents and vinyl chloride monomer), oxygenates (ketones, esters, and aldehydes), and odorous compounds (amines, mercaptans, and others from the effluent treatment or food processing) (see Table 3.7.8) (Spivey 1987). The oxidation products are carbon dioxide, water, and an acidic component (HX [X = Cl,Br,I]) from halocarbons; SO_x from sulfur compounds; NO_x from amines, nitriles, and nitrogen heterocycles; and P_2O_5 from phosphorus compounds. Excess oxygen ensures

TABLE 3.7.8 EXAMPLES OF VOC DESTRUCTION APPLICATIONS

Sources of Pollution	Types of Pollution	Oxidation Catalysts
Chemicals Manufacture -Plant operations -Petrochemicals -Storage	**Solvents** -Hydrocarbons -Aromatics -Ketones -Esters -Alcohols	**Metals/Supported Metals** Metal =Ag Pt Pd Ru Rh Ni
Chemicals Use -Coating processes -Electronics industry -Furniture manufacture -Painting -Dry cleaning -Paper industry -Wood coating -Printing	**Odor Control** -Amines -Thiols/thioethers -Heterocyclics -Aldehydes -Acids -Food by-products -Smelter off-gases	Support =SiO_2 Al_2O_3 Zeolite ThO_2 ZrO_2 Polymers
Environment Management -Landfill -Hazardous waste handling	**Other Effluent** -Monomers -Plasticizers -CO -Formaldehyde	**Metal Oxides** MgO V_2O_5 Cr_2O_3 MoO_3
Food/Beverage -Breweries -Bakeries -Food processing -Ovens		Pr_2O_3 Fe_2O_3 MnOx NiO TiO_2 Co_3O_4
Energy -Power generation -Stationary engines -Diesel trucks		La_2MO_4 (M = (Cu,Ni)) $CoMoO_4$ Mixed oxides

Notes: For specific applications, one or more catalyst types can be effective. The selection of a particular catalyst system is based on the combination of the VOC to be destroyed and the technology selected.

TABLE 3.7.9.   DESTRUCTIBILITY OF VOC BY CATALYTIC OXIDATION

VOC	Relative Destructibility	Catalytic Ignition Temperature[a] (0°C)
Formaldehyde	High	<30
Methanol	↑	<30
Acetaldehyde		100
Trimethylamine		100
Butanone (Methyl ethyl ketone)		100
n-Hexane		120
Phenol		150
Toluene		150–180
Acetic acid		200
Acetone		200
Propane	Low	250–280
(Chlorinated hydrocarbons)		(400)

Source: Data from F. Nakajima, 1991, Air pollution control with catalysis—Past, present, and future, *Catalysis Today* 10:1.

Note: [a]Catalyst: Pd/activated alumina.

that oxidation is complete and that partial oxidation products (which can be pollutants) are not formed. Scrubbing with a mild basic aqueous solution removes the acidic component in the tail streams.

The oxidation of a VOC is exothermic. However, because a VOC is present only as a dilute component, the reaction heat is insufficient to maintain the temperature required for oxidation. Therefore, additional heating is required, or a catalyst must be used that enables total oxidation at lower temperatures (see Table 3.7.8).

Most catalytic operations also require elevated temperatures to ensure a complete reaction at a suitably fast rate. Therefore, either the catalyst bed is heated, or the feed stream is preheated. Inlet gas streams are kept 50–150°C higher than the ignition temperatures (Nakajima 1991). When the preheating uses an internal flame, the process is complex, involving both the products of combustion in the flame and processes occurring at the catalyst. The operating parameters (temperature and space velocity) for an application depend on the destructability of the VOC (see Table 3.7.9).

In a typical application, more than one VOC is destroyed. Frequently, oxidation at a catalyst is competitive under stoichiometric or partial oxidation conditions, or one VOC inhibits catalytic oxidation of another. Using excess oxygen in air overcomes these problems. A range of parameters applies in catalytic oxidation for VOC destruction. These parameters depend on the application (Spivey 1987).

Mechanisms

The oxidation of a VOC (S) at a catalyst can involve a species at the surface and in the vapor phase. The Langmuir–Hinshelwood mechanism (Scheme 1) requires the adsorption of each species at nearby sites and subsequent reaction and desorption.

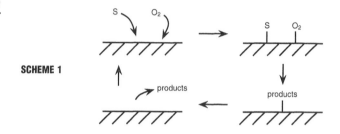

SCHEME 1

The Mars–van Krevelen mechanism of catalytic oxidations (Scheme 2) explicitly requires a redox process in which oxygen is consumed from the catalyst surface by reaction with the VOC and then is replenished by oxygen from the vapor phase.

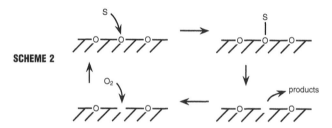

SCHEME 2

The Eley–Rideal mechanism (Scheme 3) is similar to the Mars–van Krevelen mechanism except that the products are formed from adsorbed oxygen and the VOC in the gas phase.

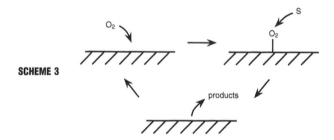

SCHEME 3

As these schemes show, for metal or nonreducible oxide catalysts, excess oxygen in the gas phase means that the catalyst surface is well-covered with oxygen and that little if any VOC is adsorbed. Thus, the Eley–Rideal mechanism is expected to be important. For metal oxide catalysts containing readily reducible metals, the Mars–van Krevelen mechanism is important.

Metal oxides that are n-type semiconductors are rich in electrons and are generally not highly active as oxidation catalysts. Vanadium pentoxide is the notable exception. In contrast, p-type semiconductors are conductive because of the electron flow into positive holes. The electron-deficient surfaces of such metal oxides readily adsorb oxygen, and if the adsorption is not too strong, they are active catalysts. Insulators, which are inexpensive and not friable or thermally unsuitable, have values as supports for more expensive, catalytically active metal oxides or noble metals.

For all mechanisms, a key factor is the strength of the interaction between the surface and the oxygen (atom, molecule, or ion) required for oxidation of the VOC. If the oxygen is too tightly bound to a surface, that surface is not highly active as a catalyst. Similarly, if the interaction is too weak, the surface coverage with oxygen is low, and the catalytic activity is consequently diminished. Various thermodynamic properties are considered as the parameter that best represents the strength of adsorption. For metals, the initial heat of oxygen adsorption is a reasonable choice (Bond 1987). For metal oxides, the reaction enthalpy for reoxidation of the used catalyst (Mars–van Krevelen mechanism) is considered the most representative (Satterfield 1991). The maximum rate for VOC oxidation over an oxide catalyst is estimated to occur when the reaction enthalpy for reoxidation of the catalyst is one-half of the reaction enthalpy for total oxidation of the VOC. Thus, the selection of a catalyst for an application depends strongly on the nature of the VOC to be destroyed, and the conditions depend on the destructabilities of the VOC (see Table 3.7.9).

Complete oxidation of each VOC in a multicomponent stream is ensured with high temperatures and excess oxygen, and mixed-oxide catalysts are frequently used. The mixed oxides, especially when promoted with alkali or alkaline earth metal oxides, frequently have activities that are different from the combination of properties of the components, and in general the activities are higher. This phenomenon probably arises from two factors: the availability and mobility of the different forms of available oxygen and the accessibility of different binding sites with various energy levels.

More than one type of surface oxygen species can be involved: adsorbed dioxygen (O_2), ions (O^{2-}; O_2^{2-}), or radical ions (O^-; O_2^-) on the surface or incorporated into the lattice of the catalyst. Sachtler (1970) and Sokolovskii (1990) review the roles of various forms of adsorbed oxygen.

Kinetics

The following equations summarize the steps required for destruction of a VOC by an Eley–Rideal mechanism:

$$O_2 + [\ ] \longrightarrow [O_2] \qquad 3.7(22)$$

$$[O_2] + [\ ] \longrightarrow 2[O] \qquad 3.7(23)$$

$$VOC + [O] \longrightarrow [S_a]_i + H_2O \qquad 3.7(24)$$

$$[S_a]_i + [O] \longrightarrow CO_2 + [\ ] \qquad 3.7(25)$$

where [ ] represents a surface site and $[S_a]_i$ represents the ith in a series of partially oxidized species S_a at the surface of the catalyst. The following rate expression is derived from the preceding mechanism:

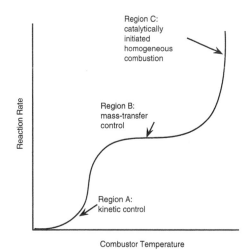

FIG. 3.7.16 Overall reaction rate. (Reprinted, with permission, from D.L. Trimm, 1991, Catalytic combustion, Chap. 3 in *Studies in inorganic chemistry 1991* 11:60.)

$$r = \frac{k_a P_{O_2} P_{VOC}}{k_b P_{O_2} + v k_c P_{VOC}} \qquad 3.7(26)$$

where k_a, k_b, and k_c are constants and v is the stoichiometric coefficient of oxygen in the overall oxidation reaction of the VOC. Under conditions of excess oxygen where $k_b P_{O_2} \gg v k_c P_{VOC}$, this equation reduces to the following approximate form:

$$r \cong k P_{VOC} \qquad 3.7(27)$$

For most applications, the kinetics are described by the following fractional power expression:

$$r = k_b P_{O_2}^a P_{VOC}^b \qquad 3.7(28)$$

in which a and b are fractional coefficients with values close to zero and unity, respectively.

The oxidation of a VOC can occur at the catalyst surface and in the gas phase. The overall reaction rate is the sum of these two components and is a strong function of temperature (see Figure 3.7.16) (Prasad 1984, Trimm 1991). The catalytic oxidation of hydrocarbons over supported metal catalysts is thought to occur via dissociative chemisorption of the VOC, followed by reaction with co-adsorbed oxygen at the surface and then desorption of the combustion products (Chu and Windawi 1996). The rate determining step in this Langmuir–Hinshelwood type of mechanism is hydrogen abstraction from the VOC. Thus the ease of oxidation of the VOC is directly related to the strength of the C–H bond. Methane is more difficult to oxidize than other paraffins, aromatics, or olefins, and oxygenates are relatively easier to oxidize.

Reactors

The reactor's heat requirement heat arises mainly from the need to preheat the inlet gases or to heat the catalyst bed. For efficient operation of a catalytic oxidation system, the exhaust heat must be recovered and used to preheat the feed, as shown in Figure 3.7.17 for a system manufactured

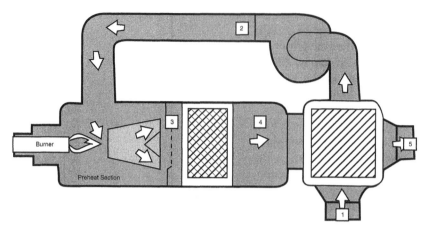

FIG. 3.7.17 Schematic diagram of catalytic process with burner. (Reprinted, with permission, from Salem Engelhard.)

by Salem Engelhard (South Lyon, Mich.). The residual heat is recovered by a secondary heat exchanger and used for area heating or other purposes requiring low-grade energy.

In some cases, destruction of all VOCs and intermediates requires a higher operating temperature than that required to combust a single VOC (see Figure 3.7.18).

The waste gases from several chemical, printing, or related industries contain mixtures of halogenated and non-halogenated VOCs. Converting each VOC requires a combination of catalysts. Further, scrubbing the effluent from the reactor is necessary to remove the acidic components generated. As an example of such a system, the catalytic solvent abatement (CSA) process designed and marketed by Tebodin V.B. is shown in Figure 3.7.19.

The heat for a catalytic incinerator can be applied directly to the catalyst bed, rather than from a burner. The swingtherm system (see Figure 3.7.20) manufactured by Mo-Do Chemetics, Ltd. (Vancouver, Canada and Ornskolsvik, Sweden) uses dual beds at a temperature of 300–350°C. The air stream is fed at 60°C and is heated by contacting a ceramic at 320°C. It then passes through a platinum-based catalyst. The exothermal VOC oxidation raises the temperature of the gases to 350°C. The gases then pass through the other bed to heat the ceramic rings; the effluent gases are cooled to 90°C. When the second reactor is warmed to 320°C, the inflow is reversed. The flow reversal occurs every 2–5 min. For VOC concentrations below 300 ppm, auxiliary heating is required.

HYDROGEN SULFIDE

Large sources of hydrogen sulfide (H_2S) are treated as a resource from which sulfur is recovered. The normal recovery is to oxidize a part of the H_2S to sulfur dioxide (SO_2) and then react H_2S and SO_2 over an alumina-based catalyst in the Claus process as follows:

$$2H_2S + SO_2 \longrightarrow 3S + 2H_2O \qquad 3.7(29)$$

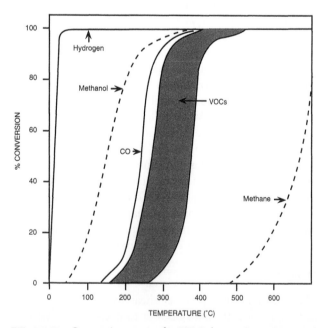

FIG. 3.7.18 Conversion curves for VOC destruction using a catalytic incinerator. (Data from Brown Engineering, Seattle, Wash; Johnson Matthey, Catalytic Systems Division, Wayne, Pa.)

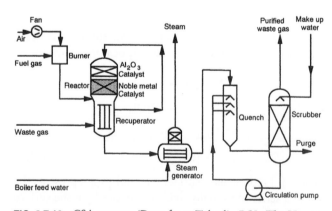

FIG. 3.7.19 CSA system. (Data from Tebodin B.V., The Hague, the Netherlands.)

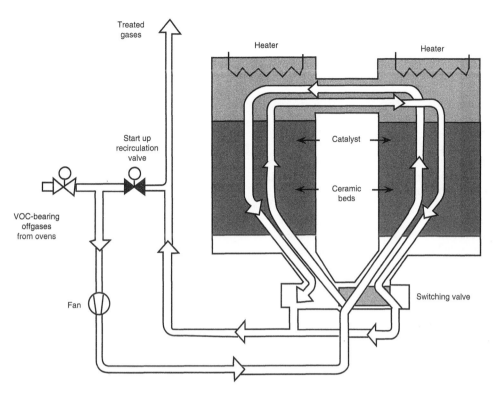

FIG. 3.7.20 Swingtherm process. (Reprinted, with permission, from *Chemical Engineering*, 1993 [October]:153.)

H_2S is initially removed from a source, such as sour natural gas, by dissolution in an alkanolamine solvent. However, for low concentrations of H_2S in air, this method is not sufficiently effective for total removal of H_2S (Kohl and Riesenfeld 1985). Instead, the H_2S is either oxidized completely to SO_2, and then the system removes the SO_2 by scrubbing or capture, or it is captured by reaction with solid or liquid-phase adsorbents.

For stationary units, any liquid or solid base reacts with H_2S. Magnetite (Fe_3O_4), limestone ($CaCO_3$), lime (CaO), zinc carbonate ($ZnCO_3$), and zinc oxide (ZnO) are each effective, and Fe_3O_4, CaO, and ZnO are used commercially.

Table 3.7.10 lists the major adsorbents for H_2S. The main advantage of using these solutions is to reduce high H_2S concentrations in gas streams. Table 3.7.11 lists the advantages and disadvantages of these adsorbents. Using solid adsorbents or dry conversion technology is frequently necessary to remove H_2S at the ppm level. Table 3.7.10 also shows examples of dry oxidation technologies. The use of solid adsorbents is expensive; therefore, for air streams containing large amounts of H_2S, both scrubbing and adsorption are used in sequence.

The market for H_2S is small compared to that for sulfur or sulfuric acid. Thus, regeneration of H_2S removed by adsorption from air streams is performed only occasionally. Several proven technologies are available for the dry oxidation of H_2S to sulfur.

Dry Oxidation

The dry oxidation process using iron oxide is inexpensive but not pleasant or simple to operate. The following equation empirically gives the suitable box sizes (Steere Engineering Co.):

$$A = \frac{GS}{3000(D + C)} \qquad 3.7(30)$$

where:

A   = cross-section through which gas passes on its way through any one box in the series
G   = maximum amount of gas to be purified, scf/hr
S   = a correction factor for the hydrogen sulfide content of the inlet gas
D   = the total depth of the oxide, ft through which the gas passes consecutively in the purifier set. In boxes with split flow where half the gas passes through each layer, the area exposed to the gas is twice the cross-sectional area of the box, while D is the depth of one layer of oxide.
C   = the factor: 4 for 2 boxes, 8 for 3 boxes, and 10 for 4 boxes, respectively
3000 = a constant

From Equation 3.7(30), the following equation gives the maximum amount of gas which can be purified by a unit:

$$G = \frac{3000(D + C)A}{S} \qquad 3.7(31)$$

TABLE 3.7.10 ADSORBENTS FOR HYDROGEN SULFIDE

Liquid Adsorbents	Trade or Generic Names	Source	Comments
Alkanolamines:			
Monoethanolamine	MEA	Dow, U.S.A.	Moderately selective
Diethanolamine	DEA	Dow, U.S.A.	Moderately selective
Triethanolamine	TEA	Dow, U.S.A.	Selective
N-methydiethanolamine	MDEA	Dow, U.S.A.	Selective
Diisopropanolamine	DIPA	Dow, U.S.A.	Selective
Aqueous potassium carbonate	Hot Carbonate		Moderately selective
Aqueous tripotassium phosphate			Selective
Alkazid process			
Dialkylglycine sodium salt	DIK Solution		Selective
Sodium alanine solution	M Solution		Moderately selective
Sodium phenolate solution	S Solution		Not selective
Aqueous Ammonia			
+Air/hydroquinone	Perox		Recovers as sulfur
+Air	Am-gas	Am-gas, Canada	Recovers as $(NH_4)_2SO_4$
Aqueous Sodium Carbonate			
+Quinoline sulfonate salt and sodium metavanadate	Stretford	W.C. Holmes and Company, U.K.	Recovers as sulfur
+Naphthaquinone sulfonate salt	Takahax	Tokyo Gas Co. and Nittetsu Chemical Engineers Co.	Recovers as sulfur
Aqueous zinc acetate			Precipitates ZnS
Aqueous calcium acetate			$Ca(SH)_2$ soluble;
+Heat to decompose product			CaS precipitates
Iron Oxide	Dry Box		Sulfur recovery, for
+Controlled reoxidation			ppm H_2S streams
Zinc Oxide, ZnO			Useful for ppm H_2S; only mod. selective
Calcium oxide, CaO	Lime		Product releases H_2S on decomp. oxidizes to sulfate
Copper, Cu (or other metals)			Reagent, not adsorbent; forms CuS and H_2, but requires expensive regeneration

Table 3.7.12 tabulates the values of S. The values for the factor C are based on the number of boxes in the series being used at the time of operation and on the assumption that the flow is reversed during purification.

The designer must use the appropriate size of iron oxide particles for the required pressure drop across the bed. The empirical relationship was developed for one system (Prasad 1984) and can be used as an approximation for similar systems:

$$\Delta P = 3.14d^{-0.61} \qquad 3.7(32)$$

Perry (1984) recommends an allowable pressure drop of 1–2 psi/ft at 800–1000 psig and linear gas velocities of 5–10 ft/min. Typical operating conditions are shown in Table 3.7.13.

Activated Carbon Process

Activated carbon is an effective catalyst for the oxidative conversion of H_2S in hydrocarbon gas streams to elemental sulfur under mild conditions. I.G. Farbenindustrie developed the first process to exploit this capability. Such a unit may process up to 200,000 cu ft/hr of water–gas, with a pressure drop of 25 in of water.

Reactions of H_2S and SO_2

The Sulfreen process designed by Lurgi (Germany) and Elf Aquitaine (France) reduces residual sulfur compounds in tail gases from Claus plants for sulfur recovery from natural gas. Alumina or carbon catalysts

TABLE 3.7.11   TECHNICAL ADVANTAGES AND DISADVANTAGES FOR AMINES USED AS H₂S ADSORBENTS

Amine	Advantages	Disadvantages
MEA	Known technology Low cost (approximately $1.6/kg) Simple regeneration. 20–30% solution has low freezing point (to −50°C).	Some degradation by CO_2, COS, and CS_2 requires use of make-up. May need addition of foaming prevention agents. Some corrosion effects.
DEA	Known technology. Low cost. Simple regeneration. Better H_2S selectivity than MEA.	Some degradation. Slightly higher cost than MEA. May need foam prevention agent.
TEA	Known technology. Low cost (approximately $1.4/kg). Simple regeneration. Highly selective to H_2S. Less degradation. 40% solution has reasonable freezing point (−32°C).	Approximately twice as expensive as MEA, based on adsorption capacity. Minimum (fluid) operating temperature is higher than for MEA.
MDEA or DIPA	Highest selectivity to H_2S. Lowest regeneration temperature; hence energy savings. Lower heat of reaction Lower corrosion effects. Low vaporization losses. Not degraded by COS or CS_2. Higher initial costs are offset by reduced make-up costs.	Higher initial cost for amine. If an MDEA or DIPA regeneration plant is not nearby, a regeneration unit is required.

are used, and efficiencies as high as 90% can be attained for streams containing 1.50% H_2S and 0.75% SO_2. Similar technologies are used in the cold-bed adsorption process, developed by AMOCO Canada Petroleum Company Ltd., and the MCRC sulfur recovery process, licensed by Delta Engineering Corporation (USA). Sulfur recoveries up to 99% are attainable for each of these processes.

TABLE 3.7.12   CORRECTION FACTOR S FOR THE HYDROGEN SULFIDE CONTENT OF THE INLET GAS

Grains H₂S/100 scf of Unpurified Gas	Factor
1000 or more[a]	720
900	700
800	675
700	640
600	600
500	560
400	525
300	500
200 or less	480

[a]1000 grains/100 scf is 22.9 grams/cu m

Oxidation to Oxides of Sulfur

An alternative strategy in sulfur recovery is the oxidation to oxides of sulfur. For treating air with low concentrations of H_2S, this strategy does not require finding a market for small amounts of sulfur produced, and it can remove any adsorbed products from the catalyst bed thermally or by flushing with water.

The Katasulf process (Germany) for cleaning gas streams for domestic use can operate in various configurations. Depending on the pollutants and their concentrations in the feed gas, a prewasher may be necessary in addition to the main gas washer for SO_2 removal from gases exiting the catalyst chamber. Other applications use side-stream or split-stream arrangements. In each case, the catalyst is activated carbon, alumina, or a combination of two of three metals (iron, nickel, and copper) and operates close to 400°C. The Katasulf process is effective for H_2S and NH_3 removal and can be operated to reduce HCN and organic sulfur compounds.

NITROGEN OXIDES

The major stationary sources of nitrogen oxide (NO_x) emissions come from the combustion of fossil fuels. The

TABLE 3.7.13 TYPICAL OPERATING CONDITIONS OF IRON OXIDE PURIFIERS

Variable Condition	Type of Purifier				
	Conventional Boxes	Deep Boxes	High-Pressure	Tower Purifiers	Continuous
Gas volume treated, millions of cubic feet (mmcf)/day	6.0	4.3	15.0	24.0	2.0
Hydrogen sulfide content, gr/100 scf	1000	740	10	500–950	1000
Pressure, psig	low	40*	325	low	
Number of units in series (boxes or towers)	5	6	4†	6§	3
Cross-sectional area per unit, sq ft	960	1200	24	776	71
Number of layers per unit	4	1	1	28	1
Depth of layer, ft	2.25; 1.50	4	10	1.4	40
Temperature: °F:					
In	60	73		85	
Out	70	93		100	
Space velocity, cu ft/(hr) (cu ft)	7.15	6.66	37.4‡	5.38	9.4
R ratio	35.6	40	112‡	32	28

Source: Data from A.L. Kohl and F.C. Risenfeld 1985, *Gas purification,* 4th ed. (Houston: Gulf Publishing Company).
Notes: *Inches of water.
†Two series of four units, three units in operation in each series.
‡At 325 psig.
§Two series of six towers.

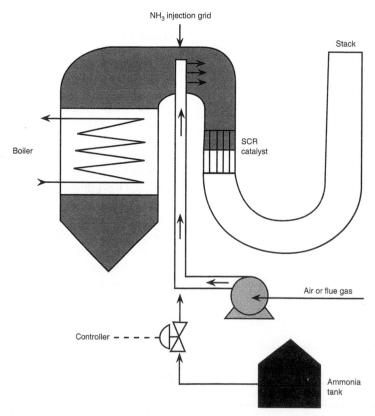

FIG. 3.7.21 SCR of NO_x.

TABLE 3.7.14 TECHNOLOGY MANUFACTURERS AND SUPPLIERS

Company	Process	Application	Number of Installations	Date of Last Installation
VOC Abatement				
AWD Technologies, Inc., subsidiary of Dow Chemical Co.	Modified PSA	Modified PSA process captures VOCs from variable concentration vent emission streams.	20	1993
Calgon Carbon Corp.	Granular activated-carbon adsorption	Granular activated-carbon adsorbs VOCs from air and other vapor streams.	100	NA
Callidus Technology, Inc.	Vacuum-regenerated activated-carbon vapor recovery	Vacuum-regenerated activated-carbon vapor recovery system controls VOC vapors from loading operations.	2	NA
Jaeger Products, Inc.	Vent gas adsorption	Process removes small amounts of acid, basic, or organic vapors into a liquid by absorption with and without chemical reaction.	100	1993
Membrane Technology and Research Inc.	Membrane vapor separation for VOC recovery	Membrane separation with proven condensation and compression techniques removes select organic compounds.	14	1993
Chlorinated Hydrocarbon Abatement				
Tebodin B.V., Consultants and Engineers	Catalytic solvent abatement process	Catalytic solvent abatement process treats exhaust gases containing chlorinated hydrocarbons and is suitable for PVC facility.	1	1991
Thermatrix	Flameless thermal oxidation	A packed-bed reactor destroys hazardous organic fumes by oxidation. This process can be used on chlorinated hydrocarbons also.	10	1992
Vara International, division of Calgon Carbon Corp.	VOC abatement process	Combination of fixed-carbon-bed adsorption and thermal oxidation concentrates and oxidizes low-VOC concentration streams.	6	NA
Catalytic VOC Oxidation				
CSM Environmental Systems, Inc.	Catalytic oxidation	Catalytic oxidation system handles phthalic anhydride and pure terephthalic acid exhaust streams.	20	NA
Haldor Topsøe Inc.	Catalytic solvent and VOC abatement	This catalytic process removes VOCs and solvents from exhaust air. The catalyst is poison-resistant and trouble-free.	134	1993
Catalytic NO$_x$ Reduction				
Engelhard Corp.	SCR of NO$_x$	SCR process controls NO$_x$ emissions with ammonia injection. Cost: \$20–90/kw. Temperature: 290–595°C.	25	NA
Lurgi AG	Catalytic conversion of NO$_x$	Catalytic conversion of NO$_x$ and SO$_x$ uses ammonia injection. Highly efficient removal of SO$_2$.	NA	NA
Research–Cottrell Co.	SCR	Ammonia injection controls and reduces NO$_x$ emissions.	NA	NA

Continued on next page

TABLE 3.7.14 *Continued*

Company	Process	Application	Number of Installations	Date of Last Installation
SNCR of NO$_x$				
Shell	Catalytic NO$_x$ reduction	Low temperature catalyst can achieve 90% reduction of NO$_x$.	4	1993
Exxon Research and Engineering Co.	Thermal NO$_x$ reduction	Reduces NO$_x$ emissions in flue gas streams with aqueous and anhydrous ammonia and has 80% efficiency. Temperature: 700–1100°C.	130	1993
Lurgi AG	SNCR of NO$_x$	Noncatalytic process reduces NO$_x$ compounds using aqueous ammonia or urea.	NA	NA
Nalco Fuel Tech	SNCR of NO$_x$	SNCR of NO$_x$ from stationary combustion sources uses stabilized aqueous urea. Cost: $500–1500/tn NO$_x$. Temperature: 815–1100°C.	70	NA
Research–Cottrell/Nalco Fuel Tech	SNCR of NO$_x$	Process uses controlled urea injection and chemicals to reduce NO$_x$ emissions.	70	NA
Thermal NO$_x$ Reduction				
ABB Stal	Dry low NO$_x$ combustion	System reduces NO$_x$ emissions for gas turbines.	4	NA
Catalytic SO$_x$ Reduction				
Haldor Topsøe A/S	Catalytic SO$_x$ and NO$_x$ removal	Catalytic process removes SO$_2$ and NO$_x$ from flue gases and offgases. NO$_x$ is reduced by ammonia to N$_2$ and H$_2$O.	29	1993
Noncatalytic SO$_x$ Reduction				
Exxon Research and Engineering Co.	Wet gas scrubbing process	Process removes particulates and SO$_x$ from FCC unit. Simple system has economic and operating advantages over other control systems.	14	1992
Lurgi AG	Wet gas scrubbing	Dust, noxious gases, and heavy metals are scrubbed from flue gas and recovered as a slurry form.	NA	NA
Incineration				
NAO Inc.	Thermal incineration of VOCs	Thermal oxidizer destroys VOC emissions with 99.9% efficiency and can recover valuable thermal energy.	NA	NA
Praxair, Inc.	Oxygen combustion	Process uses high-velocity oxygen jets to replace air that recirculates organics within the incinerator.	10	1993
Other				
Edwards Engineering Corp.	Hydrocarbon/solvent vapor recovery system	Hydrocarbon and solvent vapor recovery system, based on the Rankine refrigeration cycle, condenses vapors to liquids.	340	NA
Institut Français du Pétrole/Babcock Enterprises	Clean combustion of heavy fuel oil	AUDE boiler burns heavy fuel oil and petroleum residues cleanly and meets European directives on atmospheric discharges.	2	1993

Source: Data from *Hydrocarbon Processing*, 1993 (August):75, 102.
Notes: NA = not applicable.

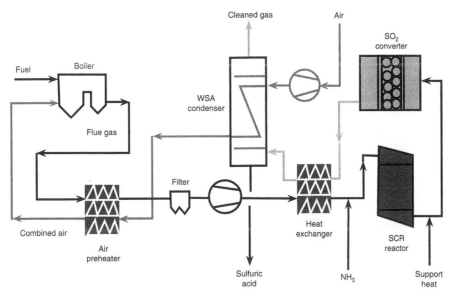

FIG. 3.7.22 SNOX process for combined treatment of NO_x and SO_x.

flue gas usually contains 2–6% oxygen and a few hundred ppm of NO_x, which consists of 90–95% NO and 5–10% NO_2. The only viable postcombustion process for NO_x emission control is chemical conversion to N_2 by a reducing agent. Several gases can be used for this purpose including methane, hydrogen, carbon monoxide, and ammonia. However, because the flue gas contains much more than O_2 than NO_x, the reducing agent must selectively react with NO_x rather than O_2 to minimize its consumption (Environment Canada: Task Force Report 1989; Nakatsuji 1991). To date, many commercial installations based on the selective reduction of NO_x by NH_3 can remove about 80% of NO_x in the flue gas. The governing equations for the ammonia-based technology are as follows:

$$4NO + 4NH_3 + O_2 \longrightarrow 4N_2 + 6H_2O \qquad 3.7(33)$$

$$2NO_2 + 4NH_3 + O_2 \longrightarrow 3N_2 + 6H_2O \qquad 3.7(34)$$

$$4NH_3 + 5O_2 \longrightarrow 4NO + 6H_2O \qquad 3.7(35)$$

The first two reactions dominate when the flue gas is heated to about 1000°C. Above 1100°C, Equation 3.7(35) becomes significant leading to the unwanted formation of NO. If the temperature is below 800°C, the reaction rate is too low for practical use. Thus, the process is tempera-

TABLE 3.7.15 CATALYST SUPPLIERS

Supplier	Materials (partial list)
Corning Corp. Corning, NY	Ceramic supports
Johnson Matthey, Catalytic Systems Div. Wayne, PA	Noble metals
W.R. Grace & Co., TEC Systems Div. DePere, WI.	Supports, metals/compounds, supported metals, other
W.R. Grace & Co., Davison Chemical Div. Baltimore, MD	Oxides, supported oxides, other
UOP Des Plaines, IL	Zeolites, adsorbents
Engelhard Corporation Iselin, NJ	Noble metals, supported metals/oxides
Norton Chemical Process Products Corp. Akron, OH	NO_x reduction catalysts

Notes: This list is not comprehensive. The authors have attempted to provide current information but recognize that data may have changed. Inclusion of a company in this list is not an endorsement of that company's products and should not be construed as such.

ture sensitive. With the use of a suitable catalyst, the NO_x reduction can be carried out at 300–400°C, a temperature normally available in a flue gas system. This process is called Selective Catalytic Reduction (SCR). Figure 3.7.21 is a schematic diagram for the SCR process. In this process, at least 1% O_2 should be present in the flue gas, thus it is suitable for boiler and furnace applications. Several simple catalytic and noncatalytic systems are in commercial operation at several sites (see Table 3.7.14).

Under ideal reaction conditions, one mole of NH_3 is required to convert one mole of NO; however, in practice, one mole of NH_3 reduces about 0.8–0.97 moles of NO_x. The excess NH_3 is required because of the side reaction with O_2 and incomplete mixing of the ammonia with the flue gas. Much of the ammonia slip ends up in the fly ash, and the odor can become a problem when the ash is sent to a landfill or sold to cement plants.

The SCR processes are simple, requiring only a proper catalyst and an ammonia injection system. The catalysts currently in use are TiO_2-based which can be mixed with vanadium or molybdenum and tungsten oxides. These catalysts eliminate the formation of ammonium bisulfate which can plug the downstream equipment, a problem in earlier SCR systems.

A new Shell de-NO_x catalyst comprises vanadium and titanium, in high oxidation states, impregnated onto silica with a high surface area (300 m²/g). The data for a commercial and semicommercial operation indicate consistent performance for periods up to one year (Groeneveld et al. 1988). The catalyst is deactivated by high concentrations of SO_2, but the poisoning is reversible with heating. A version of this process is installed to control dust-containing flue gas. This process employs a parallel passage system in which the flue gas permeates through the catalyst separating into passages for feed and purified streams.

The Shell technology is being developed for other applications. In particular, a lateral flow reactor is being developed which demonstrates a low pressure drop. The design of this reactor should allow for convenient installation and maintenance (Groeneveld et al. 1988).

Flue gases, especially from coal-burning boiler units or power generation, contain both NO_x and SO_x, with fly ash and metal-containing particulates. Processes have been developed that convert the NO_x to nitrogen, which is vented, and the SO_x to sulfuric acid, which is removed by scrubbing. Figure 3.7.22 is a schematic of the SNOX process (Haldor Topsøe A/S, Denmark). This process recovers up to 95% of the sulfur in the SO_x as sulfuric acid and reduces 95% of the NO_x to free nitrogen. All fly ash and metals are essentially captured.

Table 3.7.15 is a recent list of catalyst suppliers, and Table 3.7.14 lists the manufacturers of environmental technologies (Environmental Processes '93 1993).

—Karl T. Chuang
Alan R. Sanger

References

Bond, G.C. 1987. *Heterogeneous catalysis: Principles and application.* 2d ed. Oxford: Oxford University Press.

Chu, W., and H. Windawi. 1996. *Chem. Eng. Progress.* (March): 37.

Environment Canada: Task Force Report 1989. *Development of a national nitrogen oxide (NO_x) and volatile organic compounds (VOC) management plan for Canada.* (July).

Environmental processes '93. 1993. *Hydrocarbon Processing* 72, no. 8 (August):67.

Fair, J.R. 1987. Distillation. In *Handbook of separation process technology.* New York: John Wiley.

Groeneveld, M.J., G. Boxhoorn, H.P.C.E. Kuiper, P.F.A. van Grinsven, H. Gierman, and P.L. Zuideveld. 1988. Preparation, characterization and testing of new V/Ti/SiO$_2$ catalysts for DeNO$_x$ing and evaluation of shell catalyst S-995. *Proc. 9th Int. Congr. Catalysis,* edited by M. Ternan and M.J. Phillips, vol. 4: 1743. Ottawa: Chemical Institute of Canada.

Kast, W. 1981. Adsorption from the gas phase—Fundamentals and processes. *Ger. Chem. Eng.* 4:265.

Kohl, A.L. 1987. Absorption and stripping. In *Handbook of separation process technology.* New York: John Wiley.

Kohl, A.L., and F.C. Riesenfeld 1985. *Gas purification.* 4th ed. Houston, Gulf.

Langmuir, I. 1921. The mechanism of the catalytic action of platinum in the reactions $2CO + O_2 = 2CO_2$ and $2H_2 + O_2 = 2H_2O$. *Trans. Faraday Soc.* 17:621.

McCabe, W.L., J.C. Smith, and P. Harriott. 1993. Unit operations of chemical engineering. 5th ed. New York: McGraw-Hill.

Nakajima, F. 1991. Air pollution control with catalysis—Past, present and future. *Catalysis Today.* 10:1.

Nakatsuji, T., and A. Miyamoto. 1991. Removal technology for nitrogen oxides and sulfur oxides from exhaust gas. *Catalysis Today* 10:21.

Onda, K.H., H. Takeuchi, and Y. Okumoto. 1968. Mass transfer coefficients between gas and liquid phases in packed columns. *J. Chem. Eng.* (Japan) Vol. 1, no. 1:56.

Perry, R.H., and D.W. Green. 1984. *Perry's chemical engineers' handbook.* 6th ed. New York: McGraw-Hill.

Prasad, R., L.A. Kennedy, and E. Ruckenstein. 1984. *Catalytic combustion. Catal. Rev.—Sci. Eng.* 26:1.

Sachtler, W.H.M. 1970. The mechanism of catalytic oxidation of some organic molecules. *Catal. Rev.—Sci. Eng.* 4:27.

Satterfield, C.N. 1991. Heterogeneous catalysis in practice. 2d ed. New York: McGraw-Hill.

Sokolovskii, V.D. 1990. Principles of oxidative catalysis on solid oxides. *Catal. Rev.—Sci. Eng.* 32:1.

Spivey, J.J. 1987. Complete catalytic oxidation of volatile organics. *Ind. Eng. Chem. Res.* 26:2165.

Strigle, R.F. 1987. *Random packings and packed towers.* Houston: Gulf.

Trimm, D.L. 1991. Catalytic combustion. Chap. 3 in *Studies in inorganic chemistry 1991* 11:60.

Yang, R.T. 1987. *Gas separation by adsorption processes.* Boston: Butterworths.

3.8
GASEOUS EMISSION CONTROL: THERMAL DESTRUCTION

This section addresses the thermal destruction of gaseous wastes. These wastes are predominantly VOC-containing. The principal methods are thermal combustion and incineration and flaring, although a number of new technologies are emerging. First, this section provides an overview of these methods including major technological and some cost considerations. Then, it cites several examples of industries that employ these technologies to achieve destruction and removal efficiencies (DREs) in excess of 98% and low products of incomplete combustion (PIC) emissions.

Next this section reviews the thermodynamic and kinetic fundamentals that form the basis of thermal incineration and illustrates the principles with two sample calculations. Most of this section is devoted to the design considerations of thermal incinerators; the remaining section is allocated to flares and the emerging technologies.

Overview of Thermal Destruction

The principal methods for thermal destruction include thermal combustion and incineration, flaring, and other emerging technologies.

THERMAL COMBUSTION AND INCINERATION

Thermal combustion and incineration is the principal approach used for VOCs (usually expressed in concentration terms) but is equally applicable to liquids and solids with sufficient heat content. This technique is different from catalytic destruction, which is discussed in Section 3.7. From a global mass balance and energy balance point of view, the design and implementation of a thermal incinerator is straightforward. Thermal efficiencies can be estimated closely with CO_2 and H_2O as the principal products. For dilute streams, achieving the required temperature may require auxiliary fuel.

In the area of PIC-formation, global approaches become unworkable. Environmental engineers may have to use detailed measurement techniques to verify emission types and levels not only from the combustion process but also for the pollution control equipment.

Thermal destruction is simplest when (at least theoretically) CO_2 and H_2O are the expected products. The process becomes more complicated when hetero atoms (i.e., N, S, and Cl) or inorganics are involved, especially if

corrosive products (i.e., HCl and SO_2) are formed. The dividing line between gaseous and liquid wastes is not always sharp because liquid wastes can contain many species of high-vapor pressure and gaseous wastes can carry liquid residues in droplet form.

The main variables controlling the efficiency of a combustion process are temperature, time, and turbulence; the three Ts of combustion. At a constant combustion chamber temperature, the DRE, defined in terms of the following equation:

$$DRE\ \% = \frac{VOC_{in} - VOC_{out}}{VOC_{in}} \times 100 \qquad 3.8(1)$$

increases with residence time; increasing the temperature increases the DRE at a constant time. At efficient combustion temperatures, the rate can become mixing limited.

Modern thermal oxidation systems can accomplish +99% DRE for capacities ranging from 1000–500,000 cfm and VOC concentrations of 100–2000 ppmv. Typical residence times are 1 sec or less at temperatures of 1300–1800°F. Inlet VOC concentrations above 25% of the lower explosion limit (LEL) are generally avoided due to the potential explosion hazards. Temperatures near 1800°F and long residence times can lead to elevated nitrogen oxide levels, which may have to be controlled separately (if lowering the combustion temperature is not feasible).

Thermal incinerators are usually coupled to two types of thermal energy recovery systems: regenerative and recuperative. Both methods transfer the heat content of the combustion exhaust gas stream to the incoming gas stream (Ruddy and Carroll 1993). In a regenerative system, an inert material (such as a dense ceramic) removes heat from the gases exiting the furnace. Such a ceramic storage bed eventually approaches the temperature in the combustor, consequently reducing the heat transfer. Therefore, the hot exhaust stream contacts a cooler bed, while the incoming gas stream passes through the hot bed.

As shown in Figure 3.8.1, the VOC-laden gas stream enters bed #1 which warms this gas stream by transferring heat from a previous cycle. Some VOCs are destroyed here, but most of them are oxidized in the combustion chamber. The flue gases from this combustion exit through bed #2 and transfer most of their enthalpy in the process. Within seconds of a heating and cooling cycle, the beds are switched, and the incoming stream now enters bed #2. Consequently, a near steady-state operation is approached,

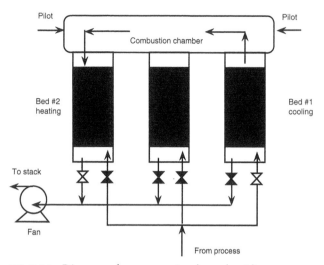

FIG. 3.8.1 Diagram of a regenerative thermal oxidizer.

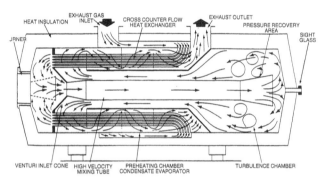

FIG. 3.8.2 Recuperation-type afterburner.

and more heat can be recovered from these systems than from a typical thermal incinerator (about 70%). Using multiple beds can lead to heat recoveries to 95%. The need for any auxiliary fuel depends on the potential thermal energy of the VOC-laden stream.

Recuperative thermal oxidation systems typically use a shell-and-tube design heat exchanger to recover heat from the flue gases for heating the incoming gases. Operating temperatures are reached quickly in these systems. Recuperative heat exchange is also more suited to cyclic operations and variations in VOC feed rates and concentrations. Figure 3.8.2 is a schematic of such a system.

Table 3.8.1 summarizes the emission sources, VOC categories, and typical operating parameters (flow rates and concentrations), as well as the cost aspects of thermal VOC control technologies.

FLARING

Flaring is the process of disposing of industrial and other combustible waste gas streams via a visible flame (flare). The flame can be enclosed in a chimney or stack.

Flares are widely used for the disposal of waste gases from several industrial processes, including process start ups, shutdowns, and emergencies. These processes are characterized by variable or intermittent flow. Flares are principally used where the heating value cannot be economically recovered. Applications include petroleum production, blast furnace and coke oven combustible gases, industrial chemical production, and landfills. Refinery

TABLE 3.8.1   THERMAL VOC CONTROL TECHNOLOGIES

Technology	Emission Source	VOC Category	Emission Rate Volumetric (scfm)	VOC (ppmv)
Thermal oxidation[a]	PV, ST TO, WW	AHC, HHC A, K	<20,000 w, w/o HR ≥20,000 w HR	20–1000 w/o HR ≥1,000 w HR
Flaring	PV, ST, TO, WW, F	AHC, A, K		

Source: E.C. Moretti and N. Mukhopadhyay, 1993, VOC control: Current practices and future trends, *Chem. Eng. Progress* 89, no. 7:20–26.
Notes: Users of VOC control technology project (1993) that about 43% of their capital expenditures will be in thermal oxidizers.
[a]Removal efficiencies   95–99+%
Capital Costs $10–200/cfm (recuperative)
    $30–450/cfm (regenerative)
Annual Operating Costs $15–90/cfm (recuperative)
    $20–150/cfm (regenerative)
Up to 95% energy recovery is possible
Not recommended for batch operations
Key:
PV = process vents
ST = storage tanks
TO = transfer operations
WW = wastewater operations
F = fugitive
AHC = aliphatic and aromatic HCs
HHC = halogenated hydrocarbons
A = alcohols, glycol ethers, ethers, epoxides, and phenols
K = ketones and aldehydes
HR = heat recovery

TABLE 3.8.2 EMERGING TECHNOLOGIES

Molten salt oxidation
Molten metal catalytic extraction
Molten glass
Plasma systems
Corona discharge

flares are typically elevated and steam-assisted. Flares can be open or enclosed and burn with a type of diffusion flame. Auxiliary fuel may have to be added to support flare combustion.

EMERGING TECHNOLOGIES

Numerous emerging technologies have been used or proposed for the thermal destruction of wastes, more so for liquid and solid wastes than for gaseous wastes. The reason is clear: most gaseous wastes that are candidates for thermal destruction carry large quantities of inert diluents (CO_2, H_2O, and N_2), whereas some of these emerging technologies are energy intensive because they operate at high temperatures.

Some of these technologies have been available for decades and will probably not find widespread application. However, some technologies combine thermal destruction with the formation of a useful product. Consequently, product formation enhances the advantages of thermal incineration, in which energy recovery is the chief economic benefit.

Preconcentration (adsorption and desorption or absorption and desorption) can be a suitable pretreatment to one of these techniques, though rarely practiced. Also, numerous emerging technologies are not thermal in nature. Table 3.8.2 lists several emerging technologies based primarily on the availability of practical and theoretical information but not restricted to gaseous wastes.

Source Examples

Hundreds of processes emitting VOCs are candidates for thermal destruction. This section presents several diverse examples. The *Air Pollution Engineering Manual* (1992) provides more detail on process descriptions and emissions.

PETROLEUM INDUSTRY

The major segments of this industry are exploration and production, transportation, refining, and marketing. Each of these segments consists of facilities and processes that emit VOCs. These emissions can be handled by a variety of thermal destruction techniques, principally flares, incinerators, and boilers.

The refining step includes numerous processes and potential emission sources, such as crude separation, light hydrocarbon processing, middle and heavy distillate processing, residual hydrocarbon processing, and auxiliary processes. Emissions are classified as process point and process and area fugitive emissions (pumps, valves, fittings, and compressors). Fugitive emissions can be a substantial fraction. Table 3.8.3 summarizes estimates of major aromatic hydrocarbons and butadiene emissions from various point and fugitive sources.

Thermal incineration controls continuous VOC emissions; this method is preferred (to flaring) when hetero atoms are in the VOCs, such as Cl and S that lead to corrosive products. The application of thermal incineration to processes of varying flows and concentrations are more problematical. A large safety factor must be employed (not to exceed 25% of the LEL). On the other hand, if the concentrations are too low, auxiliary fuel is required.

CHEMICAL WOOD PULPING

The major gaseous emissions in this industry are odorous, total reduced sulfur (TRS) compounds, characteristic of Kraft pulp mills. The principal components are hydrogen sulfide (H_2S), methyl mercaptan (CH_3SH), dimethyl sulfide (($CH_3)_2S$), and dimethyl disulfide (($CH_3)_2S_2$). In addition, emissions of noncondensible gases such as acetone and methanol are common. Emissions from uncontrolled sources, such as digester and evaporator, relief, and blow

TABLE 3.8.3 ESTIMATED EMISSIONS FROM PETROLEUM REFINERY

Chemical	Point, tn/yr	Fugitive, tn/yr	Total, tn/yr
Benzene	114	29	142
Toluene	437	111	548
Xylene (total)	31	1751	1782
Butadiene	3	1	4
Trimethyl benzene (1,2,4)	310	141	452

Source: Air Pollution Management Association, 1992, *Air pollution engineering manual,* edited by A.J. Buonicore and W.T. Davis (New York: Van Nostrand Reinhold).

gases, are expressed in lb/tn of air-dried pulp. Emissions also arise from black liquor oxidation tower vents.

Kraft pulp mills that began construction or modification after September 24, 1976 are subject to the new source performance standards (NSPS) for particulate matter and TRS emissions. Also, in 1979, the U.S. EPA issued retrofit emission guidelines to control TRS emissions at existing facilities not subject to the NSPS.

Today, combustion controls most major and minor sources of TRS emissions (National Council of the Paper Industry for Air and Stream Improvement 1985). Most commonly, existing combustors, such as power boilers and lime kilns, are used as well as specifically dedicated incinerators. The principal oxidation product is sulfur dioxide; a caustic scrubber is often installed after the incinerator to neutralize that gas. Two types of noncondensible gases are produced by the Kraft pulping process: low volume, high concentration (LVHC) and high volume, low concentration (HVLC). The latter can be burnt only in a boiler capable of accepting such large gas volumes without disrupting the unit's efficiency.

LANDFILL GAS EMISSIONS

The major component of landfill gas is methane; less than 1% (by volume) consists of nonmethane organic compounds. Air toxics detected in landfills include such compounds as benzene, chlorobenzene, chloroform, TSRs, tetrachloroethylene and toluene, and xylenes (Air and Waste Management Association 1992). Landfill gas is generated by chemical and biological processes on municipal solid waste (MSW). The gas generation rate is affected by parameters such as the type and composition of the waste, the fraction of biodegradable materials, the age of the waste, the moisture content and pH, and the temperature. Anaerobic decomposition can produce internal temperatures to 37°C (98.6°F); gas production rates are highest for moisture contents of 60–78%.

Control measures are usually based on containment combined with venting and collection systems. Low permeability solids for cover and slurry walls reduce the landfill gas movement. Capping is the process that uses a cover soil of low permeability and low porosity.

Gases can be vented or collected. Collection systems consist of several vertical and/or horizontal recovery wells that collect and convey gas within the landfill via a piping header system to a thermal oxidizer such as a flare. Blowers or compressors are used for this purpose. Gas collection efficiencies for active landfills with capping and gas collection can reach 90%.

The principal thermal destruction method for landfill gas is flaring. Emissions from landfill gas flares are a function of the gas flow rates, the concentration of the combustible component (which determines the temperature), and the residence time required. These emissions are esti-

mated from combustion calculations; they are generally not measured.

RENDERING PLANTS

Thermal destruction methods are often applied to odorous emissions. Rendering plants are prime examples of an odorous emission. In these plants, animal and poultry byproducts are processed to produce fallow, grease, and protein meals. Batch and continuous processes are used. Other sources (National Council of the Paper Industry for Air and Stream Improvement 1985; Prokop 1985) describe these processes in detail.

In a batch cooker system (the basic rendering process), cookers are charged with raw material; a cook is made under controlled time, temperature, and pressure conditions; the cooked material is discharged; and the cycle is repeated. Under continuous conditions, the raw material is charged to the cooker.

The principal odorous emissions from these processes are N- and S-containing organic compounds, which are listed in Table 3.8.4. Additional compounds include higher molecular weight organic acids, pyrazines, alcohols, and ketones. These compounds are mostly noncondensibles and arise under the cooking conditions ($\sim$220°F). The type of raw material and its age have a significant effect on the odor intensity. Continuous systems tend to be enclosed and have a greater capability of confining odors. In the batch process, the steam rate varies between 450–900 ft^3 min^{-1}. Steam is subsequently condensed and cooled to below 120°F. Odorous noncondensibles range in odor intensity from 5000–10^6 odor units/scf; the volumetric emission rate of noncondensibles varies between 25–75 ft^3 min^{-1}.

Since rendering plants use boilers for steam generation and drying, odor control by incineration usually uses the existing boilers. The following factors should be considered when boiler incineration of odor-intense effluents is implemented:

Excess air at odor pickup points should be avoided
If possible, the odor-containing stream should be used as primary combustion air
Moisture and particle concentrations should be low
High-intensity odors must contact the furnace flame
Sufficient residence time at T > 1200°F must be provided in existing boilers.

In addition, the use of an existing boiler for odor incineration must conform to engineering, safety, and insurance requirements. Sections 5.1 and 5.2 provide more information on odor and its control.

Combustion Chemistry

Thermal incineration is based on combustion chemistry including stoichiometry and kinetics. These fundamentals and sample calculations are described next.

TABLE 3.8.4 ODOROUS COMPOUNDS IN RENDERING PLANT EMISSIONS

Compound Name	Formula	Molecular Weight	Detection Threshold (ppm, v/v)	Recognition (ppm, v/v)
Acetaldehyde	CH_3CHO	44	0.067	0.21
Ammonia	NH_3	17	17	37
Butyric acid	C_3H_7COOH	88	0.0005	0.001
Dimethyl amine	$(CH_3)_2NH$	45	0.34	—
Dimethyl sulfide	$(CH_3)_2S$	62	0.001	0.001
Dimethyl disulfide	CH_3SSCH_3	94	0.008	0.008
Ethyl amine	$C_2H_5NH_2$	45	0.27	1.7
Ethyl mercaptan	C_2H_5SH	62	0.0003	0.001
Hydrogen sulfide	H_2S	34	0.0005	0.0047
Indole	$C_6H_4(CH)_2NH$	117	0.0001	—
Methyl amine	CH_3NH_2	31	4.7	—
Methyl mercaptan	CH_3SH	48	0.0005	0.0010
Skatole	C_9H_9N	131	0.001	0.050
Trimethyl amine	$(CH_3)_3N$	59	0.0004	—

Source: Air Pollution Management Association, 1992.

STOICHIOMETRY

The starting basis of thermal incineration is the complete combustion of a hydrocarbon to carbon dioxide and water in air as follows:

$$C_xH_y + \left(x + \frac{y}{4}\right)O_2 + \left(x + \frac{y}{4}\right)3.78\ N_2 \longrightarrow$$

$$x\ CO_2 + \frac{y}{2}H_2O + \left(x + \frac{y}{4}\right)3.78\ N_2 \quad 3.8(2)$$

This equation accounts for all major atoms. If the VOC contains Cl, S, or N in appreciable amounts, these components must be accounted for in the stoichiometry; usually HCl, SO_2, and NO are combustion products. If the components of a mixture are known, equations can be written for each species; if the components are not known, an apparent chemical formula can be based on the weight percents of each combustible element (e.g., C, H, N, and S).

Any oxygen in VOCs (alcohols and ketones) is subtracted from the stoichiometric oxygen requirements. The formation of thermal NO (from nitrogen in the air) is not accounted for in stoichiometric combustion equations; only when nitrogen is present in the fuel–VOC mixture is NO formation (from fuel or VOC N) accounted for stoichiometrically.

For dilute gas streams, achieving high-DRE and low-PIC emissions (not accounted for in the combustion stoichiometry) requires auxiliary fuel to maintain a minimum temperature and residence time.

Using excess air is common practice. This use is expressed in terms of the air/fuel ratio (mass based) or the equivalence ratio, defined as ϕ as follows:

$$\phi = \frac{(\text{VOC/oxygen})_{\text{actual}}}{(\text{VOC/oxygen})_{\text{stoichio.}}} \quad 3.8(3)$$

SAMPLE CALCULATION #1

A hazardous chlorinated hydrocarbon is combusted in air according to the following stoichiometry:

$$C_xH_yCl_z + \left(x + \frac{y-z}{4}\right)O_2$$

$$\longrightarrow x\ CO_2 + \frac{y-z}{2}H_2O + z\ HCl$$

Calculate the stoichiometric HCl mole fraction for $x = 6$, $y = 3$, and $z = 2$. If the combustion is conducted in a way that leads to 4 oxygen mole % in the stack gas, what is the equivalence ratio; if, under these conditions, the DRE is 99.99%, what is the emitted $C_6H_3Cl_2$ concentration?

SOLUTION

$$C_6H_3Cl_2 + \left(6 + \frac{3-2}{4}\right)O_2 + \left(6 + \frac{3-2}{4}\right)3.78\ N_2 \longrightarrow$$

$$6CO_2 + \frac{3-2}{2}H_2O + 2HCl + 6.25 \times 3.78\ N_2$$

Total moles in stack gas: 32.125; mole fraction HCl = 0.062

If E = excess oxygen, then by stoichiometry, 6.25E oxygen is in the product gases. Therefore, the following equation applies:

$$0.04 = \frac{6.25E}{32.125 + 6.25E + 6.25 \times 3.78E}$$

Solving, E = 0.254; then

$$(\text{oxygen})_{\text{actual}} = 6.25 + 6.25 \times 0.254 = 7.84$$

$$\phi = \frac{1/7.84}{1/6} = 0.765$$

1 mole $C_6H_3Cl_2$ in $\rightarrow$ 0.0001 mole $C_6H_3Cl_2$
Total moles out = 32.125 + 7.589 = 39.714

$$\frac{0.0001}{39.714} \times 10^6 = 2.52 \text{ ppm}$$

KINETICS

Stoichiometry cannot account for finite DRE and PIC concentrations. Global kinetics can address only the former. Finite DRE can be expressed as a fractional conversion as follows:

$$\ln \frac{C_{vocf}}{C_{voci}} = -kt = \ln(1 - DRE) \qquad 3.8(4)$$

where k is a (pseudo) first-order rate coefficient s^{-1}. The product of rate and residence time determines the conversion (i.e., a high DRE is achieved only by a sufficiently high rate [high temperature] and sufficient residence time).

This presentation implies a plug-flow reactor and irreversibility; the latter is usually a good assumption in combustion reactions. Some evidence exists that a modified model incorporating an ignition-delay time gives a better fit to experimental data (Lee et al. 1982).

In Equation 3.8(4), the dimensionless group $-kt$ presents the natural log of the unreacted fraction. Conversion increases rapidly with an increase in kt. This product implies a batch reactor. For a flow reactor (plug flow), t is replaced by $\bar{t}$, the mean residence time, defined as V/Q (reactor volume divided by the volumetric flow rate at a constant density) or $L/\bar{u}$ for tubular reactors (L = reactor/combustor length; $\bar{u}$ = average velocity).

Some reported data shows evidence that continuously stirred tank reactors (CSTR), or perfect mixer, behavior occurs at high conversion (Hemsath and Suhey 1974). The following equation expresses this behavior:

$$\frac{C_{vocf}}{C_{voci}} = \left(\frac{1}{1 + k\bar{t}}\right) \qquad 3.8(5)$$

Composite behavior is sometimes an explanation for the observed results, i.e., a model consisting of a plug-flow reactor followed by one or more CSTRs.

In all combustion reactions, some CO is always formed. The following two-step global model accounts for this formation:

$$VOC \xrightarrow{k_1} CO \xrightarrow{k_2} CO_2 \qquad 3.8(6)$$

where, assuming excess oxygen, the following equations apply:

$$r_{VOC} = -k_1 (VOC) \qquad 3.8(7)$$

$$r_{CO} = -k_1 (VOC) - k_2 (CO) \qquad 3.8(8)$$

$$r_{CO_2} = k_2 (CO) \qquad 3.8(9)$$

Qualitatively, the resulting concentration dependence is of the form shown in Figure 3.8.3. Under typical incineration conditions, the preceding chemical sequence is irre-

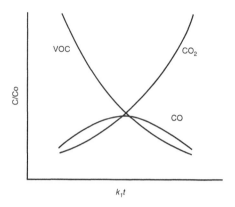

key:
c = Concentration at time t or exit concentration
C_0 = Initial concentration
k_1t = Product of rate coefficient × time, where t can be taken as the residence time at a given temperature

FIG. 3.8.3 Qualitative relationship between VOC, CO, and CO_2 concentration as a function of k_1t.

versible. Using C_{VOC} (t = 0) = $C_{VOC,i}$ and $C_{CO,i} = C_{CO_2,i}$ = 0, the solution of Equations 3.8(7) to 3.8(9) is as follows:

$$\frac{C_{VOC,t}}{C_{VOC,i}} = \exp(-k_1t) \qquad 3.8(10)$$

$$\frac{C_{CO,t}}{C_{VOC,i}} = \frac{1}{1 - \frac{k_2}{k_1}} (\exp(-k_2t) - \exp(-k_1t)) \qquad 3.8(11)$$

$$\frac{C_{CO_2,t}}{C_{VOC,i}} = 1 - \exp(-k_1t)$$

$$- \frac{1}{1 - \frac{k_2}{k_1}} (\exp(-k_2t) - \exp(-k_1t)) \qquad 3.8(12)$$

This scheme can account for variable levels of CO in the exit gases, depending on the VOCs, temperature, and residence time. This analysis does not account for other PICs; usually when CO is low, other PICs are low also. Since the rate coefficients k_1 and k_2 are the Arrhenius-type and are strongly temperature-dependent, temperature is the critical variable.

The incineration of VOCs requires a minimum residence time at a specified temperature to achieve a certain DRE. Given the global rate expression of Equation 3.8(2) and a pseudo-first-order rate coefficient $k_1 = A \exp(-E/RT)$, where E = 45 Kcal $mole^{-1}$, A = 1.5×10^{11} s^{-1}, R = gas constant = 2×10^{-3} Kcal $mole^{-1}$ K^{-1}, and T = temperature K, would 1 sec at 1000 K (1341°F) be sufficient to achieve a DRE of 99.99%?

$$k_1 = 1.5 \times 10^{11} \exp\left(-\frac{45}{2 \times 10^{-3} \times 10^3}\right) = 25.4 \text{ s}^{-1}$$

SOLUTION

For 1 sec, $k_1 t = 25.4$ and $\exp(-25.4) << 0.0001$; therefore, 1 sec is sufficient to achieve a DRE of 99.99%.

Design Considerations

This section discusses the design considerations for thermal incinerators, flares, and other emerging technologies.

THERMAL INCINERATORS

Katari et al. (1987a and b) summarize incineration techniques for VOC emissions. Particularly useful are a table on the categorization of waste gas streams (see Table 3.8.5) and a flow chart that determines the suitability of a waste gas stream for incineration and the need for auxiliaries. The categorization includes the % oxygen, VOC content vis-a-vis LEL, and heat content. Thus, a mixture of VOC and inert gas with zero or a negligible amount of oxygen (air) and a heat content >100 Btu scf^{-1} (3.7 MJ m^{-3}) can be used as a fuel mixed with sufficient oxygen for combustion (see category 5 in Table 3.8.5).

Figure 3.8.4 is a schematic of the incineration system. Waste gas from the process, auxiliary fuel (if needed), and combustion air (if needed) are combined in the combustion chamber under conditions (i.e., time, temperature, and turbulence) to achieve minimum conversion (DRE >99%). The temperature inside the combustion chamber should be well above the ignition temperature (1000–1400°F) of most VOCs; residence times of 0.3–1.0 sec may be sufficient. If higher DREs are required (>99.99%), both residence time and temperature may have to be increased depending on the VOC composition. This increase is also necessary the more nonuniform (in terms of VOC components) the waste gas stream is.

The majority of industrial waste gases for thermal destruction fall under category 1 of Table 3.8.5. The following parameters must be quantified (see Figure 3.8.4):

Heat content of the waste gas and Tp from process (Tp 5 Tw, if no heat exchanger is used)
The % LEL (VOC content and types)
The waste gas volumetric flow rate T_w
TE required based on the necessary DRE

The nature of the VOCs in the waste gas stream determine the temperature (T_E) at the exit of the combustor (not equal to the adiabatic flame temperature).

The composition of the waste gases determines the combustion air requirements; required temperatures and flow rates determine the auxiliary fuel requirements, furnace chamber size, and heat exchanger capacity. The suggested temperatures are 1800°F for 99% DREs at approximately 1 sec residence time. Applicable incinerator types include liquid injection, rotary kiln, fixed-hearth, and fluidized-beds (of which several variants exist). Dempsey and Oppelt (1993) describe these units in more detail. Although these authors principally address hazardous waste, their article

TABLE 3.8.5 CATEGORIZATION OF WASTE GAS STREAMS

Category	Waste Gas Composition	Auxiliaries and Other Requirements
1	Mixture of VOC, air, and inert gas with >16% O$_2$ and a VOC content <25% LEL (i.e., heat content <13 Btu/ft^3)	Auxiliary fuel is required. No auxiliary air is required.
2	Mixture of VOC, air, and inert gas with >16% O$_2$ and a VOC content between 25 and 50% LEL (i.e., heat content between 13 and 26 Btu/ft^3)	Dilution air is required to lower the heat content to <13 Btu/ft^3. (Alternative to dilution air is installation of LEL monitors.)
3	Mixture of VOC, air, and inert gas with <16% O$_2$	This waste stream requires the same treatment as categories 1 and 2 except the portions of the waste gas used for fuel burning must be augmented with outside air to bring its O$_2$ content to about 16%.
4	Mixture of VOC and inert gas with zero to negligible amount of O$_2$ (air) and <100 Btu/ft^3 heat content	This waste stream requires direct oxidation with a sufficient amount of air.
5	Mixture of VOC and inert gas with zero to negligible amount of O$_2$ (air) and >100 Btu/ft^3 heat	This waste stream requires premixing and use as a fuel.
6	Mixture of VOC and inert gas with zero to negligible amount of O$_2$ and heat content insufficient to raise the waste gas to the combustion temperature	Auxiliary fuel and combustion air for both the waste gas VOC and fuel are required.

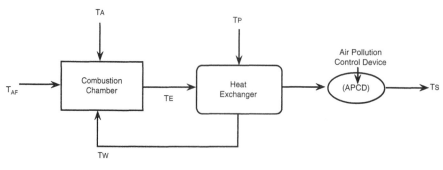

Temperature:

T_P = Waste gas from process
T_W = Waste gas leaving heat exchanger or entering combustion chamber
T_{AF} = Auxiliary fuel
T_A = Combustion air
T_E = Flue gas
T_S = Flue gas leaving heat exchanger or stack temperature

FIG. 3.8.4 Schematic diagram of incinerator system.

is a comprehensive review that includes regulatory aspects, current practice, technology, emissions and their measurements, control parameters, performance indicators, and risk assessment.

In thermal incinerators, T_w is usually below 1000–1100°F (to avoid preignition); as T_w increases, auxiliary fuel requirements can decrease. For a system with a recuperative heat exchanger, T_w can be calculated as follows:

$$T_w = T_p + \eta \, (T_E - T_p) \qquad 3.8(13)$$

where η represents the efficiency of the heat exchanger (see Figure 3.8.2). Standard heat transfer texts provide the equations for estimating η as well as the requisite material and energy balances.

If the wastes to be destroyed contain chlorine, higher temperatures may be required, and APCD needs increase, specifically the necessity to control HCl emissions. In addition, the PIC mix is more complex when organochlorine compounds are present and may require the application of Appendix VIII compound sampling techniques. Dempsey and Oppelt summarize these methods in a table.

CEMs are often used (or are required) for combustion gas components such as CO, CO_2, O_2, NO_x, and THC, one or more of which serves as a performance indicator. Again, if chlorinated compounds are combusted, continuous monitoring of HCl can be necessary. The methodology is the same as for CO and CO_2, namely NDIR. Nitrogen-containing compounds can form NO_x, S-containing compounds lead to SO_2 and perhaps some SO_3, P-containing compounds lead to P_2O_5 (a highly corrosive compound), and Br-, F-, and I-compounds form the corresponding acids.

Gaseous hazardous waste can lead to higher molecular weight PICs; for example, the combustion of methyl chloride (while yielding mainly CO_2 and HCl) can also lead to species such as chloroethanes and chlorobenzenes.

FLARES

Figure 3.8.5 shows an example of an enclosed flare, as used in landfill gas disposal. A multiple-head, gas burner is mounted at ground level inside a refractory-lined, combustion chamber that is open at the top. Enclosed flares have better combustion efficiencies than elevated flares, and emission testing is more readily performed. Typical minimum performance parameters include 0.3 sec residence time at 1000°F and CO < 100 ppm. Most refinery flares are elevated and steam assisted (see Figure 3.8.6). The steam promotes turbulence, and the induction of air into the flare improves combustion. The amount of steam required depends partly on the C/H ratio of the VOCs to be destroyed; a high ratio requires more steam to prevent a smoking flare.

U.S. EPA studies (Joseph et al. 1983; McCrillis 1988) identify several parameters important to flare design: flare head design, flare exit velocity, VOC heating value, and whether the flame is assisted by steam or air. They acquired the data from a specially constructed flare test facility. For a given flare head design, they found that the flame stability (the limit of flame stability is reached when the flame propagation speed is exceeded by the gas velocity) is a function of the fuel (VOC), gas velocity at the flare tip, and the lower heating value (LHV) of the fuel. For a given velocity, a minimum LHV is required for stable combustion. A combustion efficiency of >98% is maintained as long as the ratio of the LHV to the minimum LHV required for stable combustion is >1.2. Below that value, the DRE drops rapidly.

Gas (VOC) composition has a major influence on the stability limit. Figure 3.8.7 illustrates this influence with the results from the EPA flare test facility. Particularly interesting is the difference between methyl chloride (MeCl), propane (Prop), and butadiene (But) at a given exit velocity. The figure shows that with most gases, an increase in the flare exit velocity must be accompanied by an in-

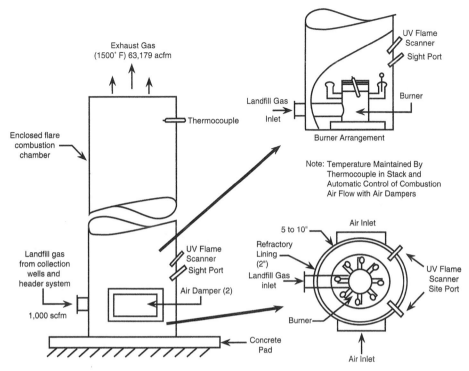

FIG. 3.8.5 Typical enclosed flare. (Adapted from John Zinc Inc., General arrangement drawing model ZTOF enclosed flare.)

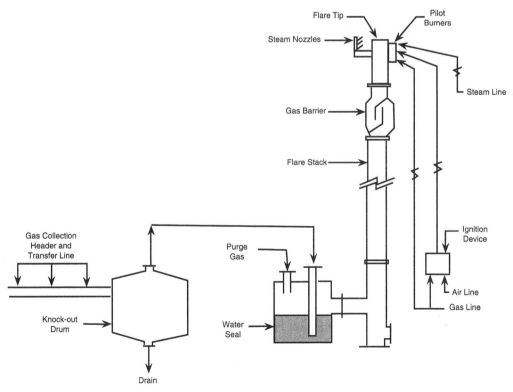

FIG. 3.8.6 Steam-assisted elevated flare system.

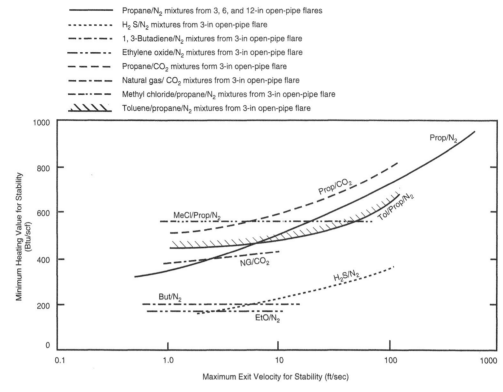

FIG. 3.8.7 Flame stability limits for different gas mixtures flared without pilot assist from a 3-in open-pipe nozzle.

crease in the heating value of the waste to maintain a stable flame. For Cl-containing compounds, the flame stability should correlate with the H/Cl ratio.

Stone et al. (1992) outline design procedures for flares. The U.S. EPA requirements for flares are specified in 40 *CFR* Section 60.18. The requirements are for steam-assisted, air-assisted, and nonassisted flares. For steam-assisted, elevated flares, the following points must be addressed:

An exit velocity at the flare tip <60 ft/sec (for 300 Btu/scf gas streams) and <40 ft/sec (for >1000 Btu/scf gas streams). Between 300–1000 Btu/scf, the following equation applies:

$$\log V_{max} = \frac{NHV + 1214}{852} \qquad 3.8(14)$$

where V_{max} = the maximum permitted velocity and NHV = the net heating value (BTU/scf)

Absence of visible emissions (5 min exception period for any 2 consecutive hr)

Presence of a flame at all times when venting occurs

NHV not less than 300 Btu/scf.

EMERGING TECHNOLOGIES

This section discusses the design considerations for emerging technologies including molten salt oxidation, molten metal reactions, molten glass, plasma systems, and corona destruction.

Molten Salt Oxidation

Molten salt technology is an old technology. The process combines combustible wastes and air in a molten salt batch (which can be a single component such as sodium carbonate or a mixture) within a molten salt reactor, usually constructed of ceramic or steel. Typical temperatures are 1500–1900°F and several seconds of residence time for the gas phase.

An attractive feature is the heating value of the fuel which should be sufficient to maintain the salt medium in the molten state. Consequently, this technology is best applied to combustible liquid and solid wastes, rather than gaseous wastes, as the latter are usually too dilute. Another attractive aspect is the neutralization of acidic species, such as HCl and SO_2, which form from Cl- or S-containing wastes.

Molten Metal Reactions

Much higher temperatures are possible for a molten metal bath (primarily iron), up to 3000°F. This technology works best for wastes that are low in oxygen, leading to an off-gas composed primarily of CO and H_2. A molten metal bath requires large heat input by induction; the resulting temperatures are high enough to break all chemical bounds, leading to extremely high DREs. Any downstream emissions are usually the result of recombination reactions rather than incomplete destruction.

This process is most advantageous for concentrated liquid and solid wastes. Dilute gaseous wastes place high demands on the energy inputs.

A potential advantage for both molten salt and molten metal technology is their application to wastes containing quantities of toxic metals. However, both technologies require extensive downstream particle control.

Molten Glass

This technology has been championed by Penberthy Eletromelt International, Seattle, Washington. It is based on charging waste (combustible or noncombustible) continuously to a pool of molten glass in an electric furnace. This approach is primarily attractive for ash-forming wastes; these ashes are melted into the glass, and the resulting composites are stable (resistant to leaching).

Although gaseous wastes do not contribute ash, the temperatures in these furnaces are sufficiently high (>2300°F) to achieve high DRE and low PIC levels. The residence time for the molten glass phase is long (hours), but the residence time for the gas phase is much shorter.

Plasma Systems

A plasma incinerator burns waste in a pressurized stream of preheated oxygen. Because temperatures can reach 5000°F, applying this technology to dilute streams is prohibitive. Such a system consists of a refractory-lined, water-cooled preheater, where a fuel is burned to heat oxygen to about 1800°F. This preheater is followed by the combustion chamber, where the waste and oxygen are mixed and auto-ignition occurs. From here, the gases pass into a residence chamber, where the destruction is completed at maximum temperature, followed by a quench chamber, a scrubber to remove acid gases, and a stack.

At plasma temperatures, the degree of dissociation of molecules is high, consequently reactivity increases; reaction rates are much higher than at normal incinerator temperatures. The high combustion efficiencies that are achievable lead to a compact design. High heat recovery is also possible. Thus, these systems lend themselves to a portable design.

Corona Destruction

The EPA has been researching VOC and air toxic destruction since 1988 (Nunez et al. 1993). This work features a fixed-bed packed with high dielectric-constant pellets, such as barium titanate, that are energized by an AC voltage applied through stainless steel plates at each end of the bed. The destruction efficiencies for VOCs such as

benzene, cyclohexane, ethanol, hexane, hexene, methane, methylene chloride, methyl ethyl ketone, styrene, and toluene is predicted from the ionization potential and bond type for these compounds. The EPA studies did not report PICs; DREs ranged from 15% (methane) to ~100% (toluene) for concentrations of 50–250 ppmv. One advantage of this process is that it operates at ambient temperatures and appears not to be sensitive to poisoning by S and Cl compounds, as catalytic systems are.

A cost comparison with carbon adsorption and catalytic and thermal incineration is favorable. However, the results presented were based on a bench-scale system only; scaleup is clearly required for further engineering evaluation.

Conclusions

For the foreseeable future, thermal oxidation remains the principal methodology whereby emissions of gaseous and particle-bound pollutants are minimized and heat is recovered. The combustion and incineration approach is primarily viable for dilute streams where selective recovery is difficult and economically infeasible. Proposed expenditures in this technology continue to be high (see Table 3.8.1).

Where wastes are well characterized and contain no chlorine, few problems are encountered. Thermal destruction is a mature technology. Future improvements include enhanced energy recovery, smaller sizes, and continuous monitors for compliance. The presence of chlorine and other hetero atoms that lead to corrosive products (e.g., HCl) impose constrains on the material choices, gas cleanup, and monitoring requirements. The chlorine level entering the unit should be closely controlled so that downstream cleanup is efficient. In some cases, monitoring for PICs, including polychlorinated dibenzo-p-dioxins and dibenzofurans (PCDD/F), is required.

Flares will come under increasing scrutiny in the future, and performance improvements will be required. No routine methods are available to measure emissions from flares. PIC inventories are inadequate. However, in regions where flaring is common, VOC emissions are expected to impact the photochemical smog potential of the atmosphere; hence, these emissions must be described in more detail.

—*Elmar R. Altwicker*

References

Air and Waste Management Association. 1992. *Air pollution engineering manual.* Edited by A.J. Buonicore and W.T. Davis. New York: Van Nostrand Reinhold.

Dempsey, C.R., and E.T. Oppelt. 1993. Incineration of hazardous waste: A critical review update. *J. Air Waste Mgt. Assoc.* 43, no. 1:25–73.

Hemsath, K.H., and P.E. Suhey. 1974. *Fume incineration kinetics and its applications.* Am. Inst. of Chemical Engineers Symp. Series No. 137, 70, 439.

Joseph, D., J. Lee, C. McKinnon, R. Payne, and J. Pohl. 1983. *Evaluation of the efficiency of industrial flares: Background—Experimental design facility*. EPA-600/2-83-070, NTIS No. PB83-263723. U.S. EPA.

Katari, V.S., W.M. Vatavuk, and A.H. Wehe. 1987a. Incineration techniques for control of volatile organic compound emissions, Part I, Fundamentals and process design considerations. *J. Air Pollut. Control Assoc.* 37, no. 1:91–99.

———. 1987b. Incineration techniques for control of volatile organic compound emissions, Part II, Capital and annual operating costs. *J. Air Pollut. Control Assoc.* 37, no. 1:100–104.

Lee, K.C., N. Morgan, J.L. Hanson, and G. Whipple. 1982. Revised predictive model for thermal destruction of dilute organic vapors and some theoretical explanations. Paper No. 82-5.3, *Air Pollut. Control Assoc. Annual Mtg.* New Orleans, La.

McCrillis, R.C. 1988. Flares as a means of destroying volatile organic and toxic compounds. Paper presented at EPA/STAPPA/ALAPCO Workshop on Hazardous and Toxic Air Pollution Control

Technologies and Permitting Issues, Raleigh, NC and San Francisco, CA.

National Council of the Paper Industry for Air and Stream Improvement. 1985. *Collection and burning of Kraft non-condensible gases—Current practices, operating experience and important aspects of design and operations*. Technical Bulletin No. 469.

Nunez, C.M., G.H. Ramsey, W.H. Ponder, J.H. Abbott, L.E. Hamel, and P.H. Kariher. 1993. Corona destruction: An innovative control technology for VOCs and air toxics. *J. Air Waste Mgt. Assoc.* 43, no. 2:242–247.

Prokop, W.H. 1985. Rendering systems for processing animal by-product materials. *J. Am. Oil Chem. Soc.* 62, no. 4:805–811.

Ruddy, E.N., and L.A. Carroll. 1993. Select the best VOC control strategy. *Chem. Eng. Progress* 89, no. 7:28–35.

Stone, D.K., S.K. Lynch, R.F. Pandullo, L.B. Evans, and W.M. Vatavuk. 1992. Flares. Part I: Flaring technologies for controlling VOC containing waste streams. *J. Air Waste Mgt. Assoc.* 42, no. 3:333–340.

3.9
GASEOUS EMISSION CONTROL: BIOFILTRATION

The biological treatment of VOCs and other pollutants has received increasing attention in recent years. Biofiltration involves the removal and oxidation of organic compounds from contaminated air by beds of compost, peat, or soil. This treatment often offers an inexpensive alternative to conventional air treatment technologies such as carbon adsorption and incineration.

The simplest biofiltration system is a soil bed, where a horizontal network of perforated pipe is placed about 2 to 3 ft below the ground (Bohn 1992) (see Figure 3.9.1). Air contaminants are pumped through the soil pores, adsorbed on the surface of the moist soil particles, and oxidized by microorganisms in the soil.

Mechanisms

Biofiltration combines the mechanism of adsorption, the washing effect of water (for scrubbing), and oxidation. Soils and compost have porosity and surface areas similar to those of activated carbon and other synthetic adsorbents. Soil and compost also have a microbial population of more than 1 billion antiomycetes (microorganism resembling bacteria and fungi) per gram (Alexander 1977). These microbes oxidize organic compounds to carbon dioxide and water. The oxidation continuously renews the soil beds adsorption capacity (see Figure 3.9.2).

Another distinction is that the moisture in the waste gas stream increases the adsorption capacity of water-soluble gases and is beneficial for the microbial oxidation on which the removal efficiency of biofilters depends. Conversely, the moisture adsorbed by synthetic adsorbents reduces

their air contaminant adsorption capacity and removal efficiency. In addition, biofilter beds also adsorb and oxidize volatile inorganic compounds (VICs) to form calcium salts.

Gases in air flowing through soil pores adsorb onto or, as in GC, partition out on the pore surfaces so that VOCs remain in the soil longer than the carrier air. Soil–gas partition coefficients indicate the relative strengths of retention. The coefficients increase with VOC molecular weight and the number of oxygen, nitrogen, and sulfur functional groups in the VOC molecules. In dry soils, the coefficients for VOCs have been reported from 1 for methane to 100,000 for octane. However, under moist conditions, because of the water-soluble nature, the soil–gas partition coefficient for octane is probably a thousand, and acetaldehyde is around several thousand (Bohn 1992).

The biofilter's capacity to control air contaminants depends on the simultaneous operation of both adsorption and regeneration processes. Thus, overloading the system through excessive air flow rates can affect the biofilter's removal efficiency so that adsorption rates are lower than the rates at which chemicals pass through the filter. Once all adsorption sites are occupied, removal efficiency diminishes rapidly.

A second limiting factor is the microbial regeneration rate of the adsorbed chemical, which must equal or exceed the adsorption rate. Toxic chemicals can interfere with microbial processes until a bacterial population develops that can metabolize the toxic chemical. Biofilter bed acidity also reduces the removal efficiency because the environment for soil bacterial is inhospitable. In most cases

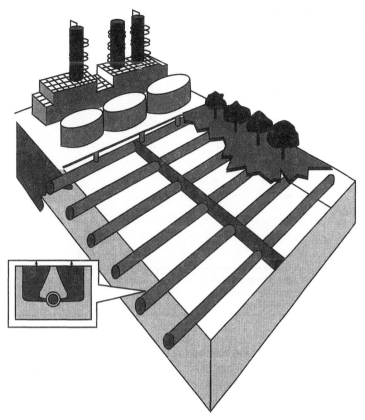

FIG. 3.9.1 Soil bed. A biofilter consists of a bed of soil or compost, beneath which is a network of perforated pipe. Contaminated air flows through the pipe and out the many holes in the sides of the pipe (enlarged detail), thereby being distributed throughout the bed.

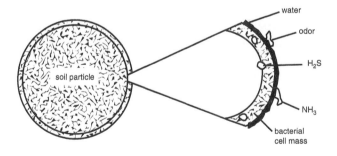

Adsorption of ammonia (NH₃), hydrogen sulfide (H₂S), and other odors on the soil particle surface.

water

odor

H₂S

NH₃

soil particle

bacterial cell mass

REGENERATION REACTIONS
Odorous hydrogen sulfide is oxidized to odorless sulfate
$$H_2S + 2O_2 \rightarrow SO_4^{2-} + 2H^+$$
Ammonia is dissolved in water and oxidized to odorless nitrate
$$NH_3 + H_2O \rightarrow NH_4^+ + OH_2^-$$
$$2NH_4^+ + 3O_2 \rightarrow 2\,NO_3^- + 8H^+$$
Bacteria oxidize odorous volatile organics to odorless carbon dioxide and water
$$\text{volatile organics} + O_2 \rightarrow CO_2 + H_2O$$

FIG. 3.9.2 Adsorption of odors and regeneration of active sites.

of biofilter failure, the limiting factor is filter overloading rather than microbiological processes because of the great diversity and number of soil bacteria.

After start up, biofilter beds require an adaptation time for the microbes to adapt to a new air contaminant input and to reach steady state. For rapidly biodegradable compounds, the adaptation time is no more than several hours. As the biodegradability decreases, the adaptation time can take weeks. After start up, the bed is resistant to shock load effects. Table 3.9.1 summarizes the biodegradability of various gases.

Fixed-Film Biotreatment Systems

Most biological air treatment technologies are fixed-film systems that rely on the growth of a biofilm layer on an inert organic support such as compost or peat (biofilters) or on an inorganic support such as ceramic or plastic (biotrickling filters). For both systems, solid particles must be removed from waste gases before the gases enter the system; particulates plug the pores. Both systems are best suited for treating vapor streams containing one or two major compounds. When properly designed, biofilters are well suited for treating streams that vary in concentration from minute to minute.

TABLE 3.9.1 GASES CLASSIFIED ACCORDING TO DEGRADABILITY

Rapidly Degradable VOCs	Rapidly Reactive VICs	Slowly Degradable VOCs	Very Slowly Degradable VOCs
Alcohols	H_2S	Hydrocarbons*	Halogenated hydrocarbons†
Aldehydes	NO_x	Phenols	Polyaromatic hydrocarbons
Ketones	(but not N_2O)	Methyl chloride	CS_2
Ethers	SO_2		
Esters	HCl		
Organic acids	NH_3		
Amines	PH_3		
Thiols	SiH_4		
Other molecules containing O, N, or S functional groups	HF		

Source: H. Bohn, 1992, Consider biofiltration for decominating gases, *Chem. Eng. Prog.* (April).
Notes: *Aliphatics degrade faster than aromatics such as xylene, toluene, benzene, and styrene.
†Such as TCE, TCA, carbon tetrachloride, and pentachlorophenol.

Figure 3.9.3 is a schematic diagram of a biofiltration system. A biofiltration system uses microorganisms immobilized in the form of a biofilm layer on an adsorptive filter substrate such as compost, peat, or soil. As a contaminated vapor stream passes through the filtered bed, pollutants transfer from the vapor to the liquid biolayer and oxidize. More sophisticated enclosed units allow for the control of temperature, bed moisture content, and pH to optimize degradation efficiency. At an economically viable vapor residence time (1 to 1.5 min), biofilters can be used for treating vapor containing about 1500 μg/l of biodegradable VOCs.

Figure 3.9.4 is a schematic diagram of a biotrickling filter for treating VOCs (Hartman and Tramper 1991). Biotrickling filters are similar to biofilters but contain conventional packing instead of compost and operate with recirculating liquid flowing over the packing. Only the recirculating liquid is initially inoculated with a microorganism, but a biofilm layer establishes itself after start up. The automatic addition of acid or base monitors and controls the pH of the recirculating liquid.

The pH within a biofilter is controlled only by the addition of a solid buffer agent to the packing material at the start of the operation. Once this buffering capacity is exhausted, the filtered bed is removed and replaced with fresh material. For the biodegradation of halogenated contaminants, biofilter bed replacement can be frequent. Therefore, biotrickling filters are more effective than biofilters for the treatment of readily biodegradable halogenated contaminants such as methylene chloride.

Biotrickling filters, possibly because of higher internal biomass concentrations, offer greater performance than biofilters at higher contaminant loadings. At a 0.5-min vapor residence time, the maximum concentration of styrene that can be degraded with 90% efficiency using biotrickling filters is two times greater than what can be degraded using biofilters.

Applicability and Limitations

Biofiltration has been used for many years for odor control at slaughter houses in Germany, the Netherlands, the United Kingdom, and Japan and to a limited extent in the United States. Recently, many wastewater and sludge treatment facilities have used biofilters for odor control purposes. The use of biofilters to degrade more complex air contaminants from chemical plants has occurred only

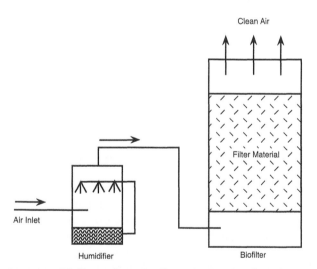

FIG. 3.9.3 Biofilter schematic diagram. (Reprinted, with permission, from A.P. Togna and B.R. Folsom, 1992, Removal of styrene from air using bench-scale biofilter and biotrickling filter reactors, Paper No. 92-116.04, *85th Annual Air & Waste Management Association Meeting and Exhibition, Kansas City, June 21–26.*)

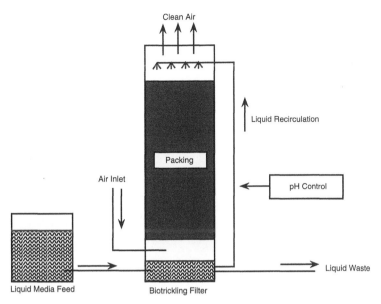

FIG. 3.9.4 Biotrickling filter schematic diagram. (Reprinted, with permission from Togna and Folsom, 1992.)

TABLE 3.9.2 CPI APPLICATIONS OF BIOFILTRATION

Company	Location	Application
S.C. Johnson & Son, Inc.	Racine, Wis.	Propane and butane removal from room air; 90% removal efficiency about 3000 cfm
Monsanto Chemical Co.	Springfield, Mass.	Ethanol and butyraldehyde removal from dryer air; 99% removal, 28,000 cfm; styrene removal from production gases
Dow Chemical Co.	Midland, Mich.	Chemical process gases
Hoechst Celanese Corp.	Coventry, R.I.	Process gases
Sandoz	Basel, Switzerland	Chemical process gases
Esso of Canada	Sarnia, Ontario	Hydrocarbon vapors from fuel storage tanks (proposed)
Mobil Chemical Co.	Canandaigua, N.Y.	Pentane from polystyrene–foam molding (proposed)
Upjohn Co.	Kalamazoo, Mich.	Pharmaceutical production odors; 60,000 cfm (proposed)

Source: H. Bohn, 1992, Consider biofiltration for decominating gases, *Chem. Eng. Prog.* (April).

within the last few years. Table 3.9.2 lists some chemical process industry CPI applications.

Biotrickling filters are more effective than biofilters for the treatment of readily biodegradable halogenated contaminants such as methylene chloride. Biofiltration can be unsuitable for highly halogenated compounds such as TCE, TCA, and carbon tetrachloride because they degrade slowly aerobically.

The limiting factors of soil-bed treatments are biodegradability of the waste and the permeability and chemistry of the soil. Because these factors vary, the design of soil beds is site-specific. For VIC removal, the lifetime of the bed depends on the soil's capacity to neutralize the acids produced. Any complex mixture with widely

different chemical, physical, and biodegradive properties, such as petroleum hydrocarbon vapors, can require more than one optimized biofilter or biotrickling filter in series.

—*Samuel S. Cha*

References

Alexander, M. 1977. *Introduction to soil microbiology.* 2d ed. New York: John Wiley.

Bohn, H. 1992. Consider biofiltration for decontaminating gases. *Chem. Eng. Prog.* (April).

Hartman, S., and J. Tramper. 1991. Dichloromethane removal from waste gases with trickling-bed bioreactor. *Bioprocess Eng.* (June).

Fugitive Emission: Sources and Controls

David H.F. Liu

4.1
FUGITIVE INDUSTRIAL PARTICULATE EMISSIONS

This section is concerned with the dust generated in processing operations that is not collected through a primary exhaust or control system. Such emissions normally occur within buildings and are discharged to the atmosphere through forced- or natural-draft ventilation systems.

Industrial fugitive emissions contribute more than 50% of the total suspended and inhalable emissions (Cowherd and Kinsey 1986). In addition, these particulates frequently contain toxic or hazardous substances.

Sources

Most dry processing operations generate dust. Generation points include the following:

Dumping of materials
Filling of materials
Drying of materials
Feeding or weighing of materials
Mechanical conveying of materials
Mixing or blending of materials
Pneumatic conveying of materials
Screening or classifying of materials
Size reduction of materials
Bulk storage of materials
Emptying of bulk bags

Generally, these materials are finely divided products that are easily airborne, resulting in sanitation problems and possibly fire or explosions.

Emission Control Options

Fugitive particulates can be controlled by three basic techniques: process modification, preventive measures, and add-on capture and removal equipment.

PROCESS MODIFICATION

Changing a process or operation to reduce emissions can be more practical than trying to control the emissions. For example, a pneumatic conveyance system eliminates the emission problems of a conveyor belt, which can require any or all preventive procedures (i.e., enclosures, wet suppression, housekeeping of spilled materials, and stabilization of materials) and add-on equipment. Another solution is to use new intermediate bulk containers (IBCs), such as shown in Figure 4.1.1.

PREVENTIVE MEASURES

Housekeeping provides numerous beneficial results of good operation and maintenance (O&M). Good O&M includes the prevention of process upsets and defective equipment, prompt cleanup of spillage before dust becomes airborne, the partial or complete enclosure or shielding of dust sources.

Water, a water solution of a chemical agent, or a micron-sized foam can be applied to the surface of a particulate-generating material. This measure prevents (or suppresses) the fine particles contained in the material from leaving the surface and becoming airborne. The chemical agents used in wet suspension can be either surfactants or foaming agents. Figure 4.1.2 shows a suppression system at a crusher discharge point.

The use of foam injection to control dust from material handling and processing operations is a recently developed method to augment wet suppression techniques. The foam is generated by a proprietary surfactant compound added to a small quantity of water, which is then vigorously mixed to produce a small-bubble, high-energy foam in the 100 to 200 μm size range. The foam uses little liq-

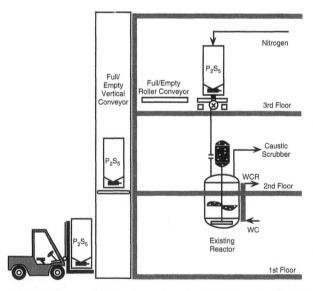

FIG. 4.1.1 Inverted IBCs elevated to the third floor for charging reactors, eliminating worker exposure to hazardous P_2S_5. (Reprinted, with permission, from B.O. Paul, 1994, Material handling system designed for 1996 regulations, *Chemical Processing* [July].)

uid volume and when applied to the surface of a bulk material, wets the fines more effectively than untreated water does. Foam has been successfully used in controlling the emissions from belt transfer points, crushers, and storage-pile load-ins.

CAPTURE AND REMOVAL

Most industrial process, fugitive particulate emissions are controlled by capture and collection or industrial ventilation systems. These systems have three primary components:

A hood or enclosure to capture emissions that escape from the process

A dry dust collector that separates entrained particulates from the captured gas stream

A ducting or ventilation system to transport the gas stream from the hood or enclosure to the APCD

Before designing a dust collection system, the environmental engineer must thoroughly understand the process and its operations and the requirements of operating personnel. The engineer must evaluate the materials in the dust in addition to the particle size and define the characteristics of the material including the auto-ignition temperature, explosive limits, and the potential for electrostatic buildup in moving these materials. The engineer must locate the pickup and select the components to be used in the system.

Each operation generates various amounts of dust. The environmental engineer uses the amount of dust, the particle size, and the density of the dust in determining the

capture velocity of the dust, which affects the pickup or hood design. Normal capture velocity is a function of particle size and ranges from 6 to 15 ft/sec (Opila 1993). In practice, the closer the hood is to the dust source, the better the collection efficiency is and the less air is needed for the collection process. The engineer should design enclosures and hoods at suitable control velocities to smooth the air flow. Examples of dust collection follow (Kashdan, et al. 1986):

A bag tube packer (see Figure 4.1.3) where displaced air is treated from the feed hopper supplying the packer, from the bag itself, and from the spill hopper below the bagger for any leakage during bag filling

Open-mouth bag filling (see Figure 4.1.4) where the dust pickup hood collects the air displaced from the bag

Barrel or drum filling (see Figure 4.1.5) that uses a dust pickup design with the same contour of the drum be-

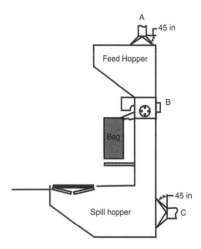

FIG. 4.1.3 A bag tube packer. (Reprinted, with permission, from Robert L. Opila, 1993, Carefully plan dust collection systems, *Chem. Eng. Prog.* [May].)

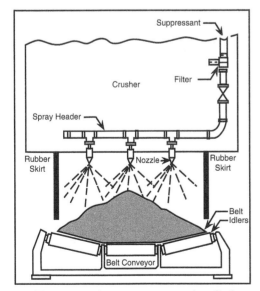

FIG. 4.1.2 Wet suppression system at a crusher discharge point. (Reprinted from C. Cowherd, Jr. and J.S. Kinsey, 1986, *Identification, assessment, and control of fugitive particulate emissions,* EPA-600/8-86-023, Research Triangle Park, N.C.: U.S. EPA.)

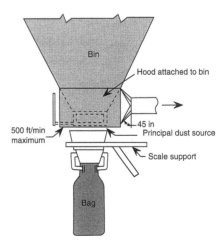

FIG. 4.1.4 Open-mouth bag packer. (Reprinted, with permission, from Opila, 1993.)

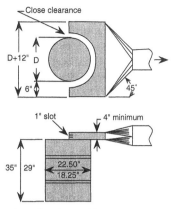

FIG. 4.1.5 Dust pickup for drum filling. (Reprinted, with permission, from Opila, 1993.)

ing filled. The pickup is designed for a single drum diameter; the design becomes complicated if 15, 30, and 55 gal drums are filled at the same filling station. The height and diameter of these drums vary affecting the capture velocity of the air.

The pickup of dust created when material flows down a chute onto a conveyor belt (see Figure 4.1.6)

The pickup points required for dust-free operation of a flatdeck screen (see Figure 4.1.7)

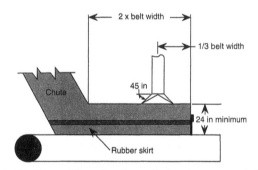

FIG. 4.1.6 Dust pickup for chute discharging onto a belt conveyor. (Reprinted, with permission, from Opila, 1993.)

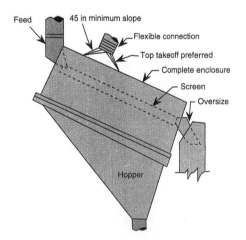

FIG. 4.1.7 Dust pickup on a flatdeck screen. (Reprinted, with permission, from Opila, 1993.)

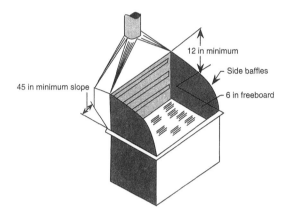

FIG. 4.1.8 Slot type hood mounted on a ribbon mixer. (Reprinted, with permission, from Opila, 1993.)

TABLE 4.1.1   MINIMUM AIR VELOCITIES IN DUCTS TO PREVENT DUST SETTLING

Type of Dust	Velocity, ft/min
Low density (gases, vapors, smoke, flour, lint)	2000
Medium-low density (grain, sawdust, plastic, rubber)	3000
Medium-high density (cement, sandblast, grinding)	4000
High-density (metal turnings, lead dust)	5000

Source: J.A. Danielson, ed., 1973, *Air pollution engineering manual*, 2d ed., AP–40. (Washington, D.C.: U.S. EPA).

A slot-type hood used with a ribbon mixer (see Figure 4.1.8)

Maintaining the conveying velocity throughout the collection system is important. If conveying velocities are too low, saltation (deposition of particles within a duct) occurs until the duct is reduced to a point where conveying velocity is obtained. This material buildup can be troublesome for a sanitary plant operation. Table 4.1.1 provides guidelines for choosing a velocity. In addition, dry and wet cleaning of the dust collection system must be provided.

—David H.F. Liu

References

Cowherd, C., Jr., and J.S. Kinsey. 1986. *Identification, assessment, and control of fugitive particulate emissions.* EPA-600/8-86-023, Research Triangle Park, N.C.: U.S. EPA.

Kashdan, E.R., et al. 1986. Hood system design for capture of process fugitive particulate emissions. *Heating/Piping/Particulate Emissions* 58, no. 2 (February):47–54.

Opila, Robert L. 1993. Carefully plan dust collection systems. *Chem. Eng. Prog.* (May).

4.2
FUGITIVE INDUSTRIAL CHEMICAL EMISSIONS

This section is concerned with unintentional equipment leaks of VOCs from industrial plants. The quantity of these fugitive emissions is hard to measure, but in some cases they account for 70 to 90% of the air emissions from chemical manufacturing operations (U.S. EPA 1984). Many of these emissions contain HAPs.

Sources

Fugitive emissions are unintentional releases (leaks) from sources such as valves, pumps, compressors, pressure-relief valves, sampling connection systems, open-ended lines, and flanges as opposed to point-source emissions from stacks, vents, and flares.

Table 4.2.1 compares the fugitive emissions from these sources in terms of source emissions factors and total source contributions of VOCs. A count of all equipment multiplied by the emission factors estimates the total emissions.

Fugitive emissions do not occur as part of normal plant operations but result from the following:

Malfunctions
Wear and tear
Lack of proper maintenance
Operator error
Improper equipment specifications

Improper installation
Use of inferior technology
Externally caused damage

Fugitive emissions can be significantly reduced with the adoption of improved technology, maintenance, and operating procedures.

Source Controls

Good O&M has a significant influence on lowering the fugitive emissions and includes the following:

Daily inspection for leaks by plant personnel
Immediate leak repair
Installation of gas detectors in strategic plant locations, with sample analysis performed regularly
Monitoring of vibration in rotating machinery
Minimization of pipe and connector stresses caused by the vibration of pumps and compressors
Inspection and testing of relief-valves and rupture disks for leaks
Reduction in the number and volume of samples collected for control purposes
Inspection and periodic replacement of seals and gaskets

Because the emission control of equipment-related sources is unit-specific, this section discusses the control techniques by source category.

TABLE 4.2.1 COMPARISON OF FUGITIVE EMISSIONS FROM EQUIPMENT TYPES

Equipment Type	Process Fluid	Emission Factor, kg/hr	Percent of Total VOC Fugitive Emissions
Valve	Gas or vapor	0.0056	47
	Light liquid	0.0071	
	Heavy liquid	0.00023	
Pump	Light liquid	0.0494	16
	Heavy liquid	0.0214	
Compressor	—	0.228	4
Pressure-relief valve	Gas or vapor	0.1040	9
Sampling connection	—	0.0150	3
Open-ended line	—	0.0017	6
Flange	—	0.00083	15
			100

Source: U.S. Environmental Protection Agency (EPA), 1982, *Fugitive emission sources of organic compounds,* EPA 450/3-80/010, Research Triangle Park, N.C.

VALVES

Except for check and pressure-relief valves, industrial valves need a stem to operate. An inadequately sealed stem is a source of fugitive gas emissions. Valves with emissions greater than 500 ppmv are considered *leakers*. The 500-ppmv-limit threshold can place excessive demands on many low-end valves. Figure 4.2.1 shows the primary maintenance points for a packed, stemmed valve. The cost of keeping these valves in compliance can be expensive over time compared to another valve of greater initial cost.

The following special valves are designed to control fugitive emissions:

Bellows-type stems for both rising stem and quarter-turn valves show almost zero leakage and require no maintenance during their service life (Gumstrup 1992). However, because bellows seals are more costly than packed seals, they are typically used in lethal and hazardous service. Figure 4.2.2 shows a typical design of a bellows-sealed valve.

Diaphragm valves and magnetically actuated, hermetically sealed control valves are two other valves least prone to leaking. Figure 4.2.3 shows a typical diaphragm valve.

For services not requiring zero leakage on rising and rotary valve stem seals, a new group of improved packed seals is available. Most new seals claim to reduce fugitive emissions to a maximum leakage of less than 100 ppmv (Ritz 1993). Gardner (1991) discusses other types of valves for emission control.

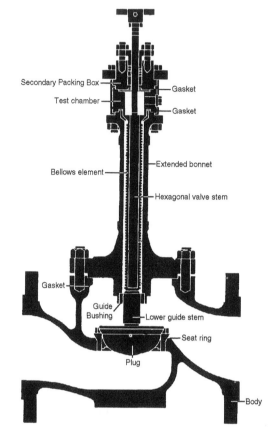

FIG. 4.2.2 Typical design of a bellows-sealed valve.

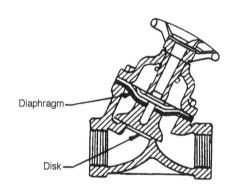

FIG. 4.2.3 Typical design of diaphragm valves. (Reprinted from U.S. EPA, 1984.)

PUMPS

The failure of a sealant where a moving shaft meets a stationary casting is a source of fugitive gas emissions. Figure 4.2.4 shows the possible leak area. Lubricants between the rotating shaft and the stationary packing control the heat generated by friction between the two materials.

The following techniques can hold fugitive emissions to a minimum:

Equipping pumps with double, mechanical seals that have liquid buffer zones and alarms or automatic pump shut-

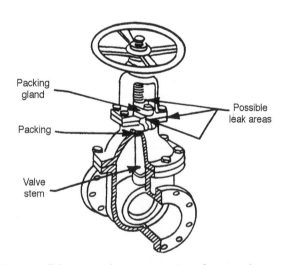

FIG. 4.2.1 Primary maintenance points for a valve stem. (Reprinted from U.S. Environmental Protection Agency, 1984, *Fugitive VOC emissions in the synthetic organic chemicals manufacturing industry*, EPA 625/10-84/004, Research Triangle Park, N.C.: U.S. EPA, Office of Air Quality and Standards.)

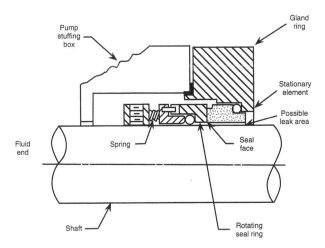

FIG. 4.2.4 Typical design of a packed, sealed pump shaft. (Reprinted from U.S. EPA, 1984.)

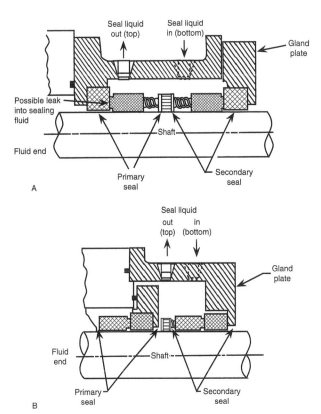

FIG. 4.2.5 Two arrangements of dual mechanical pump seals. (*A*) Back-to-back arrangement; (*B*) tandem arrangement (Reprinted from U.S. EPA, 1984).

offs for seal failures (see Figure 4.2.5). The effective use of mechanical seals requires closed tolerances. Shaft vibration and misalignment induce radial forces and movement, which can damage both seals and bearings. The primary causes of radial shaft motion include poor alignment of the shaft with the motor connected to the pump, poor baseplate installation, incorrect operating conditions, and loose or failed bearings (Clark and Littlefield 1994). Continuously coating bearings with lubricants keeps the surfaces free of debris.

Using canned-motor pumps (see Figure 4.2.6). These units are closed-couple designs in which the cavity housing the motor rotor and the pump casing are interconnected. As a result, the pump bearings run in the process liquid, and all seals are eliminated.

Adopting diaphragm and magnetic-drive pumps. These two pumps do not require a sealant to control leakage.

Adams (1992) discusses the selection and operation of mechanical seals.

COMPRESSORS

The standard for compressors requires (1) using mechanical seals equipped with a barrier fluid system and control degassing vents or (2) enclosing the compressor seal area and venting emissions through a closed-vent system to a control device. These systems provide control efficiencies approaching 100% (Colyer and Mayer 1991).

Most concepts to control fugitive emissions from pumps apply to compressors. Likewise, analyzing the vibration characteristics anticipates pending problems.

PRESSURE-RELIEF DEVICES

Figure 4.2.7 shows a pressure-relief valve and rupture device to control fugitive emissions. This device does not al-

low any emissions until the pressure is large enough to rupture the disk. The pressure-relief valve opens at a set pressure and then reseats when the pressure returns to below the set value.

Reseating the pressure-relief device after a discharge is often a source of a fugitive emission. The source can vary from a single improper reseating to a continuous failure because of a degraded seating element. Like other valves, pressure-relief valves with emissions greater than 500 ppmv above background are considered leakers.

An alternative to rupture disks and other techniques that achieve less than 500 ppmv above background is for plants to vent pressure-relief devices to a closed-vent system connected to a control device.

SAMPLING CONNECTION SYSTEMS

MACT consists of closed-purge sampling, closed-loop sampling, and closed-vent vacuum systems in the rules for sampling connections. These systems are described as follows:

A closed-purge sampling system eliminates emissions due to purging either by returning the purge material directly to the process or by collecting the purge in a system that is not open to the atmosphere for recycling or disposal.

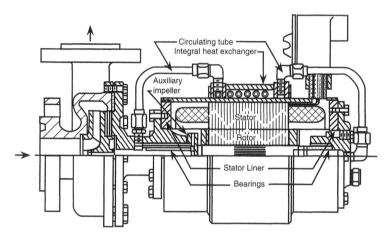

FIG. 4.2.6 Chempump canned-motor pump.

A closed-loop sampling system also eliminates emissions due to purging by returning process fluid to the process through an enclosed system that is not directly vented to the atmosphere.

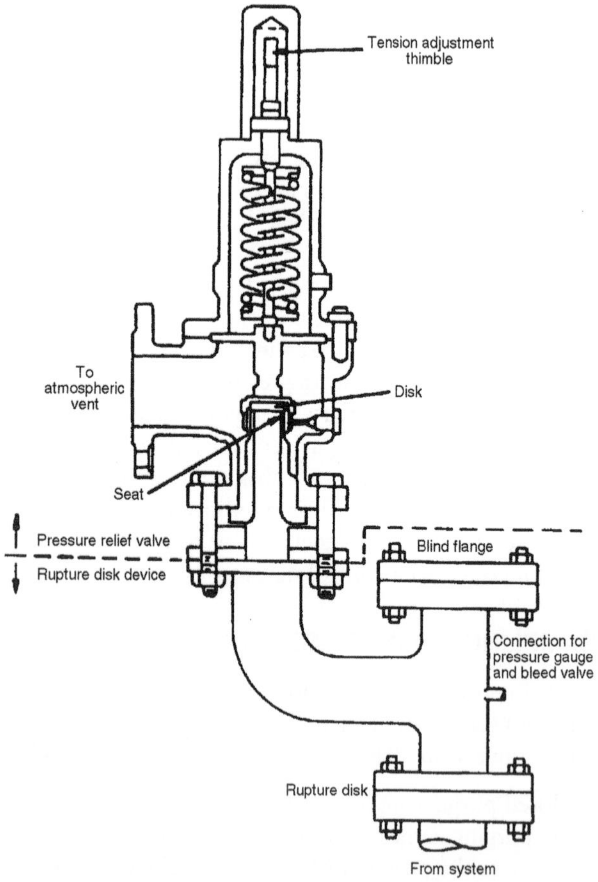

A closed-vent system captures and transports the purged process fluid to a control device.

Figure 4.2.8 shows two such systems.

OPEN-ENDED LINES

Enclosing the open end of a valve or line with a cap, plug, or a second valve eliminates emissions except when the line is used for draining, venting, or sampling operations. The control efficiency associated with these techniques is approximately 100% (Colyer and Mayer 1991).

FLANGES AND CONNECTORS

Flanges are significant sources of fugitive emissions, even at well-controlled plants, due to the large number of flanges and connectors. In most cases, tightening the flange bolts on the flanged connectors, replacing a gasket, or correcting faulty alignment of a surface eliminates a leak. The use of all welded construction minimizes the number of flange joints and screwed connections.

Unsafe-to-monitor connectors can expose personnel to imminent hazards from temperature, pressure, or explosive conditions. During safe-to-monitor periods, plant personnel should monitor these connectors, especially the critical ones, as frequently as possible using leak detectors.

AGITATORS

Limited screening data indicate that agitators are a significant source of emissions. Agitators are technologically similar to pumps so emissions are controlled using seal technology. However, agitators have longer and larger diameter shafts than pumps and produce greater tangential loading. Therefore, the performance of pump seal systems cannot estimate agitator seal performance. A leak from an

FIG. 4.2.7 Pressure-relief valve mounted on a rupture disk device.

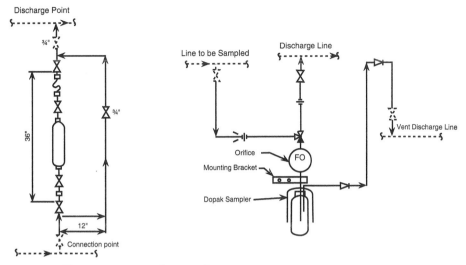

FIG. 4.2.8 Two arrangements for sampling systems.

agitator is defined as a concentration of 10,000 ppmv or greater.

—David H.F. Liu

References

Adams, W.V. 1992. Controlling fugitive emissions from mechanical seals. *Hydrocarbon Processing* (March):99–104.

Clark, E., and D. Littlefield. 1994. Maximize centrifugal pump reliability. *Chemical Engineering* (February).

Colyer, R.S., and J. Mayer. 1991. Understanding the regulations governing equipment leaks. *Chem. Eng Prog.* (August).

Gardner, J.F. 1991. Selecting valves for reduced emissions. *Hydrocarbon Processing* (August).

Gumstrup, B. 1992. Bellows seal valves. *Chemical Engineering* (April).

Ritz, G. 1993. Advances in control valve technology. *Control* (March).

U.S. Environmental Protection Agency (EPA). 1984. Fugitive VOC emissions in the synthetic organic chemicals manufacturing industry. EPA 625/10-84/004. Research Triangle Park, N.C.: U.S. EPA, Office of Air Quality Planning and Standards.

4.3
FUGITIVE DUST

Fugitive dust that deposits large particles on buildings, vehicles, and materials is a nuisance. Inhalation of dust can be debilitating, especially when the dust contains toxic elements and minerals. In addition, fugitive dust reduces visibility.

Fugitive dust consists of geological material suspended into the atmosphere by wind action and by human activities. Most of this dust soon deposits within a short distance of its origin, yet a portion of it can be carried many miles by atmospheric winds.

Sources

Table 4.3.1 shows the approximate emissions of fugitive dust from manmade sources in the United States for 1990.

TABLE 4.3.1 FUGITIVE DUST EMISSIONS IN THE UNITED STATES BY SOURCE CATEGORY DURING 1990

Source Category	Particulate matter (million tn/yr)	Total (%)
Mining and quarrying	0.37	1
Erosion	4.1	9
Agricultural tilling	7.0	16
Paved roads	8.0	18
Construction	10.0	22
Unpaved roads	15.5	34

Source: Air and Waste Management Association, 1992, Standards and non-traditional particulate source control (Pittsburgh, Pa.).

TABLE 4.3.2 CONTROL TECHNIQUES FOR VARIOUS SOURCES

Fugitive Emission Source	Chemical Stabilizers	Vegetative Cover	Watering	Windscreens	Wind Barriers/Berms	Plantings	Pile Shaping and Orientation	Paving and Gravel	Sweeping and Cleaning	Reduced Speed	Curbing and Stabilizing Shoulders	Operations Change	Reduced Drop Distance	Water Sprays and Foggers	Electrostatic Curtains	Partial or Complete Enclosure	Hooding and Ducting	Covers	Wheel Washes	Foams
Paved roads			X	X	X	X			X	X	X									
Unpaved roads	X		X	X	X	X		X		X	X									
Unpaved parking lots	X		X	X	X	X		X		X	X									
Active storage piles			X	X	X	X	X					X				X		X		
Inactive storage piles	X	X	X	X	X	X	X					X				X		X		
Exposed areas	X	X	X	X	X	X		X				X								
Construction sites			X	X	X			X				X				X				
Conveyor transfer				X		X						X	X	X	X	X	X			X
Drop points				X								X	X	X	X	X	X			X
Loading and unloading				X	X							X	X	X	X	X	X			X
Vehicle carryout								X	X										X	
Truck and rail spills									X	X	X							X		
Crushing and screening			X	X	X							X	X	X	X	X	X	X		
Waste sites	X	X		X		X		X				X								
Tilling operations			X									X								
Feed lots	X	X	X									X								

Source: Adapted from E.T. Brookman and D.J. Martin, 1981, A technical approach for the determination of fugitive emission source strength and control requirements, *74th Annual APCA Meeting, Philadelphia, June 1981.*

These estimates are derived from standardized U.S. EPA emission factors that relate soil characteristics, meteorological conditions, and surface activities to emission rates. These estimates do not include all potential emitters (for example, natural dust emissions, such as wind devils, are omitted as are many from industrial activities). Mass emissions are restricted to particles that are less than 10 μm.

Prevention and Controls

Table 4.3.2 is a partial listing of control techniques for reducing dust emissions from various sources. Table 4.3.3 presents an approximate categorization of control system capabilities. When more than one control technique can be used in series, the combined efficiency is estimated by a series function.

Partially or completely closing off the source from the atmosphere to prevent wind from entraining the dust can effectively reduce emissions. Frequently, confinement plus suppression is an optimum combination for effective, economical dust control.

WIND CONTROL

Preventing the wind from entraining dust particles is accomplished by keeping the wind from blowing over materials. This prevention involves confinement or wind control. Tarps that cover piles are a wind control procedure for storage piles, truck, railcars, and other sources (see Figure 4.3.1).

A variation of the source enclosure method to control fugitive dust emissions involves applying wind fences (also referred to as windscreens). Porous wind fences significantly reduce emissions from active storage piles and exposed ground areas. The principle of a windscreen is to provide a sheltered region behind the fenceline where the mechanical turbulence generated by ambient winds is significantly reduced. The downwind extent of the protected region is many times the physical height of the fence. This sheltered region reduces the wind erosion potential of the exposed surface in addition to allowing gravitational settling of larger particles already airborne. Figure 4.3.2 is a diagram of a portable windscreen used at a coal-fired power plant.

For unpaved routes and ditches, special fabrics placed over exposed surfaces prevent dust from becoming air-

TABLE 4.3.3 APPROXIMATE CATEGORIZATION OF CONTROL SYSTEM CAPABILITIES

	Type of Control system					
	RACT		BACT		LAER	
Source	Control	Efficiency, %	Control	Efficiency, %	Control	Efficiency, %
Unpaved roads	Wetting agent (water)	50	Wetting agent (other than water)	60–80	Paving and sweeping	85–90
	Speed control	25–35	Drastic speed control	65–80		
			Soil stabilization	50		
			Apply gravel	50		
			Road carpet	80		
Active storage piles	Wetting agent (water)	50–75	Wetting agents (other than water)	70–90	Encrusting agents	90–100
	Pile orientation	50–70	Pile orientation	50–70	Tarp cover	100
	Leading slope angle	35	Wind screens	60–80		
Inactive storage piles	Vegetation	65	Chemical stabilization plus vegetation	80–90	Tarp cover	100
Transfer points	Water sprays	35	Wetting agent sprays	55	Enclosure with sprays	90–100
			Fogging sprays	80	Electrostatic-enhanced fogging sprays (EEFS)	80–95
Conveyors	Water sprays	35	Wetting agent sprays	55	Enclosure with sprays	90–100
			Fogging sprays	80	EEFS	80–95
Car dumpers	Water sprays	35	Wetting agent sprays	40	Enclosure with sprays	85–90
			Fogging sprays	75	EEFS	75–90
Construction activities	Watering	50	Chemical stabilization	80	Enclosure	90

Source: H.E. Hesketh and F.L. Cross, Jr., 1983, *Fugitive emission and controls* (Ann Arbor, Mich.: Ann Arbor Science).
RACT-Reasonable available control technology
BACT-Best available control technology
LAER-Lowest available emission rate

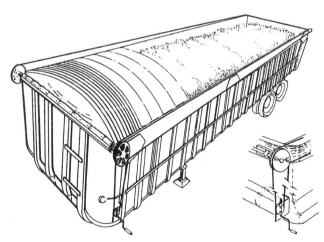

FIG. 4.3.1 Complete truck tarp; tarp movement and hold-down system. (Reprinted, with permission, from Aero Industries.)

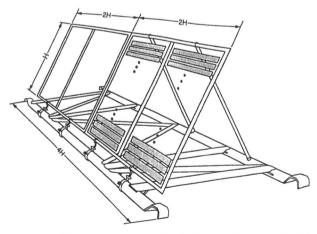

FIG. 4.3.2 Diagram of a portable windscreen. (Reprinted, with permission, from C. Cowherd, Jr. and J.S. Kinsey, 1986, *Identification, assessment, and control of fugitive particulate emissions*, EPA 600/8-86-023, Research Triangle Park, N.C.: U.S. EPA.)

TABLE 4.3.4   CLASSIFICATION OF TESTED CHEMICAL SUPPRESSANTS

Dust Suppressant Category	Trade Name
Salts	Peladow
	LiquiDow
	Dustgard
	Oil well brine
Lignosulfonates	Lignosite
	Trex
Surfactants	Biocat
Petroleum-based	Petro Tac
	Coherex
	Arco 2200
	Arco 2400
	Generic 2 (QS)
Mixtures	Arcote 220/Flambinder
	Soil Sement

Source: U.S. Environmental Protection Agency (EPA), 1987, *Emission control technologies and emission factors for unpaved road fugitive emission,* EPA 625/5-87/002 (Cincinnati: U.S. EPA, Center for Environmental Research Information).

borne from wind or machinery action. These fabrics are referred to as *road carpets,* and alternatively they prevent erosion by rain runoff.

WET SUPPRESSION

Wet suppression systems apply either water, a water solution of a chemical agent, or a micron-sized foam to the surface of a particulate-generating material. This measure prevents (or suppresses) the fine particles in the material from leaving the surface and becoming airborne.

The chemical agents used in wet suppression systems are either surfactants or foaming agents for materials handling and processing operations or various dust palliatives applied to unpaved roads. In either case, the chemical agent agglomerates and binds the fines to the aggregate surface, thus eliminating or reducing its emission potential.

In agricultural fields, electrostatic foggers provide electrostatically charged water droplets that agglomerate suspended particles, thereby increasing the particle size and deposition velocity.

VEGETATIVE COVER

Vegetation reduces wind velocity at the surface and binds soil particles to the surfaces. Of course, an area with vegetative cover should not be disturbed after planting (U.S. EPA 1977). Likewise, tilling implements, orientations, and frequencies should aim to limit the suspension of surface dust in agricultural fields.

TABLE 4.3.5   IMPLEMENTATION ALTERNATIVES FOR CHEMICAL STABILIZATION OF AN UNPAVED ROAD

Cost Elements	Implementation Alternatives
Purchase and ship chemical	Ship in railcar tanker (11,000–22,000 gal/tanker)
	Ship in truck tanker (4000–6000 gal/tanker)
	Ship in drums via truck (55 gal/drum)
Store chemical	Store on plant property
	In new storage tank
	In existing storage tank
	Needs refurbishing
	Needs no refurbishing
	In railcar tanker
	Own railcar
	Pay demurrage
	In truck tanker
	Own truck
	Pay demurrage
	In drums
	Store in contractor tanks
Prepare road	Use plant-owned grader to minimize ruts and low spots
	Rent contractor grader
	Perform no road preparation
Mix chemical and water in application truck	Put chemical in spray truck
	Pump chemical from storage tank or drums into application truck
	Pour chemical from drums into application truck, generally using forklift
	Put water in application truck
	Pump from river or lake
	Take from city water line
Apply chemical solution via surface spraying	Use plant-owned application truck
	Rent contractor application truck

Source: U.S. EPA, 1987.

CHEMICAL STABILIZATION

Dust from unpaved roads or uncovered areas can be reduced or prevented by chemical stabilization. Chemical suppressants can be categorized as salts, lignin sulfonates, wetting agents, petroleum derivatives, and special mixtures. Manufacturers generally provide information for typical applications, dilution and application rates, and costs. Table 4.3.4 is a partial list of chemical stabilizers.

Table 4.3.5 lists alternatives for implementing chemical stabilization of an unpaved road.

—David H.F. Liu

Reference

U.S. Environmental Protection Agency (EPA). 1977. *Guideline for development of control strategies in areas with fugitive dust problems.* EPA 450/2-77/029. Washington, D.C.: U.S. EPA.

Odor Control

Samuel S. Cha ∣ Amos Turk

5.1
PERCEPTION, EFFECT, AND CHARACTERIZATION

Odor Terminology

An odor is a sensation produced by chemical stimulation of the chemoreceptors in the olfactory epithelium in the nose. The chemicals that stimulate the olfactory sense are called *odorants* although people frequently refer to them as *odors*.

Several dimensions of human responses to the odor sensation can be scientifically characterized. These dimensions are threshold, intensity, character, and hedonic tone.

THRESHOLD

Threshold, or detectability, refers to the theoretical minimum concentration of odorant stimulus necessary for perception in a specified percentage of the population, usually the mean. A threshold value is not a fixed physiological fact or a physical constant but is a statistical point representing the best estimate from a group of individual scores. Two types of thresholds can be evaluated: detection and recognition.

The *detection threshold* is the lowest concentration of odorant that elicits a sensory response in the olfactory receptors of a specified percentage, usually 50%, of the population being tested (see Figure 5.1.1). The detection threshold is the awareness of an odor without necessarily recognizing it.

The *recognition threshold* is the minimum concentration recognized as having a characteristic odor quality by a specified percentage of the population, usually 50%. It differs from the detection threshold because it is the point at which people can describe a specific odor character to the sensory response.

In the measurement of environmental odors, which are generally complex mixtures of compounds, the threshold is not expressed as a concentration level. Instead, the threshold is expressed as dilution-to-threshold ratios (D/T); it is dimensionless. A D/T ratio of 100 means that a given volume (i.e., 1 cu ft) of odorous air requires 100 volumes (i.e., 100 cu ft) of odor-free air to dilute it to threshold, or a barely detectable odor. This concept is shown in Figure 5.1.1 where a 50% panel response occurs at 235 D/T for one sample and 344 D/T for the other sample.

Over the years, many terms have been used to express the concentration of odor including the following:

ODOR UNIT—one volume of odorous air at the odor threshold; often the volume is defined in terms of cubic feet as follows:

$$\frac{\text{odor units}}{\text{cu ft}} = \frac{\text{volume of sample diluted to threshold (cu ft)}}{\text{original volume of sample (cu ft)}}$$

5.1(1)

ODORANT QUOTIENT—Expressed by the following equation. The Z is for Zwaardermaker who was the earliest investigator to use dilution ratios for odor measurement (ASTM 1993).

$$Z_t = \frac{C_o}{C_t} = \frac{\text{odorant concentration of a sample}}{\text{odorant concentration at threshold}}$$

5.1(2)

Both odor units/cu ft and Z_t are dimensionless and are synonyms of the D/T; the odor unit is a unit of volume and is not a synonym of D/T (ASTM 1993; Turk 1973).

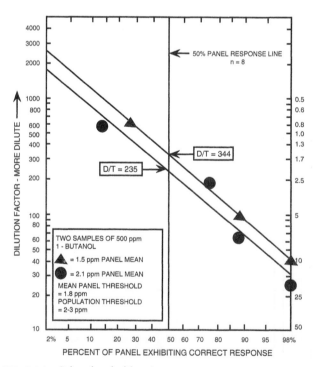

FIG. 5.1.1 Odor threshold ratio.

Another term, the *odor emission rate,* is often used to describe the severeness of the downwind impact, i.e., the problem a typical odor source creates at downwind locations. The odor emission rate, expressed in unit volume per minute, is the product of the D/T value of the odorous air and the air flowrate (cfm or its equivalent).

Listings of threshold values for pure chemicals are common in many environmental resource references. However, the reported threshold values are limited and vary as much as several orders of magnitude, suggesting that thresholds are limited in usefulness. However, a review of threshold value methodology finds that when threshold values are subjected to basic methodological scrutiny, the range of experimentally acceptable values is considerably reduced. Two recent documents of odor threshold reference guides are available. One published by the American Industrial Hygiene Association (1989) is based on a review for 183 chemicals with occupational health standards. The other document is published by the U.S. EPA (1992). It compiles odor thresholds for the 189 chemicals listed in the CAAAs.

INTENSITY

Odor intensity refers to the perceived strength of the odor sensation. Intensity increases as a function of concentration. The relationship of the perceived strength (intensity) and concentration is expressed by Stevens (1961) as a psychophysical power function as follows:

$$S = K I^n \qquad 5.1(3)$$

where:

    S = perceived intensity of the odor sensation
    I = physical intensity of stimulus (odorant concentration)
    n = constant
    K = constant

This equation can be expressed in logarithm as follows:

$$\log S = \log K + n \log I \qquad 5.1(4)$$

where:

    n = slope
    K = y-intercept

Figure 5.1.2 shows an intensity function on logarithm coordinates of the standard odorant 1-butanol. The slope of the function, also called the *dose-response function,* varies with the type of odorant. Odor pollution control is concerned with the dose-response function, or the degree of dilution necessary to decrease the intensity. This function can be described by the slope. A low slope value indicates that the odor requires greater dilution for it to dissipate; a high slope value indicates that the intensity can be reduced by dilution more quickly. Compounds with low slope values are hydrogen sulfide, butyl acetate, and amines. Compounds with high slope values are ammonia and aldehydes. This function explains why hydrogen sulfide, butyl acetate, and amines can be detected far away from the odor origin. On the other hand, ammonia and aldehydes cause odor problems at locations near the origin.

Category Scales

Category scales were the first technique developed to measure odor intensity. One widely used scale was developed for a 1930 study of odor used as gas alarms. The scale has the following six simple categories:

0 No odor
1 Very faint odor
2 Faint odor
3 Easily noticeable odor
4 Strong odor
5 Very strong odor

The limitations of category scaling are that the number of categories to choose from is finite and they are open to bias through subjective number preferences or aversions.

Magnitude Estimation

In magnitude estimation, observers create their own scales based on a specific reference point. The data from several observers is then normalized to the reference point. Magnitude estimation is a form of ratio scaling and has many advantages over category scaling (i.e., an unlimited range and greater sensitivity); however, it does require a more sophisticated observer and statistical analysis.

Reference Scales

The reference scale for measuring odor intensity is standardized as ASTM Standard Practice E 544 (ASTM 1988).

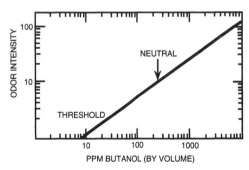

$S = kI^n$ on log–log coordinates

S = PERCEIVED INTENSITY
k = Y–INTERCEPT
I = ODORANT CONCENTRATION
n = SLOPE OF PSYCHOPHYSICAL FUNCTION

FIG. 5.1.2 Butanol reference scale.

The method uses a standard reference odorant, 1-butanol, set in a series of known concentrations. The advantages of this method are that it 1) allows the comparison of subjective odor intensities between laboratories, 2) allows for odor control regulations to be expressed in terms of perceived intensity rather than odor thresholds, and 3) allows cross modality comparisons (i.e., sound and odor). One disadvantage is that some people find it difficult to compare odors that have a different odor character than the reference standard.

CHARACTER

Another dimension of odor is its *character,* or what the odor smells like. ASTM publication DS 61 (Dravnieks 1985) presents character profiles for 180 chemicals using a 146 descriptor scale. The scale uses terms such as fishy, nutty, moldy, rancid, sewer, and ammonia. This dimension is useful in air pollution control for describing the source or process responsible for community odors because different odor characters are associated with various processes and industries. Table 5.1.1 provides a short list of the industrial odor character descriptions.

HEDONIC TONE

Hedonic tone, also referred to as acceptability, is a category judgment of the relative pleasantness or unpleasantness of an odor. In the context of air pollution field work, the hedonic tone is often irrelevant. Perception of an odor is based on the combination of frequency of occurrence, odor character, and odor intensity. Even pleasant odors can become objectionable if they persist long enough.

Human Response to Odors and Odor Perception

Olfactory acuity in the population follows a normal bell curve distribution of sensitivity, ranging from hypersensitive to insensitive and anosmic (unable to smell). Individual odor threshold scores can be distributed around the mean value to several orders of magnitude.

Olfactory studies have revealed interesting information regarding odor perception. Two sensory channels are responsible for human detection of inhaled chemical substances in the environment. These are *odor perception* and the *common chemical sense* (CCS). Odor perception is a function of cranial nerve I, whereas the CCS is a function of cranial nerve V. The CCS is described as pungency and associated sensations such as irritation, prickliness, burning, tingling, and stinging. Unlike the olfactory structure, the CCS lacks morphological receptor structures and is not restricted to the nose or oral cavity (Cometto-Muñiz and Cain 1991).

Human responses to odor perception follow patterns associated with both the olfactory and the CCS functions.

TABLE 5.1.1   INDUSTRIAL ODOR DESCRIPTORS

Odor Character Descriptor	Potential Sources
Nail polish	Painting, varnishing, coating
Fishy	Fish operation, rendering, tanning
Asphalt	Asphalt plant
Plastic	Plastics plant
Damp earth	Sewerage
Garbage	Landfill, resource recovery facility
Weed killer	Pesticide, chemical manufacturer
Gasoline	Refinery
Airplane glue	Chemical manufacturer
Household gas	Gas leak
Rotten egg	Sewerage, refinery
Rotten cabbage	Pulp mill, sewage sludge
Cat urine	Vegetation

These are discussed next to facilitate the understanding of what prompts odor complaints and the difficulties associated with odor identification and measurement.

SENSITIZATION, DESENSITIZATION, AND TOLERANCE OF ODORS

Repeated exposure to an odor can result in either an enhanced reaction described as sensitization or a diminishing reaction defined as tolerance. When people become sensitized to an odor, the complaints regarding the odor can increase. On the other hand, tolerance of an odor can attribute to a person's unawareness of the continuous exposure to a potentially harmful substance. As an illustration of the complexity of odor perception, pungency stimulation shows an increased response to odors from a continuous or quickly repetitive stimulus during a short term exposure. Also, a person can be more sensitive to one odorant than another. Such differences are often caused by repeated exposure to an odor. For example, tolerance is not uncommon for chemists or manufacturers who are exposed to an odorant daily over a period of years. Fatigue also affects odor perception; repeated exposure to an odorant can result in a desensitization to the odorant, where an observer can no longer detect an odor although it is strongly detectable by another.

ODOR MIXTURES

An additional problem in odor identification is that ambient odors are generally mixtures of compounds in different concentration levels. A comparison of the perceived intensity of two chemicals presented alone at a concentration and the two chemicals in a mixture at the same concentrations usually shows that the perceived intensity of either chemical in the mixture is lower than it is alone. This response is known as hypoadditivity. On the other hand, the CCS responses show that the perception of the odor is mainly additive, i.e., equal in the mixture to the perception of either chemical alone (Cometto-Muñiz and Cain 1991).

Questions of safety with chemical mixtures are more successfully answered by experimentally determining the odor threshold of the mixture and then relating it to the chemical composition of the mixture. Some mixtures can contain a highly odorous but relatively nontoxic chemical together with a nonodorous but highly toxic chemical. The chemicals in a mixture can also disassociate through aging or different chemical processes, and the judgment of whether a toxic component is present can be incorrect if it is based solely on the detection of the odorous component. Assumptions about the relationship between odor and risk can only be made for specific cases for chemical mixtures.

OTHER FACTORS AFFECTING ODOR PERCEPTION

Human response to odor is also based on the community, meteorology, and topography interactions. An odor with an acceptable hedonic tone and low intensity can generate complaints when the odor character is unfamiliar, causing concern with the toxicity. Conversely, some high intensities of hedonically unpleasant odors are tolerated because they are considered socially acceptable. Even pleasant odors, such as perfumes or roasting coffee, are considered objectionable if they persist for a long time or are frequently present.

Demographically, awareness of air pollution problems increases with education, income and occupational level, and age. Some of this awareness is also based on economic factors, such as property values.

ODOR AND HEALTH EFFECTS

Odors can warn of potentially dangerous materials. The warning property of odor has long been recognized. In the Middle Ages, odors were held responsible for disease, rather than being the result of disease and poor hygiene. Physicians protected themselves by theoretically purifying the air in a sick room with perfumed water. In this century, the property of odor as a warning agent has been used on a worldwide scale through the addition of pungent odorants such as ethyl mercaptan to odorless natural gas. The perceived connection between odors and disease persists. With national attention focused on waste and chemical spills, in the absence of specific information to the contrary, the average person concludes that a bad smell is unhealthy.

The relationship between odor and health effects should be clarified. Many odorants are perceptible at concentrations far below harmful concentration levels. For exam-

ple, hydrogen sulfide (H_2S) has been detected at concentrations as low as 0.15 ppb, whereas the acceptable exposure limit or the threshold limit value–time-weighted average (TLV–TWA) recommended by the American Conference of Governmental Industrial Hygienists (ACGIH) for this compound is 10 ppm, 6 orders of magnitude greater. Thus, the detection of the "rotten egg" character of H_2S is not necessarily an indicator of a potential health effect. Indeed, at concentrations near the TLV, the odor intensity of H_2S is unbearable. Similarly, the TLVs for creosol and ethyl acrylate are 4 orders of magnitude greater than their threshold values.

The nose is not a suitable screening device to determine the presence or the absence of a health risk. Although detecting an odor indicates that a chemical exposure has occurred, a more detailed sampling investigation should be conducted.

Studies that have reviewed community odor and health problems reveal that a variety of common ailments are related to chemical exposure. However, in most cases, the identified chemicals were well below the thresholds for toxicity. This evidence suggests that detecting unpleasant odors can cause adverse physiological and neurogenic responses such as nausea, stress, and low concentration levels and that these effects are a result of chemical exposure. Therefore, further studies are necessary to define allowable exposures and how they relate to odor detection.

By definition, chemicals that are hazardous to health are considered toxic whether odorous or not. They are therefore controlled under existing laws such as the Toxic Substances Act and the CAA and their related regulations.

—*Amos Turk*
Samuel S. Cha

References

American Industrial Hygiene Association (AIHA). 1989. *Odor Thresholds for chemicals with established occupational health standards.* AIHA.

American Society for Testing and Materials (ASTM). 1988. Standard practice for referencing suprathreshold odor intensity. E 544-75 (Reapproved 1988). ASTM.

———. 1993. *Standard terminology relating to sensory evaluation of materials and products.* E 253-93a. ASTM.

Cometto-Muñiz, J.E. and W.S. Cain. 1991. Influence of airborne contaminants on olfaction and the common chemical sense. Chap. 49 in *Smell and taste in health and disease,* edited by T.V. Getchell et al. New York: Raven Press.

Dravnieks, A. 1985. Atlas of odor character profiles. *ASTM Data Series 61,* ASTM.

Stevens, S.S. 1961. The psychophysics of sensory function. In *Sensory communications,* edited by Rosenblith. Cambridge, Mass.: MIT Press.

Turk, A. 1973. Expressions of gaseous concentration and dilution ratios. *Atmospheric Environment,* vol. 7:967–972.

U.S. Environmental Protection Agency (EPA). 1992. *Reference guide to odor thresholds for hazardous and pollutants listed in the Clean Air Amendments of 1990.* EPA 600/R-92/047.

5.2
ODOR CONTROL STRATEGY

Odor control is similar to any other air pollution control problem. Choosing a new material with less odor potential, changing the process, using an add-on pollution control system such as a scrubber or an afterburner, or raising the stack height all achieve community odor control. However, since odor is a community perception problem, the control strategies must be more flexible. When the tolerance level of a community changes, the odor control requirements also change. In addition, unique local topography and weather mean that each odor control problem usually requires its own study. An adopted control strategy for a similar problem at another place may not be suitable under different circumstances.

As far as control strategy is concerned, eliminating the odor-causing source is always better than using add-on control equipment because the equipment can become an odor source itself. For example, when a fume incinerator controls solvent odor, the incinerator, if not properly operated, can produce incomplete combustion by-products with a more offensive odor and lower odor threshold than the original odor. When multiple sources are involved, the control effort should be targeted to sources with higher odor emission rates, lower slopes of the dose-response function, and unpleasant odor characters.

Odor control techniques fall into the following categories:

Activated carbon adsorption
Adsorption with chemical reaction
Biofiltration
Wet scrubbing
Combustion
Dispersion

This section discusses the advantages and disadvantages of each technique.

Activated Carbon Adsorption

Activated carbon adsorption is a viable method in many odor control problems. Due to the nonpolar nature of its surface, activated carbon is effective in adsorbing organic and some inorganic materials. In general, organics having molecular weights over 45 and boiling points over 0°C are readily adsorbed.

The service life of activated carbon is limited by its capacity and the contaminating load. Therefore, provisions must be made for periodic renewal of the activated carbon. The renewal frequency is determined on the basis of performance deterioration (breakthrough) or according to a time schedule based on previous history or calculations of expected saturation. The exhausted carbon bed can be discarded or reactivated. Reactivation involves passing superheated steam through the carbon bed until sufficient material is desorbed. Recondensation of the material can recover it for reuse. This process is most suitable when only single compounds are involved. (Turk 1977).

Adsorption with Chemical Reaction

Impregnating filter media such as granular activated carbon or activated alumina with a reactive chemical or a catalyst can convert the adsorbed contaminants to less odorous compounds. For example, an air filter medium consisting of activated alumina and potassium permanganate can remove odorous hydrogen sulfide, mercaptans, and other sulfur contaminants. Another approach, patented by Turk and Brassey, can remove hydrogen sulfide from oxygen containing gas. The process involves adding ammonia to the air stream containing hydrogen sulfide prior to its passage through activated carbon. The ammonia catalyzes the oxidation of hydrogen sulfide to elemental sulfur. Impregnating activated carbon with sodium or potassium hydroxide can also neutralize hydrogen sulfide and form elemental sulfur. Both the ammonia/granular activated carbon system and the caustic/granular activated carbon system perform better than unimpregnated carbon.

Biofiltration

Biofiltration is an odor control technology that uses a biologically active filter bed to treat odorous chemical compounds. Materials such as soil, leaf compost, peat, and wood chips can be used for the filter bed. The filter bed provides an environment for microorganisms to degrade and ultimately remove the odorous chemical compounds.

Ideal operating requirements for a filter bed include high adsorption capacity, low pressure drop, high void fraction, high nutrient content, neutral to slightly alkaline pH, moderate temperatures, and adequate moisture content. Hydrogen sulfide, ammonia, and most organic components can be broken down in a biological filter, while most inorganic components cannot.

Wet Scrubbing

Wet scrubbing is another widely used technology for odor control. Most wet scrubbers for odor control employ reactive chemicals. Reactive scrubbing involves the removal of odorous materials by neutralization, oxidation, or other chemical reactions. Odor removal by any liquid–gas process is a function of the solubility of the chemical compound in the liquid phase, the total effective gas–liquid contact area, the concentration of the odorous chemical compound in the gas stream, and the residence time of the gas stream in the scrubber.

Atomized mist scrubbers and packed scrubbers are both capable of providing large gas–liquid contact areas for gas absorption. In a mist scrubber, pneumatic nozzles use a high-velocity compressed air stream to atomize a chemical solution into 5- to 20-μm-diameter drops, providing a large contact area. Unlike packed scrubbers, mist scrubbers do not clog even if the gas stream contains large particles. In a packed scrubber system, the chemical solution is often recirculated. As a result, the water and chemical consumption is significantly reduced.

For more complex odor problems caused by a mixture of odorants, multiple stages of scrubbers utilizing different chemicals in each stage are often used. Figure 5.2.1 shows a multiple-stage scrubber installed to control odorous emissions from a wastewater sludge treatment facility. The system uses sulfuric acid in the first-stage, cocurrent, coarse-packing, packed-tower scrubber to remove ammonia and particulates; a caustic/hypochlorite solution in the second-stage, cocurrent, horizontal-mist scrubber to

remove or reduce sulfur compounds and VOCs; and a weak caustic solution in the third-stage, cross-flow, packed-bed scrubber to remove the remaining odorants, especially those chlorine compounds generated in the second-stage scrubber. The control system also uses a 50-ft-tall stack to enhance the dilution of the remaining odorants during plume dispersion (Ponte and Aiello 1993).

Combustion

Combustion is an effective technique for odor control. The odor control efficiency of a combustion system depends largely on the level of complete combustion. Incomplete combustion can actually increase an odor problem by forming more odorous chemical compounds. Three major factors influence the design principles for odor emission control: temperature, residence time, and mixing. All three factors are interrelated to each other, and variation in one can cause changes in the others. Generally, temperatures of 1200 to 1400°F and residence times of 0.3 to 0.5 sec with good mixing conditions effectively destroy the odorous chemical compounds in a combustion chamber. If moisture and corrosion are not of concern, facilities often use the boiler onsite for odorant destruction.

Dispersion

The use of elevated emission points can reduce the odorant concentration at downwind ground level because of dispersion or dilution. As shown in Figure 5.2.2, aerodynamic building wake effects often downwash emissions from rooftop sources. When the stack height is increased, emissions avoid the building wake, disperse at a higher

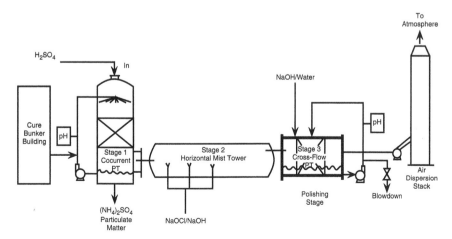

FIG. 5.2.1 Multiple-stage scrubber system schematic diagram.

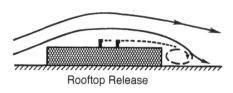
Rooftop Release

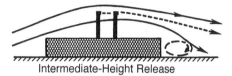
Intermediate-Height Release

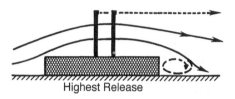
Highest Release

FIG. 5.2.2 Wake effect and stack height.

level, and are diluted during transport before reaching the ground.

—*Amos Turk*
Samuel S. Cha

References

Ponte, M., and K. Aiello. 1993. Odor emission and control at the world's largest chemical fixation facility. Presented at the New York Water Pollution Control Association, Inc., 65th Annual Winter Meeting.

Turk, A. 1977. Adsorption. In *Air Pollution,* 3rd ed. Edited by A.C. Stern. New York: Academic Press.

Turk, A., and J.M. Brassey. 1986. Removal of hydrogen sulfide from air streams. U.S. Patent 4,615,714.

6

Indoor Air Pollution

David H.F. Liu

6.1
RADON AND OTHER POLLUTANTS

PARTIAL LIST OF SUPPLIERS

Aircheck; Amersham Corporation; Electro-Mechanical Concepts, Inc.; Health Physics Associates, Ltd.; R.A.D. Service and Instruments Ltd.; Radon Inspection Service; Radon Testing Corporation of America; Ross Systems, Inc.; Scientific Analysis, Inc.; Teledyne Isotopes, Inc.; Terradex Corp.; University of Pittsburgh.

Growing scientific evidence indicates that the air within homes and other buildings can be more seriously polluted than outdoor air even in the largest and the most industrialized cities. For many people who spend 90% of their time indoors that indoor pollution can be unhealthy.

This section describes the source and effects of radon and other indoor pollutants and discusses techniques to improve the quality of indoor air.

Radon

This section is concerned with radon including its source and effects and control techniques.

SOURCE AND EFFECTS

Radon gas is produced by the decay of naturally occurring uranium found in almost all soils and rocks. Figure 6.1.1 shows the decay chain that transforms uranium into radon and its progeny. Radon is also found in soils contaminated with certain types of industrial waste, such as the by-products of uranium mining. Phosphate rock is a source of radon because deposits of phosphate often contain high levels of uranium, approximately 50 to 150 ppm.

A significant amount of radon is present in wells and soil in many parts of this country. Radon is commonly associated with granite bedrock and is also present in the natural gas and coal deposits in this rock. In its natural state, radon rises through airspace in the soil and enters a house through its basement, is released from agitated or boiled water, or escapes during natural gas use.

The most common pathways through which radon gas seeps in from the soil include cracks in concrete floors and walls, drain pipes, floor drains, sumps, and cracks or pores in hollow block walls. Radon is drawn in by reduced air pressure, which results when the interior pressure drops below the pressure in the ground. This pressure drop is commonly caused by a warmer indoor climate; kitchen or attic exhaust fans; or consumption of interior air by furnaces, clothes dryers, or other appliances.

Radon is a colorless, odorless, almost chemically inert, radioactive gas. It is soluble in cold water, and its solubility decreases with increasing temperature. This characteristic of radon causes it to be released during water-related activities, such as taking showers, flushing toilets, and general cleaning.

Scientists and health officials express fears that the reduced infiltration of fresh air from the outside to increase energy efficiency is eliminating the escape route for radon and making a bad indoor pollution problem worse. Other factors to consider include the inflow rate of radon which depends on the strength of the radon source beneath the house and the permeability of the soil.

Since radon is naturally radioactive, it is unstable, giving off radiation as it decays. The radon decay products,

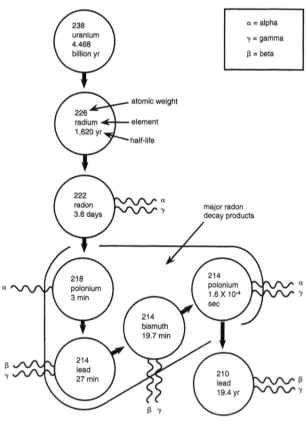

FIG. 6.1.1 Radium decay chart. About halfway through its decay sequence, uranium becomes a gas, radon, which as it disintegrates gives off radioactive particles of polonium, bismuth, and lead. Also noted is the half-life of each material—the time required for half of its radioactivity to dissipate.

radon progeny, or radon daughters, which are formed, cling to dust. If inhaled, the dust can become trapped in the lung's sensitive airways. As the decay products break down further, more radiation is released which can damage lung tissue and lead to lung cancer after a period of ten to thirty years. Outside, radon dissipates quickly. However, in an enclosed space, such as a house, it can accumulate and cause lung cancer. Scientists believe that smoking increases any cancer risk from radon.

Figure 6.1.2 shows how the lung cancer risks of radon exposure compare to other causes of the disease. Scientists at the U.S. EPA estimate that if 100 individuals are exposed to a level of 4 picocuries per liter (pCi/l) over seventy years, between one and five of them will contract lung cancer. If these same individuals live in houses with levels of 200 pCi/l for only ten years, the number of anticipated lung cancer deaths would rise to between four and forty-two out of 100.

Working Levels (WL) and pCi/l

By definition, a curie is the decay rate of one gram of radium—37 billion decays per sec. Radioactivity in the environment is usually measured in trillionth of a curie or a picocurie (pCi). With a concentration of 1 pCi/l of air, about 2 alpha particles are emitted per minute from radon atoms in each liter of air.

Another unit, the WL derived from mine regulations, is sometimes used. Actually, the WL is a measure of the concentration of radon daughters rather than of radon itself. Approximately, 1 pCi/l of radon gas is equivalent to 0.005 WL, or 1 WL is equivalent to about 200 pCi/l of radon gas.

RADON DETECTION

The two most common radon testing devices are the charcoal canister and the alpha-check detector. Alpha-check detectors are ideal for making long-term measurements but are not suited for quick results (Lafavore 1987). A charcoal-adsorbent detector is the most practical approach for most. This test method is low-cost; the price for a single unit ranges from $10 to 50 (Cohen 1987). However, a disadvantage of charcoal is its sensitivity to temperature and humidity.

Charcoal-adsorbent detectors consist of granules of activated charcoal that adsorb gases (including radon) to which they are exposed. When the charcoal is exposed long enough to become saturated, the canister is resealed and shipped back to a laboratory where the radioactivity is measured. The laboratory calculates the level of radon to which the device was exposed.

Ideally, testing should be done in the late fall or early spring when the house is closed and the heating system is

not turned on. Less natural ventilation does not dilute the radon concentrations at these times. The next best times are during winter months. Detectors that measure for three days or less require a closed-house procedure to start at least twelve hours before the test.

The detector should be placed in the lowest finished space in a spot where it is not disturbed. If the house has a basement, it should be placed there. The detector should be kept away from any sources of air movement, such as fire places and heating vents. To avoid the diluting effect of floor drafts, the tester should place the detector at least 20 in above the floor. The recommended exposure time for a radon detector is one to seven days.

The EPA recommends 4 pCi/l as the maximum acceptable indoor radon level. When the radon levels in a home range between 4 and 20 pCi/l, homeowners should take actions to reduce the air level within one to two years. If the level is between 20 and 200 pCi/l, action should be taken within several months. A reading above 200 pCi

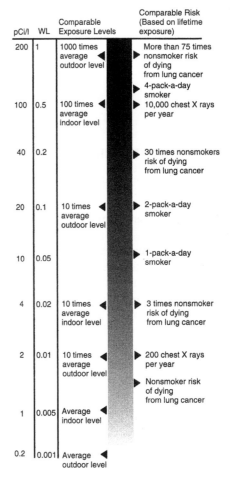

FIG. 6.1.2 Radon risk evaluation chart. This chart shows how the lung cancer risks of radon exposure compare to other causes of the disease. For example, breathing 20 pCi/l poses about the same lung cancer risk as smoking two packs of cigarettes a day. (Reprinted from U.S. Environmental Protection Agency.)

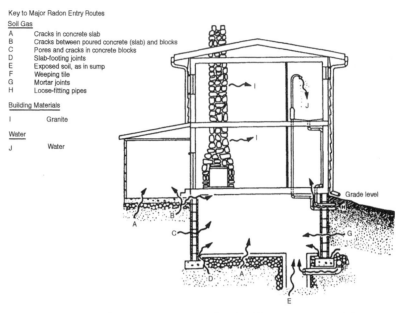

Key to Major Radon Entry Routes

Soil Gas

A   Cracks in concrete slab
B   Cracks between poured concrete (slab) and blocks
C   Pores and cracks in concrete blocks
D   Slab-footing joints
E   Exposed soil, as in sump
F   Weeping tile
G   Mortar joints
H   Loose-fitting pipes

Building Materials

I   Granite

Water

J   Water

FIG. 6.1.3 Major radon entry routes into detached houses (Reprinted from U.S. Environmental Protection Agency (EPA), 1986, *Reduction techniques for detached houses,* EPA 625/5-86/019, Research Triangle Park, N.C.: U.S. EPA.)

should be promptly reported and confirmed by the state department of environmental protection or department of public health.

RADON CONTROL TECHNIQUES

Techniques to control radon in residence can be broadly classified as follows:

Source removal (new construction considerations)
Source control
Sealing major radon source
Sealing radon entry routes (see Figure 6.1.3)
Subslab ventilation (Figure 6.1.4)
Drain-tile soil ventilation (Figure 6.1.5)
Active ventilation of hollow block basement walls (see Figure 6.1.6)
Avoidance of house depressurization
Ventilation of indoor radon concentration
Natural circulation
Forced-air ventilation
Heat recovery ventilation

However, the effectiveness of indoor air ventilation decreases with increased ventilation rates. Also, the radon source concentration and radon entry rate can reach a finite level where high ventilation rates are not productive (see Figure 6.1.7).

Table 6.1.1 summarizes the methods of reducing the radon level in a residence. No two houses have identical radon problems. Therefore, a routine method for reducing radon levels does not exist. Most homes usually re-

quire more than one of the nine methods listed to significantly reduce their radon levels. Before deciding on an approach, a homeowner should consider the unique characteristics of the house and consult a contractor specializing in the remediation of radon problems.

Other Indoor Pollutants

This section discusses the source and effects and control techniques for other indoor pollutants.

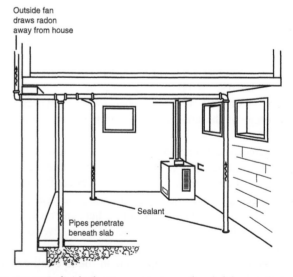

Outside fan draws radon away from house

Sealant

Pipes penetrate beneath slab

FIG. 6.1.4 Individual pipe variation of subslab ventilation. (Reprinted from U.S. EPA, 1986.)

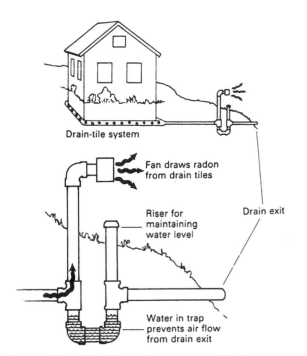

FIG. 6.1.5 Drain-tile soil ventilation system draining to remote discharge area.

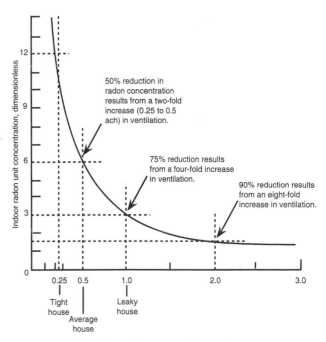

FIG. 6.1.7 Effect of ventilation on indoor radon concentrations.

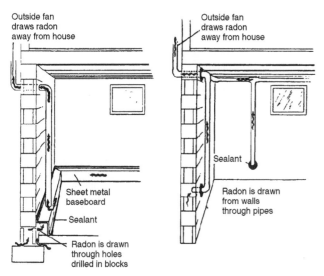

FIG. 6.1.6 Two variations of wall ventilation: the baseboard method and the single-point pipe method.

SOURCE AND EFFECTS

Other indoor pollutants include asbestos, bioaerosols, carbon dioxide and carbon monoxide, formaldehyde, nitrogen oxides, ozone, inhalable particulates, and VOCs.

Asbestos

Asbestos is a silicate mineral fiber that is flexible, durable, and incombustible and makes good electrical and thermal isolators. It has been used as insulation for heating, water, and sewage pipes; sound absorption and fireproofing materials; roof, siding, and floor tiles; corrugated paper; caulking; putty; and spackle. In short, asbestos was used extensively in all types of construction until about 1960. Once released from its binding material (by erosion, vibration, renovation, or cleaning) the fibers can remain airborne for long periods.

Bioaerosols

Biological contaminants include animal dander, cat saliva, human skin scales, insect excreta, food remnants, bacteria, viruses, mold, mildew, mites, and pollen. The sources of these contaminants include outdoor plants and trees, people, and animals. Pollens are seasonal; fungal spores and molds are prevalent at high temperatures. A central air-handling system can distribute these contaminants throughout a building.

Carbon Dioxide and Carbon Monoxide

Carbon monoxide, a colorless, odorless, toxic gas formed by the incomplete combustion of fossil fuels, is the most prevalent and dangerous indoor pollutant. It results from poorly ventilated kitchens, rooms over garages, and unvented combustion appliances (stoves, ovens, heaters, and the presence of tobacco smoke).

Formaldehyde

Formaldehyde, the simplest of aldehydes, is a colorless gas that is emitted from various building materials, household

TABLE 6.1.1 SUMMARY OF RADON REDUCTION TECHNIQUES

Method	Principle of Operation	House Types Applicable	Estimated Annual Average Concentration Reduction, %	Confidence in Effectiveness	Operating Conditions and Applicability	Estimated Installation and Annual Operating Costs
Natural ventilation	Exchanges air causing replacement and dilution of indoor air with outdoor air by uniformly opening windows and vents	All[a]	90[b]	Moderate	Opening windows and air vents uniformly around house. Air exchange rates up to 2 air changes per hour (ACH) attainable. Can require energy and comfort penalties and loss of living space use	No installation cost. Operating costs for additional heating range up to 3.4-fold increase from normal (0.25 ACH) ventilation conditions.
Forced-air ventilation	Exchanges air causing replacement and dilution of indoor air with outdoor air by fans located in windows or vent openings	All	90[b]	Moderate	Continuous operation of a central fan with fresh air makeup, window fans, or local exhaust fans. Forced air ventilation used to increase air exchange rates to 2 ACH. Can require energy and comfort penalties and loss of living space use	Installation costs range up to $150. Operating costs range up to $100 for fan energy and up to 3.4-fold increase in normal (0.25 ACH) heating energy costs[c].
Forced-air ventilation with heat recovery	Exchanges air causing replacement and dilution of indoor air with outdoor air by a fan-powered ventilation system	All	90[b]	Moderate to high	Continuous operation of units rated at 25–240 cfm. Air exchange increase from 0.25 to 2 ACH. In cold climates, recovery up to 70% of heat that would be lost through house ventilation without heat recovery	Installation costs range from $400 to 1500 for 25–240-cfm units. Operating costs range up to $100 for fan energy plus up to 1.4-fold increase in heating costs assuming a 70% efficient heat recovery[c].
Active avoidance of house depressurization	Provides clean makeup air to household appliances which exhaust or consume indoor air	All	0–10[e]	Moderate[f]	Providing outside makeup air to appliances such as furnaces. fireplaces, clothes dryers, and room exhaust fans	Installation costs of small dampered ductwork are minimal. Operating benefits can result from using outdoor air for combustion sources.
Sealing major radon sources	Uses gas-proof barriers to close off and exhaust or ventilate sources of soil-gas-borne radon	All	Local exhaust of the source can produce significant house-wide reductions.	Extremely case-specific	Sealing areas of major soil-gas entry such as cold rooms, exposed earth, sumps, or basement drains and ventilating by exhausting collected air to the outside	Most jobs are accomplished for less than $100. Operating costs for a small fan are minimal.

Continued on next page

TABLE 6.1.1 *Continued*

Method	Principle of Operation	House Types Applicable	Estimated Annual Average Concentration Reduction, %	Confidence in Effectiveness	Operating Conditions and Applicability	Estimated Installation and Annual Operating Costs
Sealing radon entry routes	Uses gas-proof sealants to prevent soil-gas-borne radon-entry	All	30–90	Extremely case-specific	Sealing all noticeable interior cracks, cold joints, openings around services, and pores in basement walls and floors with appropriate materials	Installation costs range between $300 and 500.
Drain-tile soil ventilation	Continuously collects dilutes and exhausts soil-gas-borne radon from the footing perimeter of houses	BB[a] PCB[a] S[a]	Up to 98	Moderate[g]	Continuous collection of soil-gas-borne radon using a 160-cm fan to exhaust a perimeter drain tile Applicable to houses with a complete perimeter, footing level, drain tile system and with no interior block walls resting on subslab footings	Installation cost is $1200 by contractor. Operating costs are $15 for fan energy and up to $125 for supplemental heating.
Active ventilation of hollow-block basement walls	Continuously collects, dilutes, and exhausts soil-gas-borne radon from hollow-block basement walls	BB[a]	Up to 99+	Moderate to high	Continuous collection of soil-gas-borne radon using one 250-cm fan to exhaust all hollow-block perimeter basement walls Baseboard wall collection and exhaust system used in houses with French (channel) drains	Installation costs for a single, suction-and-exhaust-point system is $2500 (contractor-installed in unfinshed basement). Installation cost for a baseboard wall collection system is $5000 (contractor-installed in unfinished basement). Operating costs are $15 for fan energy and up to $125 for supplemental heating.
Subslab soil ventilation	Continuously collects and exhausts soil-gas-borne radon from the aggregate or soil under the concrete slab	BB[a] PCB[a] S[a]	80–90, as high as 99 in some cases	Moderate to high	Continuous collection of soil-gas-borne radon using one fan (~100 cfm, ≥0.4 in H_2O suction) to exhaust aggregate or soil under slab For individual suction point approach, roughly one suction point per 500 sq ft of slab area Alternate approach of piping network under slab, permitting adequate ventilation without a power-driven fan	Installation cost for individual suction point approach is about $2000 (contractor installed). Installation costs for retrofit subslab piping network are over $5000 (contractor installed). Operating costs are $15 for fan energy (if used) and up to $125 for supplemental heating.

Source: U.S. Environmental Protection Agency (EPA), 1986, *Reduction for detached houses,* EPA 625/5-86/019 (Research Triangle Park, N.C.: U.S. EPA).
Note: [a]BB (block basement) houses with hollow-block (concrete block or cinder block) basement or partial basement, finished or unfinished
PCB (poured concrete basement) houses with full or partial, finished or unfinished poured-concrete walls
C (crawl space) houses built on a crawl space
S (slab or slab-on-grade) houses built on concrete slabs

products, or combustion processes. Indoor sources include pressed-wood products, including particle board, paneling, fiberboard, and wallboard; textiles, such as carpet backings, drapes and upholstery fabrics, linens, and clothing; urea–formaldehyde foam insulation; adhesives; paints; coatings; and carpet shampoos.

Minimal outgassing by each product can significantly increase the formaldehyde level when the ventilation rate is low. Hot and humid conditions usually cause formaldehyde to outgas at a greater rate. Product aging diminishes its emission rate although in some cases, this process can take several years.

Nitrogen Oxides

Nitrogen oxides are combustion by-products produced by the burning of natural gas or oil in oxygen-rich environments such as kitchen stoves and ovens, furnaces, and unvented gas and kerosene heaters. When a fireplace or wood stove is used, some of these pollutants enter the room. Cracks in the stovepipe, downdrafts, or wood spillage from a fireplace can exacerbate the condition. Coal burning adds sulfur dioxide to the nitrogen oxides.

Ozone

Ozone is recognizable by its strong, pungent odor. Indoors, significant ozone can be produced by electrostatic copying machines, mercury-enhanced light bulbs, and electrostatic air cleaners. Poorly ventilated offices and rooms housing photocopying machines can accumulate significant levels of ozone.

Inhalable Particulates

Particulates are not a single type of pollutant; they describe the physical state of many pollutants—that is, all suspended solid or liquid particles less than a few hundred μm in size. Among the pollutants that appear as particulates are asbestos and other fibrous building materials, radon progeny, smoke, organic compounds, infectious agents, and heavy metals, such as cadmium in cigarette smoke.

Because of the diversity of particulates and their chemical nature, considering the adverse health effects of this category as a whole is not possible.

VOCs

VOCs are chemicals that vaporize readily at room temperature. High levels of organic chemicals in homes are attributed to aerosols, cleaners, polishes, varnishes, paints, pressed-wood products, pesticides, and others.

CONTROL TECHNIQUES

The basic control techniques to improve the quality of indoor air are source removal, ventilation, isolation, and air cleaners.

Source Removal

This technique involves removing or modifying the source of pollution and replacing it with a low-pollution substitute. Asbestos pipe insulation should be encased securely or replaced by nonasbestos pipe insulation if possible. Limiting the use of formaldehyde insulation, particle boards, carpets, fabric, or furniture containing formaldehyde can limit exposure to formaldehyde. Replacing kerosene and gas space heaters with electric space heaters can eliminate exposure to carbon dioxide, carbon monoxide, and nitrogen oxide. Biological contaminants and VOCs can be controlled by source removal.

Ventilation

Increasing ventilation can remove the offending pollutants, such as VOCs, and dilute the remaining pollution to a safer concentration. For example, gas stoves can be fitted with range hoods and exhaust fans that draw the air and effluent in over the cooking surface and blow carbon dioxide, carbon monoxide, and nitrogen oxides outdoors. However, range hoods that use charcoal filters to clean the air and then revent it to the room are ineffective in controlling carbon dioxide, carbon monoxide, and nitrogen oxides. One solution is to open a window near the stove and fit it with a fan to blow out the pollutants.

Forced ventilation with window fans works the same as natural ventilation but insures a steadier and more reliable ventilation rate. Figure 6.1.8 shows heat exchangers that are the basis of a heat-recovery ventilation system.

One way to measure the success of a ventilation system is to use a monitor that detects carbon dioxide (a gas that causes drowsiness in excess amounts). If carbon dioxide levels are high, other pollutants are also likely to be present in excessive amounts. Monitors are available that trigger a ventilation system to bring in more fresh air when needed (Soviero 1992).

Isolation

Isolating certain sources of pollution and preventing their emissions from entering the indoor environment is best. Formaldehyde outgassing from urea–formaldehyde foam insulation can also be partially controlled by vapor barriers; wallpaper or low-permeability paint applied to interior walls; and plywood coated with shellac, varnish, polymeric coatings, or other low-diffusion barriers. These

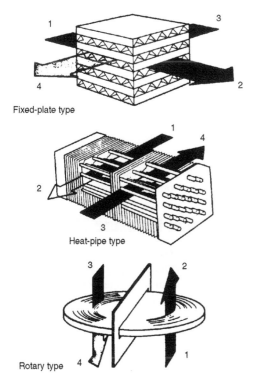

Fixed-plate type

Heat-pipe type

Rotary type

FIG. 6.1.8 Heat exchangers for heat-recovery ventilator. A heat-exchange element is the basis of a heat-recovery ventilator. Fresh outdoor air (1) is warmed as it passes through the exchanger and enters the house (2). Stale indoor air (3) leaving the house is cooled as it transfers heat to the exchanger and is vented outside (4). In the fixed-plate type, heat is transferred through plastic, metal, or paper partitions. The turning wheel of the rotary type picks up heat as it passes through the warm air path and surrenders heat to the cold air stream half a rotation later. Liquid refrigerant in the pipes of the heat-pipe type evaporates at the warm end and condenses at the cold end, transferring heat to the cold air.

barriers contain the formaldehyde outgasses, which are seemingly reabsorbed by their source rather than released into the home.

Air Cleaners

The air cleaners for residential purposes are based on filtration, adsorption, and electrostatic precipitation as follows:

Filters made of charcoal, glass fibers, and synthetic materials are used to remove particles. Pollen or lint, which

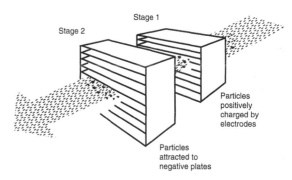

Stage 1

Stage 2

Particles positively charged by electrodes

Particles attracted to negative plates

FIG. 6.1.9 Two-step precipitator.

are relatively large particles, are easily trapped by most filters. High-efficiency particulate air (HEPA) filters can remove particles larger than 0.3 μm, which include bacteria and spores, but not viruses.

Whereas filters trap larger particles, adsorbents react with the molecules. Three common adsorbents are activated charcoal, activated alumina, and silica gel. Adsorbents remove gases, such as formaldehyde, and ammonia.

Electrostatic air cleaners work by charging airborne particles with either a negative or positive electric charge. These charged particles gravitate to oppositely charged, special collector plates within the air cleaner. The technique is effective against dust, smoke particles, and some allergens. Figure 6.1.9 shows the operation principles of a two-step precipitator.

The National Aeronautics and Space Administration (NASA) studies have long indicated another simpler way to handle pollutants: household plants which absorb some toxins. The spider plant and philodendron, for example, remove formaldehyde and carbon dioxide from indoor air (Soviero 1992).

—David H.F. Liu

References

Cohen, Bernard. 1987. Radon: *A homeowner's guide to detection and control.* (August). New York: Consumers Union.

Lafavore, Michael. 1987. Radon: *The invisible threat.* Emmaus, Pa.: Rodale Press.

Soviero, Marcelle M. 1992. Can your house make you sick? *Popular Science* (July).

6.2
AIR QUALITY IN THE WORKPLACE

PARTIAL LIST OF SUPPLIERS

Altech; American Gas & Chemical Ltd; Bacharach Inc.; CEA Instruments Inc.; Coleman & Palmer Instrument Co.; Du Pont; Enmet Analytical Corp.; Gas Tech Inc.; Gilian; Lab Safety Supply; MDA Scientific Inc.; 3 M; National Draeger Inc.; Sensidyne Inc.; Sierra Monitor Inc.

OSHA requires employers to provide a safe and healthy working environment for employees. Of greatest concern are gases and vapors in the extremely toxic category, having TLV, permissible exposure limit (PEL), short-term exposure limit (STEL), or TLV-C values of 10 ppm or less. These standards are usually developed by the National Institute of Occupational Safety and Health (NIOSH).

This section focuses on exposure limits, continuous dosage sensors, and the material safety data sheet (MSDS), which contains standardized information about the properties and hazards of toxic substances.

Exposure Limits

Table 6.2.1 lists some commonly occurring toxic components. The OSHA limits for many industrial chemicals are in 29 *CFR* 1910.1000 (OSHA 1989).

The NIOSH uses the following terms to describe exposure limits:

ACTION LEVEL—The exposure concentration at which certain provisions of the NIOSH-recommended standard must be initiated, such as periodic measurements of worker exposure, training of workers, and medical surveillance.

CA—A substance that NIOSH recommends be treated as a potential human carcinogen.

CEILING—A description usually used with a published exposure limit that refers to a concentration that should not be exceeded, even for an instant.

IMMEDIATELY DANGEROUS TO LIFE OR HEALTH (IDLH)—A level defined for respiratory protection that represents the maximum concentration from which, in the event of respiratory failure, a person can escape within 30 min without experiencing any impairing or irreversible health effects.

PEL—An exposure limit published and enforced by OSHA as a legal standard.

RECOMMENDED EXPOSURE LIMIT (REL)—The exposure recommended by NIOSH not to be exceeded.

STEL—The maximum concentration to which workers can be exposed for 15 min four times throughout the day

with at least 1 hr between exposures.

TWA—The average time, over a work period, of a person's exposure to a chemical or agent determined by sampling for the contaminant throughout the time period.

Occupational Exposure Monitoring

Table 6.2.2 lists selected workplace safety and health hazard standards. Monitoring can be accomplished through the use of color change badges, color detector tubes, and other monitoring techniques.

COLOR CHANGE BADGES

Breathing zone monitoring can be performed via a small media (monitor) fastened to the worker's collar or lapel for periods of time corresponding to the STEL and TWA exposure limits. Table 6.2.3 lists the available badges and badge holders, including a smoke detector. When the exposed badge is treated with a developing agent, a color change results, which can be interpreted by electronic monitors. Suppliers analyze the exposed badges within 24 hr and return an analysis report.

COLOR DETECTOR (DOSIMETER) TUBES

Color detector tubes (CDT) determine contaminant concentrations in work areas without pumps, charts, or training. The testor simply opens the detector tube, inserts it into the tube holder, and pulls a predetermined amount of sample air through the CDT. The color progression on the packed adsorption bed indicates the concentration of a particular toxic gas. Disposable CDTs are inexpensive, and some 250 different tubes are available. Table 6.2.4 lists some of the available tubes.

OTHER MONITORING TECHNIQUES

Membrane filters with air sampling pumps monitor nuisance dust, lead silica, zinc, mineral oil, mist, and more.

The following activities are recommended for an employee exposure control program:

Assign workers the correct monitors to determine their exposure level.

Take enough samples to get an accurate, representative sampling.

Evaluate the results to determine compliance with current OSHA standards.

TABLE 6.2.1 TOXIC GASES AND VAPORS (10 PPM AND BELOW)

Name of Element or Compound	TLV-TWA, ppm	STEL, ppm	TLV-C, ppm	Human Carcinogenity Status
Acetic acid (vinegar)	10			
Acrolein	0.1	0.3		
Acrylonitrile	2			Suspected
Aniline	2			
Arsine	0.05			
Benzene	10			Suspected
Biphenyl	0.2			
Boron trifluoride			1	
Bromine	0.1	0.3		
Bromine pentafluoride	0.1			
Bromoform	0.5			
1,3 Butadiene	10			Suspected
Carbon disulfide	10			
Carbon tetrabromide	0.1			
Carbon tetrachloride	5			Suspected
Carbonyl difluoride	2	5		
Chlorine	0.5	1		
Chlorine dioxide	0.1	0.3		
Chlorine trifluoride			0.1	
Chloroform	10			Suspected
bis (Chloromethyl)ether	0.001			Confirmed
Diborane	0.1			
Dichloroacetylene			0.1	
Dimethyl hydrazine	0.5			Suspected
Dimethyl sulfate	0.1			Suspected
Ethylene oxide	1			Suspected
Fluorine	1	2		
Formaldehyde			0.3	Suspected
Germane	0.2			
Hydrazine	0.1			Suspected
Hydrogen bromide			3	
Hydrogen chloride			5	
Hydrogen cyanide			10	
Hydrogen fluoride			3	
Hydrogen selenide	0.05			
Hydrogen sulfide	10			
Methyl isocyanate	0.02			
Methyl mercaptan	0.5			
Naphthalene	10	15		
Nickel carbonyl	0.05			
Nitric acid	2	4		
Nitrobenzene	1			
Nitrogen dioxide	3	5		
2 Nitropropane	10			Suspected
Osmium tetroxide	0.0002	0.0006		
Oxygen difluoride			0.05	
Ozone			0.1	
Pentborane	0.005			
Phenylhydrazine	0.1			Suspected
Phosgene	0.1			
Phosphine	0.3	1		
Phosphorus oxychloride	0.1			
Phosphorus pentachloride	0.1			
Phosphorus trichloride	0.2	0.5		
Silane	5			
Sodium azide			0.11	
Stibine			0.1	
Sulfur dioxide	2	5		
Sulfur tetrafluoride			0.1	
Tellurium hexafluoride	0.02			
Thionly chloride			1	
Toluidine	2			Suspected
Vinyl bromide	5			Suspected
Vinyl chloride	5			Confirmed

TABLE 6.2.2　SELECT WORKPLACE SAFETY AND HEALTH
STANDARDS

Contaminant	Concentration[a]	
	(ppm)	(mg/m³)
Ammonia	50	35
Carbon dioxide	5000	9000
Carbon monoxide	50	55
Cresol	5	22
Formaldehyde	2	3
Furfuryl alcohol	50	200
Nitric oxide	25	30
Nitrogen dioxide	5	9
Octane	500	2350
Ozone	0.1	0.2
Propane	1000	1800
Sulfur dioxide	5	13
TCA	50	240
Inert or nuisance dust, respirable fraction	—	5
Asbestos	[b]	[b]
Coal dust	—	2.4

Source: Occupational Safety and Health Administration (OSHA), 1989, Air contaminants—Permissible exposure limits standard, *Code of Federal Regulations,* Title 29, sec. 1910.1000, App. 2 (Washington, D.C.: U.S. Government Printing Office).

Notes: [a]Values are 8-hr, time-weighted averages, except values for nitrogen dioxide, which are ceiling values.

[b]Fewer than two fibers longer than 5 μm in each cubic centimeter.

If the results show overexposure, determine the cause, and adjust the process and procedures of the task involved.

If no overexposure is found, a decision should be made on the frequency of routine monitoring.

Employees should be monitored for possible new exposures any time the process or procedure changes.

Document all results.

TABLE 6.2.3　PERSONNEL-MONITORING TOXIC GAS
EXPOSURE BADGES

Description	OSHA PEL[b]
Ethylene oxide	1 ppm
Ethylene oxide, STEL[a]	5 ppm
Xylene	100 ppm
Formaldehyde	0.3 ppm
Carbon monoxide	50 ppm
Smoke-check	—

Source: Cole–Parmer Instrument Co.
Notes: [a]Short-term exposure limits for 15-min exposure; all others require 8 hr.
[b]OSHA permissible exposure levels.

MSDSs

A major area that OSHA regulations address is the Hazard Communication Standard. The overall goal of the standard is to implement risk management and safety programs by regulated employers. According to the standard, employers must instruct employees on the nature and effects of the toxic substances with which they work, either in written form or in training programs. The instruction must include the following:

The chemical and common name of the substance
The location of the substance in the workplace
Proper and safe handling practices
First aid treatment and antidotes in case of overexposure
The adverse health effects of the substance
Appropriate emergency procedures
Proper procedures for cleanup of leaks or spills
Potential for flammability, explosion, and reactivity
The rights of employees under this rule

Most of this information is available from the MSDS, 29 CFR 1910.1200, U.S. Department of Labor.

Employers must keep copies of MSDSs for each hazardous chemical in the workplace readily accessible to their employees. If the nature of the job is such that employees travel between different workplaces during a work shift,

TABLE 6.2.4 FEATURES AND CAPABILITIES OF COLOR-CHANGING DOSIMETER TUBES

Description	Measuring Range (ppm)	Typical Applications	Shelf Life (yr)
Ammonia (NH_3)	50 to 900	Process control	3
Ammonia (NH_3)	10 to 260	Industrial hygiene, leak detection	3
Benzene (C_6H_6)	5 to 200	Industrial hygiene	1
Carbon dioxide (CO_2)	0.05 to 1.0%	Air contamination, concentration control	2
Carbon monoxide (CO)	10 to 250	Blast furnace, garage, combustion	1
Hydrogen sulfide (H_2S)	100 to 2000	Industrial raw gases, metallurgy	3
Hydrogen sulfide (H_2S)	3 to 150	Metal or oil refinery, chemical lab	3
Methyl bromide (CH_3Br)	5 to 80	Insert fumigation for mills and vaults	1
Nitrogen dioxide (NO_2)	0.5 to 30.0	Arc welding, acid dipping of metal products	1
Sulfur dioxide (SO_2)	20 to 300	Metal refining, waste gas analysis	2
Smoke tubes	—	Ventilation and air flow determination	—

Source: Cole-Parmer Instrument Co.

the MSDSs can be kept at a central location at the employer's primary facility.

—*David H.F. Liu*

References

Occupational Safety and Health Administration (OSHA). 1989. Air contaminants—Permissible exposure limits standards. *Code of Federal Regulations,* Title 29, sec. 1910.1000. Washington, D.C.: U.S. Government Printing Office.

Bibliography

Adams, W.V., et al. 1990. *Guidelines for meeting emissions regulations for rotating machinery with mechanical seals.* STLE Publication SP-30 (October).

Air and Waste Management Association. 1992. *Air pollution engineering manual.* Edited by A.J. Buonicore and W.T. Davis. New York: Van Nostrand Reinhold.

American Conference of Governmental Industrial Hygienists (ACGIH). 1967. *Air sampling instruments.* Cincinnati, Ohio.

———. 1978. *Air sampling instruments for evaluation of atmospheric contaminants.* 5th ed.

———. 1993. *1993–1994 Threshold limit values.* Cincinnati, Ohio.

American Society for Testing and Materials (ASTM). 1968. *Basic principles of sensory evaluation.* STP 433. ASTM.

———. 1968. *Manual on sensory testing methods.* STP 434. ASTM.

———. 1971. *1971 Annual book of ASTM standards.* Part 23: Water: Atmospheric analysis. Philadelphia: ASTM.

———. 1981. *Guidelines for the selection and training of sensory panel members.* STP 758. ASTM.

Amoore, J.E., and E. Hautala. 1983. Odor as an aid to chemical safety: Odor thresholds compared with threshold limit values and volatiles for 214 industrial chemicals in air and water dilution. *J. Applied Toxicology* 3:272–290.

Benitez, J. 1993. *Process engineering and design for air pollution control.* PTR Prentice Hall.

Bohn, H., and R. Bohn. 1988. Soil beds weed out air pollutants. *Chem. Eng.* (25 April).

Bohn, H.I., et al. 1980. Hydrocarbon adsorption by soils as the stationary phase of gas-solid chromatography. *J. Environ. Quality* (October).

Buonicore, J. and W.T. Davis. 1992. *Air pollution engineering manual.* New York: Van Nostrand Reinhold.

Calvert, S., and H.M. Englund, eds. 1984. *Handbook of air pollution control.* John Wiley and Sons, Inc.

Cha, S.S., and K.E. Brown. 1992. Odor perception and its measurement—An approach to solving community odor problems. *Operations Forum,* vol. 9, no. 4:20–24.

Cheremisinoff, P.N., and A.G. Morressi. 1977. *Environmental assessment and impact statement handbook.* Ann Arbor Science Publishers Inc.

Cooper, C. David, and F.C. Alley. 1986. *Air pollution control—A design approach.* Prospect Heights, Ill.: Waveland Press, Inc.

Cothern, C.R., and J.E. Smith, Jr., eds. 1987. *Environmental radon.* New York: Plenum Press.

Crocker, B.B. 1974. Monitoring plant air pollution. *Chemical Engineering Progress* (January).

Downing, T.M. 1991. Current state of continuous emission monitoring technology. Presented to Municipal Utilities Conference of the American Public Power Association Meeting, Orlando, FL, February 1991.

Dravnieks, A. 1972. Odor perception and odorous air pollution. *Tappi* 55:737–742.

Eklund, B. 1992. Practical guidance for flux chamber measurements of fugitive volatile organic emission rates. *Journal of the Air and Waste Management Association.* (December).

Franz, J.J., and W.H. Prokop. 1980. Odor measurement by dynamic olfactometry. *J. Air Pollution Association* 30:1283–1297.

Gas detectors and analyzers. 1992. *Measurements and Control* (October).

Helmer, R. 1974. Desodorisierung von geruchsbeladend abluft in bodenfiltern. *Gesundheits-Ingenieur,* vol. 95, no. 1:21–26.

Hesketh, H. 1974. Fine particle collection efficiency related to pressure drop, scrubbant and particle properties, and contact mechanisms. *J. APCA* 24:939–942.

Hines, A.L., T.K. Ghosh, S.K. Loyalka, and R.C. Warder, Jr. 1993. *Indoor air: Quality and control.* Englewood Cliffs, N.J.: PRT Prentice Hall.

Hochheiser, S., F.J. Burmann, and G.B. Morgan. 1971. Atmospheric surveillance, the current state of air monitoring technology. *Environmental Science and Technology* (August).

Hodges, D.S., V.F. Medina, R.L. Islander, and J.S. Devinny. 1992. Biofiltration: Application for VOC emission control. 47th Annual Purdue Industrial Waste Conference, West Lafayette, Indiana, May 1992.

Jacobs, M.B. 1967. *The analytical toxicology of industrial inorganic poisons.* Vol. 22, *Chemical analysis.* New York: Interscience.

Laird, J.C. 1978. Unique extractive stack sampling. 1978 ISA Conference,

Houston.

Laznow, J., and T. Ponder. 1992. Monitoring and data management of fugitive hazardous air pollutants. ISA Conference, Houston, October 1992.

Licht, W. 1980. *Air pollution control engineering.* New York: Marcel Dekker, Inc.

Liptak, B. 1994. *Analytical instrumentation.* Radnor, Pa.: Chilton Book Company.

Lord, H.C., and R.V. Brown. 1991. Open-path multi-component NDIR monitoring of toxic combustible or hazardous vapors. 1991 ISA Conference, Anaheim, California, Paper #91-0401.

Meyer, Beat. 1983. *Indoor air quality.* Reading, Mass.: Addison-Wesley Publishing Co.

National Air Pollution Control Administration. 1969. *Air quality criteria for sulfur oxides.* Pub. No. AP-50. Washington, D.C.: 161–162.

National Research Council, Committee on Odors from Stationary and Mobile Sources. 1979. *Odors from stationary and mobile sources.* Washington, D.C.: National Academy of Sciences.

Pevoto, L.F., and L.J. Hawkins. 1992. Sample preparation techniques for very wet gas analysis. ISA Conference, Houston, October 1992.

Reynolds, J.P., R.R. Dupont, and L. Theodore. 1991. *Hazardous waste incinerator calculations, problems and software.* John Wiley and Sons Inc.

Sax, N.J., and R.J. Lewis. 1986. *Rapid guide to hazardous chemicals in the workplace.* New York: Van Nostrand.

Seinfeld, J.H. 1975. *Air pollution—Physical and chemical fundamentals.* McGraw-Hill Book Company.

Stephan, D.G., G.W. Walsh, and R.A. Herrick. 1960. Concepts in fab-

ric air filtration. *Am. Ind. Hyg. Assoc. J.* 21:1–14.

Stern, A.C., ed. 1968. *Air pollution.* 2d ed. Vol. II, *Analysis, monitoring, and surveying.* Chap. 16. New York: Academic Press.

Stern, A.C., H.C. Wohlers, R.W. Boubel, and W.P. Lowry. 1973. *Fundamentals of air pollution.* New York: Academic Press.

Theodore, L., and J. Reynolds. 1987. *Introduction to hazardous waste incineration.* John Wiley and Sons Inc.

Turk, A. 1977. Adsorption. In *Air pollution,* 3d ed. Edited by A.C. Stern. New York: Academic Press.

U.S. Consumer Products Safety Commission. *Status report on indoor pollution in 40 Tennessee homes.*

U.S. Environmental Protection Agency (EPA). 1977. *Technical guidance for the development of control strategies in areas with fugitive dust problems.* EPA 450/3-77/010. Washington, D.C.: U.S. EPA.

———. 1991. Burning of hazardous waste in boilers and industrial furnaces: Final rule corrections; Technical amendments. *Federal Register* 56, no. 137 (17 July):32688–32886.

———. 1992. *National air quality emissions trends report, 1991.* EPA-450-R-92-001. Research Triangle Park, N.C.: Office of Air Quality Planning and Standards (October).

Unterman, R. 1993. Biotreatment systems for air toxics. AIChE Central Jersey Section Seminar on Catalysis and Environmental Technology, May 20, 1993.

Vesilind, P.A. 1994. *Environmental engineering.* 3d ed. Butterworth Heinemann.

Weiss, M. 1994. Environmental res complicate CEM success. *Control* (March).

Wood, S.C. 1994. Select the right NO control technology. *Chem. Eng. Prog.* (January).

Index

Milton Keynes UK
Ingram Content Group UK Ltd.
UKHW050441111024
449327UK00045B/2549